Lecture Notes in Control and Information Sciences 363

Editors: M. Thoma, M. Morari

Gildas Besançon (Ed.)

Nonlinear Observers and Applications

Springer

Editor

Gildas Besançon

Département Automatique, GIPSA-lab
(ex-Laboratoire d'Automatique de Grenoble)
ENSIEG BP 46
38402 Saint-Martin d'Heres, France
Email: Gildas.Besancon@inpg.fr

Library of Congress Control Number: 2007931052

ISSN print edition: 0170-8643
ISSN electronic edition: 1610-7411
ISBN-10 3-540-73502-X Springer Berlin Heidelberg New York
ISBN-13 978-3-540-73502-1 Springer Berlin Heidelberg New York

Springer is a part of Springer Science+Business Media
springer.com

Typesetting: by the authors and SPS using a Springer LaTeX macro package

Printed on acid-free paper SPIN: 12087648 89/SPS 5 4 3 2 1 0

"Measure what is measurable and make it measurable what is not so."

Galileo Galilei.

To Alina.

Preface

This book basically gathers the material of the *28th* Grenoble International Summer School on Control, dedicated to "Nonlinear observers and applications", organized within the Control Systems Department (former *Laboratoire d'Automatique de Grenoble*) of the newly formed GIPSA-lab research center, and hosted by the "Maison Jean Kuntzmann" of the IMAG Institute, in September 2007.

The motivation for this school and the present book is twofold: on the one hand, the observer problem is undoubtedly crucial in control systems, and on the other hand there are not so many comprehensive documents on this topic. The present one does not claim to be exhaustive, but can give a good idea on the problem, on the possible tools to get solutions, and on various applications. Its spirit basically follows from the pioneering studies which took place in particular in Grenoble in the early eighties, and the subsequent developments, owing a lot to Guy Bornard and his coworkers. In particular, a general overview on observer tools for nonlinear systems is here proposed, focuses on high gain and adaptive gain techniques are given, as well as immersion and optimization-based approaches. Some applications in control and fault or parameter estimation are finally discussed.

I am grateful to all the contributors for their participation in this project.

Dr A. Voda, associate professor in the GIPSA-lab research center, and who has been in charge of the annual organization of high-level summer schools in Grenoble for quite a few years now, is to be sincerely thanked for her help and suggestions in the organization of this *28th* event. The school and the book edition are also supported by the French research center C.N.R.S, research ministry M.E.N.R.T., and polytechnical institute I.N.P. Grenoble, while the school practical organization strongly relies on the administrative staff of the GIPSA-lab control systems department. In particular, my warm thanks go here to Mrs Marie-Thérèse Descotes-Genon and Mrs Marie-Rose Alfara.

June 4th 2007

Saint-Martin d'Hères,
Gildas Besançon

Contents

List of Contributors

M. Alamir
Département d'Automatique,
GIPSA-lab, ENSIEG, BP 46,
38402 Saint-Martin d'Hères, France.
Mazen.Alamir@gipsa-lab.inpg.fr

G. Besançon
Département d'Automatique,
GIPSA-lab, ENSIEG, BP 46,
38402 Saint-Martin d'Hères, France.
Gildas.Besancon@gipsa-lab.inpg.fr

N. Boizot
LASSY,Université du Luxembourg,
6 rue Coudenhove Kalergi,
L-1359 Luxembourg
nicolas.boizot@uni.lu

E. Busvelle
LE2I, IUT d'Auxerre,
Route des plaines de l'Yonne,
89000 Auxerre, France
busvelle@u-bourgogne.fr

H. Hammouri
LAGEP, Université Claude Bernard,
Lyon 1, Villeurbanne, France.
hammouri@lagep.cpe.fr

A. Isidori
DIS and CASY-DEIS
Università Roma 1 La Sapienza
and Università di Bologna, Italy.
isidori@ese.wustl.edu

L. Marconi
CASY-DEIS, Università di Bologna,
Italy.
lmarconi@deis.unibo.it

A. Ţiclea
Universitatea Polytehnica Bucureşti,
Romania.
ticlea@indinf.pub.ro

1

An Overview on Observer Tools for Nonlinear Systems

Gildas Besançon

Control Systems Department (former *Laboratoire d'Automatique de Grenoble*), GIPSA-lab. ENSIEG BP 46, 38402 Saint-Martin d'Hères, France
Gildas.Besancon@inpg.fr

1.1 Introduction and Problem Statement

1.1.1 Context and Motivations

The problem of *observer* design naturally arises in a *system* approach, as soon as one needs some *internal* information from *external* (directly available) measurements. In general indeed, it is clear that one cannot use as many sensors as signals of interest characterizing the system behavior (for cost reasons, technological constraints, etc.), especially since such signals can come in a quite large number, and they can be of various types: they typically include time-varying signals characterizing the system (*state variables*), constant ones (*parameters*), and unmeasured external ones (*disturbances*).

This need for internal information can be motivated by various purposes: modeling (*identification*), monitoring (*fault detection*), or driving (*control*) the system. All those purposes are actually jointly required when aiming at keeping a system under control, as summarized by figure 1.1 hereafter. This makes the reconstruction - or observer - problem the heart of a general control problem.

The purpose here will thus be to give an overview on some possible tools for *observer design* (the related problems of control, fault detection or parameter identification being considered in subsequent chapters): in short, an observer relies on a model, with on-line adaptation based on available measurements, and aiming at information reconstruction, i.e. it can be characterized as a *model-based, measurement-based, closed-loop, information reconstructor.*

Usually the model is a state-space representation, and it will be assumed here that all pieces of information to be reconstructed are born by state variables. In front of this, one can try to design an explicit dynamical system whose state should give an estimate of the actual state of the considered model, or just settle the problem as an optimization one. The present chapter will focus on the first case, the second one being considered in a subsequent one.

About the considered model, it can in general be either continuous-time or discrete-time, deterministic or stochastic, finite-dimensional or infinite-dimensional, smooth or "with singularities". But in order to give a quite consistent presentation, the present chapter will be restricted to the case of smooth,

G. Besançon (Ed.): Nonlinear Observers and Applications, LNCIS 363, pp. 1–33, 2007.
springerlink.com

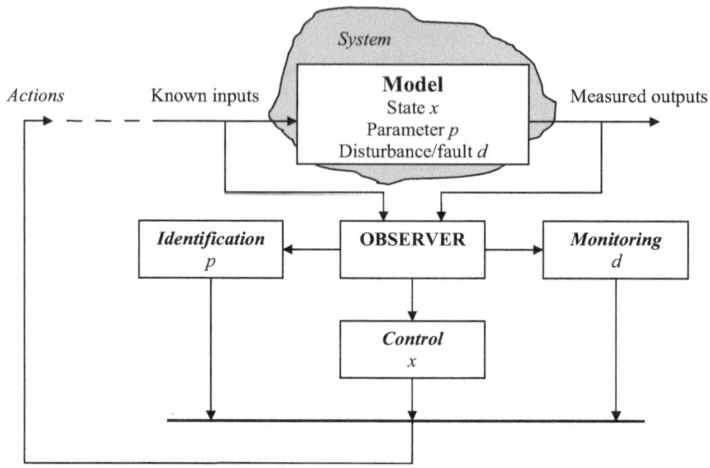

Fig. 1.1. Observer as the heart of control systems

finite-dimensional, deterministic, continuous-time state-space descriptions (even though some elements of observer design can be found for other cases in the literature).

In this framework, the subsequent subsection specifies the problem formulation, while section 1.2 presents the main related observability notions. Section 1.3 then discusses some possible techniques for observer design from a viewpoint which can be summarized by figure 1.2 below:

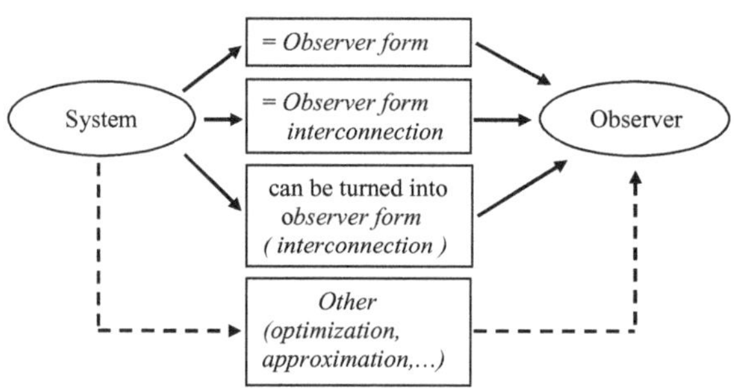

Fig. 1.2. A methodology for observer design

In short, the idea is to rely on specific structures for which observers are available (*observer forms*, in the figure), and try to bring the system under consideration to such cases. Hence, some *basic structures* for observer design are first presented, classified in two categories: those with a constant (input-independent) correction gain, and those with a time-varying (possibly input-dependent) correction gain. Then, methods for possible extensions of those designs to more general classes of systems are proposed, either by means of interconnections, or by transformations. Some conclusions and further remarks are finally given in section 1.4.

Notice that the material here presented roughly updates [7], summarizing some own research and viewpoint on the problem (in the continuity of [6] for instance) as well as various results borrowed from the quite large available ones (including overviews of [15] or [26] for instance).

In all the subsequent sections, the following notations/terminology will be used:

- Standard notations x, u, y and t respectively for state vector, input vector, output vector, and time variable (which might be omitted as an argument when not indispensable),
- I for the identity matrix of appropriate dimensions,
- v_i for the ith component of a vector v,
- $M = M^T > 0$ for a symmetric positive definite matrix M,
- *Stable matrix* for a matrix with all eigenvalues having strictly negative real parts.

while the abbreviation 'w.r.t.' will stand as usual for 'with respect to'.

1.1.2 Observer Problem Statement

Model Under Consideration

All over the chapter, the system under consideration will be considered to be described by a state-space representation generally of the following form:

$$\begin{aligned} \dot{x}(t) &= f(x(t), u(t)) \\ y(t) &= h(x(t)) \end{aligned} \tag{1.1}$$

where x denotes the state vector, taking values in X a connected manifold of dimension n, u denotes the vector of known external inputs, taking values in some open subset U of $\mathbb{R}^m$, and y denotes the vector of measured outputs taking values in some open subset Y of $\mathbb{R}^p$.

Functions f and h will in general be assumed to be $\mathcal{C}^\infty$ w.r.t. their arguments, and input functions $u(.)$ to be locally essentially bounded and measurable functions in a set $\mathcal{U}$.

The system will be assumed to be forward complete.

More generally, the dynamics might explicitly depend on time via $f(x(t), u(t), t)$, while y might further directly depend on u and even t, via $h(x(t), u(t), t)$. Such an explicitly time-dependent system is usually called 'time-varying' and generalizes (1.1) into:

$$\begin{aligned} \dot{x}(t) &= f(x(t), u(t), t) \\ y(t) &= h(x(t), u(t), t) \end{aligned} \tag{1.2}$$

However, the observer techniques discussed in this chapter are based on more specific forms of state-space representations, among which the following ones can be mentioned:

- Control-affine systems:

$$f(x, u) = f_0(x) + g(x)u$$

- State-affine systems[1]:

$$f(x,u) = A(u)x + B(u), \quad h(x) = Cx \ (or\ C(u)x + D(u))$$

- Linear Time-Varying (LTV) systems:

$$f(x,u,t) = A(t)x + B(t)u, \quad h(x,u,t) = C(t)x + D(t)u$$

- Linear Time-Invariant (LTI) systems:

$$f(x,u) = Ax + Bu, \quad h(x,u) = Cx + Du$$

Finally, the system will be said to be 'uncontrolled' whenever f and h do not depend on u.

In general, $\chi_u(t, x_{t_0})$ will denote the solution of the state equation in (1.1) under the application of input u on $[t_0, t]$ and satisfying $\chi_u(t_0, x_{t_0}) = x_{t_0}$, while u will be omitted for uncontrolled cases.

Observer Problem

Given a model (1.1), the purpose of acting on the system, or monitoring it, will in general need to know $x(t)$, while in practice one has only access to u and y. The observation problem can then be formulated as follows:

> Given a system described by a representation (1.1), find an estimate $\hat{x}(t)$ for $x(t)$ from the knowledge of $u(\tau), y(\tau)$ for $0 \leq \tau \leq t$.

Clearly this problem makes sense when one cannot invert h w.r.t. x at any time.

In front of this, one can look for a solution in terms of optimization, by looking for the best estimate $\hat{x}(0)$ of $x(0)$ which can explain the evolution $y(\tau)$ over $[0,t]$, and from this, get an estimate $\hat{x}(t)$ by integrating (1.1) from $\hat{x}(0)$ and under $u(\tau)$. In order to cope with disturbances, one should rather optimize the estimate of some initial state over a moving horizon, namely minimize some criterion of the form:

$$\int_{t-T}^{t} \|h(\chi_u(\tau, z_{t-T})) - y(\tau)\|^2 d\tau$$

w.r.t. z_{t-T} for any $t > T$, and $y(\tau)$ corresponding to the measured output over $[t-T, t]$ under the effect of the considered input u.

This is a general formulation for a solution to the problem, relying on available optimization tools and results for practical use and guarantees (see e.g. [1, 43, 47]): so it takes advantage of its systematic formulation, but suffers from usual drawbacks of nonlinear optimization (computational burden, local minima...).

Alternatively, one can use the idea of an explicit "feedback" in estimating $x(t)$, as this is done for control purposes: more precisely, noting that if one knows the initial value $x(0)$, one can get an estimate for $x(t)$ by simply integrating (1.1) from $x(0)$, the feedback-based idea is that if $x(0)$ is unknown, one can try to

---

[1] Including *bilinear systems* as a particular case, for which A, B, C, D are linear w.r.t. u.

correct on-line the integration $\hat{x}(t)$ of (1.1) from some erroneous $\hat{x}(0)$, according to the measurable error $h(\hat{x}(t)) - y(t)$, namely to look for an estimate $\hat{x}$ of x as the solution of a system:

$$\dot{\hat{x}}(t) = f(\hat{x}(t), u(t)) + k(t, h(\hat{x}(t)) - y(t)), \ \textit{with}\ k(t,0) = 0. \qquad (1.3)$$

Such an auxiliary system is what will be defined as an *observer*, and the above equation is the most common form of an observer for a system (1.1) (as in the case of linear systems [36, 42]).

More generally, an observer can be defined as follows:

Definition 1. *Observer*
Considering a system (1.1), an observer is given by an auxiliary system:

$$\begin{aligned} \dot{X}(t) &= F(X(t), u(t), y(t), t) \\ \hat{x}(t) &= H(X(t), u(t), y(t), t) \end{aligned} \qquad (1.4)$$

such that:

(i) $\hat{x}(0) = x(0) \Rightarrow \hat{x}(t) = x(t), \quad \forall t \geq 0;$
(ii) $\|\hat{x}(t) - x(t)\| \to 0$ *as* $t \to \infty;$

If (ii) holds for any $x(0), \hat{x}(0)$*, the observer is* global.
If (ii) holds with exponential convergence, the observer is exponential.
If (ii) holds with a convergence rate which can be tuned, the observer is tunable.

Notice that the overview on observer design presented in the sequel will mainly be dedicated to global exponential tunable observers.

Notice also that with notations of (1.1) and (1.4), the difference $\hat{x} - x$ will be called *observer error.*

Notice finally that with the above point of view, the observation problem turns to be a problem of observer design.

1.2 Nonlinear Observability

The purpose of this section is to discuss some conditions required on the system for possible solutions to the above mentioned observer problem. Such conditions above all correspond to what are usually called *observability* conditions. In short, they must express that there indeed is a possibility that the purpose of the observer can be achieved, namely that it might be possible to recover $x(t)$ from the only knowledge of u and y up to time t: at a first glance, this will be possible only if $y(t)$ bears the information on the full state vector when considered over some time interval: this roughly corresponds to the notion of "*observability*".

However, when restricting the definition of an observer strictly to items (i)-(ii), one can find observers yielding solutions to the observation problem even in cases when y does not bear the full information on the state vector:

Consider for instance the simple system:

$$\dot{x} = -x + u, \ y = 0$$

Clearly one cannot get any information on x from y, and yet the system:

$$\dot{\hat{x}} = -\hat{x} + u$$

satisfies (i)-(ii) and yields an estimate of x, since:

$$\dot{\overbrace{\hat{x} - x}} = -(\hat{x} - x).$$

This corresponds to a notion of "*detectability*". Notice that in that case, however, the rate of convergence cannot be tuned. Additional remarks in that respect can be found e.g. in [6].

If we restrict ourselves to the case of observers in the sense of *tunable* observers, then observability becomes a necessary condition. Such a condition can be specified in a geometric way as shown hereafter, while analytical additional conditions are discussed afterwards.

1.2.1 Geometric Conditions of Observability

For a possible design of a (tunable) observer, one must be able to recover the information on the state via the output measured from the initial time, and more particularly to recover the corresponding initial value of the state. This means that observability is characterized by the fact that from an output measurement, one must be able to distinguish between various initial states, or equivalently, one cannot admit *indistinguishable* states (following [33]):

Definition 2. *Indistinguishability*
A paire $(x_0, x'_0) \in I\!R^n \times I\!R^n$ *is indistinguishable for a system (1.1) if:*

$$\forall u \in \mathcal{U},\ \forall t \geq 0,\ h(\chi_u(t, x_0)) = h(\chi_u(t, x'_0)).$$

A state x *is indistinguishable from* x_0 *if the pair* (x, x_0) *is indistinguishable.*

From this, observability can be defined:

Definition 3. *Observability [resp. at* x_0*]*
A system (1.1) is observable [resp. at x_0*] if it does not admit any indistinguishable paire [resp. any state indistinguishable from* x_0*].*

This definition is quite general (global), and even too general for practical use, since one might be mainly interested in distinguishing states from their neighbors:

Consider for instance the case of the following system:

$$\dot{x} = u, \quad y = sin(x). \tag{1.5}$$

Clearly, y cannot help distinguishing between x_0 and $x_0 + 2k\pi$, and thus the system is not observable. It is yet clear that y allows to distinguish states of $]-\frac{\pi}{2}, \frac{\pi}{2}[$.

This brings to consider a weaker notion of observability:

Definition 4. *Weak observability [resp. at x_0]*
A system (1.1) is weakly observable [resp. at x_0] if there exists a neighborhood U of any x [resp. of x_0] such that there is no indistinguishable state from x [resp. x_0] in U.

Notice that this does not prevent from cases where the trajectories have to go far from U before one can distinguish between two states of U.
Consider for instance the case of a system:

$$\dot{x} = u; \quad y = h(x)$$

with h a $\mathcal{C}^\infty$ function as in figure 1.3 below: clearly the system is weakly observable since any state is distinguishable from any other one by applying some nonzero input u, but distinguishing two points of $[-1, 1]$ needs to wait for y to move away from 0.

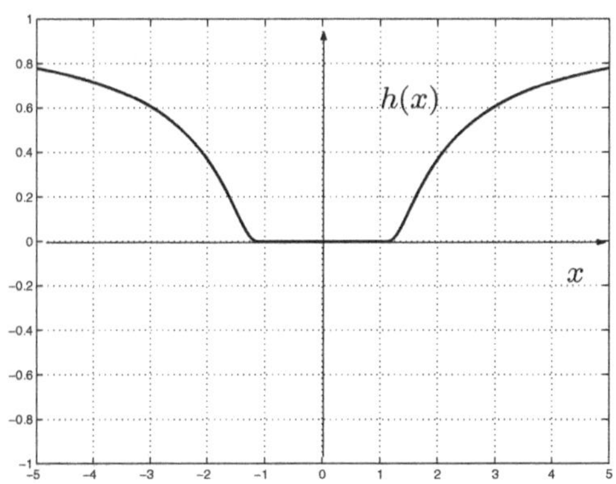

Fig. 1.3. Output function of a weakly but not locally observable system

Hence, to prevent from this situation, an even more local definition of observability can be given:

Definition 5. *Local weak observability [resp. at x_0]*
A system (1.1) is locally weakly observable [resp. at x_0] if there exists a neighborhood U of any x [resp. of x_0] such that for any neighborhood V of x [resp. x_0] contained in U, there is no indistinguishable state from x [resp. x_0] in V when considering time intervals for which trajectories remain in V.

This roughly means that one can distinguish every state from its neighbors without "going too far". This notion is of more interest in practice, and also presents the advantage of admitting some 'rank condition' characterization.

Such a condition relies on the notion of *observation space* roughly corresponding to the space of all observable states:

Definition 6. *Observation space*
The observation space for a system (1.1) is defined as the smallest real vector space (denoted by $\mathcal{O}(h)$) of C^∞ functions containing the components of h and closed under Lie derivation along $f_u := f(.,u)$ for any constant $u \in I\!R^m$ (namely such that for any $\varphi \in \mathcal{O}(h)$, $L_{f_u}\varphi \in \mathcal{O}(h)$, where $L_{f_u}\varphi(x) = \frac{\partial\varphi}{\partial x} f(x,u)$).

Definition 7. *Observability rank condition [resp. at x_0]*
A system (1.1) is said to satisfy the observability rank condition [resp. at x_0] if:

$$\forall x, \quad dimd\mathcal{O}(h)\mid_x = n \quad \text{[resp. } dimd\mathcal{O}(h)\mid_{x_0} = n\text{]}$$

where $d\mathcal{O}(h)\mid_x$ is the set of $d\varphi(x)$ with $\varphi \in \mathcal{O}(h)$.

From this we have [33]:

Theorem 1. *A system (1.1) satisfying the observability rank condition at x_0 is locally weakly observable at x_0.*

More generally a system (1.1) satisfying the observability rank condition is locally weakly observable.

Conversely, a system (1.1) locally weakly observable satisfies the observability rank condition in an open dense subset of X.

In short this follows from the facts that:

(i) the observability rank condition at some x_0 means the existence of n elements of the observation space defining a diffeomorphism around x_0;
(ii) for any indistinguishable pair (x_0, x_0') and any element $\varphi \in \mathcal{O}(h)$, $\varphi(x_0) = \varphi(x_0')$.

As an example of application, consider again system (1.5): for this system one clearly has $d\mathcal{O}(h) = span\{cos(x)dx, sin(x)dx\}$ and thus $dimd\mathcal{O}(h)\mid_{x_0} = 1$ for any x_0, namely the system satisfies the observability rank condition.
As a second example, consider a system of the following form:

$$\begin{aligned} \dot{x} &= Ax \\ y &= Cx \quad \text{with } x \in I\!R^n. \end{aligned} \tag{1.6}$$

For this system, the observability rank condition is equivalent to local weak observability (which is itself equivalent to observability) and is characterized by the so-called *Kalman rank condition*:

Theorem 2. *For a system of the form (1.6)*

- *The observability rank condition is equivalent to $rank\mathcal{O}_m = n$ with $\mathcal{O}_m$ the so-called* observability matrix *defined by* $\mathcal{O}_m = \begin{pmatrix} C \\ CA \\ CA^2 \\ \vdots \\ CA^{n-1} \end{pmatrix}$ *the ;*
- *The observability rank condition is equivalent to the observability of the system.*

The first point results from straightforward computations (e.g. as in [34]) since here the kth Lie derivation $L_f^k h(x) = CA^k x$, while the second one results from the definition of observability (see e.g. [40]).

Notice that if system (1.6) satisfies the above observability rank condition, the pair (A, C) is usually called *observable.*

Notice also that the above result also holds for controlled systems with $\dot{x} = Ax + Bu$.

Notice finally that the above observability rank condition is also sufficient for a possible observer design for (1.6) (even necessary and sufficient for a *tunable* observer design - see later).

However, in general, the observability rank condition is not enough for a possible observer design: this is due to the fact that in general, observability depends on the inputs, namely it does not prevent from the existence of inputs for which observability vanishes.

As a simple example, consider the following system:

$$\begin{aligned} \dot{x} &= \begin{pmatrix} 0 & u \\ 0 & 0 \end{pmatrix} x \\ y &= \begin{pmatrix} 1 & 0 \end{pmatrix} x \end{aligned} \tag{1.7}$$

it is clearly observable for any constant input $u \neq 0$, but not observable for $u = 0$.

This means that the purpose of observer design requires a look at the inputs.

1.2.2 Analytic Conditions for Observability

In view of example (1.7) additional conditions to those previously presented might be required for possible observer designs, related to inputs. The purpose here is to discuss such conditions, while effective designs will be proposed later on.

More precisely, notions of *universal inputs* and *uniform observability* for systems (1.1) are first introduced (as in [15] for instance), and the stronger notions of *persistency* and *regularity* more usually defined for state affine systems [15] are then presented for the more general case of systems (1.1).

Definition 8. *Universal inputs [resp. on* $[0,t]$*]*
An input u is universal (resp. on $[0,t]$) for system (1.1) if $\forall x_0 \neq x_0'$, $\exists \tau \geq 0$ (resp. $\exists \tau \in [0,t]$) s.t. $h(\chi_u(\tau, x_0)) \neq h(\chi_u(\tau, x_0'))$.

An input u is a singular input if it is not universal.

As an example, for system (1.7), $u(t) = 0$ is a singular input.

It can be underlined here that for $\mathcal{C}^w$ systems, universal $\mathcal{C}^w$ inputs are dense in the set of $\mathcal{C}^w$ functions for the topology induced by $\mathcal{C}^\infty$ [46].

But one has to also notice that in general characterizing singular inputs is not easy. Things are easier for systems which do not admit such singular inputs:

Definition 9. *Uniformly observable systems (resp. locally)*
A system is uniformly observable (UO) if every input is universal (resp. on $[0,t]$).

Example 1. The system (1.8) below is uniformly observable [23]:

$$\dot{x} = \begin{pmatrix} 0 & 1 & 0 & \cdots & 0 \\ & & \ddots & \ddots & \\ & & & & 0 \\ \vdots & & & & 1 \\ 0 & \cdots & & & 0 \end{pmatrix} x + \begin{pmatrix} \varphi_1(x_1) \\ \varphi_2(x_1, x_2) \\ \vdots \\ \varphi_n(x_1, \ldots, x_{n-1}) \\ \varphi_{n-1}(x_1, \ldots, x_n) \end{pmatrix} u \tag{1.8}$$
$$y = x_1; \quad x = (x_1 \ldots, x_n)^T$$

This can be checked by considering any pair of distinct states $x \neq x'$: assuming indeed that their respective components x_k and x'_k coincide up to order i and that $x_{i+1} = x'_{i+1}$, then it is clear from (1.8) that $\dot{x}_{i-1} - \dot{x}'_{i-1} \neq 0$ and thus there exists t_0 such that $x_i(t) \neq x'_i(t)$ for $0 < t < t_0$. By induction, we easily end up with the existence of some time for which $x_1(t) \neq x'_1(t)$, which is true for any u.

This property actually means that observability is independent of the inputs and thus can allow an observer design also independent of the inputs, as in the case of LTI systems (see later).

For systems which are not uniformly observable, in general possible observers will depend on the inputs, and not all inputs will be admissible. Restricting the set of inputs to universal ones, as in the case of uniformly observable systems - for which *all* inputs are universal, is actually not enough:

Consider for instance the following system:

$$\dot{x} = \begin{pmatrix} 0 & u \\ -u & 0 \end{pmatrix} x; \quad y = (1 \ 0)x$$

For this system, the input defined by $u(t) = 1$ for $t < t_1$ and $u(t) = 0$ for $t \geq t_1$ is clearly universal, but if a disturbance appears after t_1, it is also clear that x cannot be correctly reconstructed.

This shows that universality must be guaranteed over the time, namely must be *persistent*. In order to characterize this persistency, notice first that we have the following property:

Proposition 1. *An input u is a universal input on $[0, t]$ for system (1.1) if and only if $\int_0^t ||h(\chi_u(\tau, x_0)) - h(\chi_u(\tau, x'_0))||^2 d\tau > 0$ for all $x_0 \neq x'_0$.*

This can be easily checked from definition 8.

Then one can define persistency as follows:

Definition 10. *Persistent inputs*
An input u is a persistent input for a system (1.1) if

$$\exists t_0, T : \forall t \geq t_0, \forall x_t \neq x'_t, \quad \int_t^{t+T} ||h(\chi_u(\tau, x_t)) - h(\chi_u(\tau, x'_t))||^2 d\tau > 0$$

Equivalently, this can be expressed as:

$$\int_{t-T}^{t} ||h(\chi_u(\tau, x_{t-T})) - h(\chi_u(\tau, x'_{t-T}))||^2 d\tau > 0, \quad \forall x_{t-T} \neq x'_{t-T} \tag{1.9}$$

which might be more suitable thinking of t as a current time and the inequality as a property on the past measurements.

This basically guarantees observability over a given time interval.

However this does not prevent observability from possibly vanishing as time goes to infinity. If this happens, effective observers would in general have to compensate this by a correction gain going to infinity:

Consider for instance the system defined by:

$$\begin{aligned} \dot{x}_1 &= x_2 \\ \dot{x}_2 &= x_2 + u \\ y &= \begin{pmatrix} \frac{x_1}{1+x_2^2} \\ x_2 \end{pmatrix} = \begin{pmatrix} y_1 \\ y_2 \end{pmatrix} \end{aligned}$$

For this system any (bounded) input can be checked to be persistent in the sense of definition 10, since whenever x_{t-T} and x'_{t-T} differ from one another in their second component, (1.9) clearly holds, while if they only differ in their first component, it can be shown that the left-hand side roughly behaves as $(e^{-4(t-T)} - e^{-4t})$ which is indeed positive. But from this it is also clear that the system is *less and less observable* as $t \to \infty$.

It can be noticed that the state variables x_1, x_2 could here be reconstructed by an auxiliary system of the form (1.3), for instance given as follows:

$$\begin{aligned} \dot{\hat{x}}_1 &= y_2 - k_1(t)\left(\frac{\hat{x}_1}{1+y_2^2} - y_1\right) \\ \dot{\hat{x}}_2 &= \hat{x}_2 + u - k_2(\hat{x}_2 - y_2) \end{aligned}$$

for any $k_2 > 0$ and some k_1 growing as y_2 (namely of the form $\kappa_1(1 + y_2^2)$ for $\kappa_1 > 0$). This system indeed clearly guarantees that $\hat{x}_2 - x_2 \to 0$ (since $\frac{d}{dt}(\hat{x}_2 - x_2) = -k_2(\hat{x}_2 - x_2)$), and also that $\hat{x}_1 - x_1 \to 0$ (from $\frac{d}{dt}(\hat{x}_1 - x_1) = -\kappa_1(\hat{x}_1 - x_1)$), but with a correction gain $k_1(t)$ growing to infinity.

In order to avoid this, one needs a *guarantee* of observability, namely some *regular persistency*:

Definition 11. *Regularly persistent inputs*
An input u is a regularly persistent input for a system (1.1) if:

$$\begin{aligned} &\exists t_0, T : \forall x_{t-T}, x'_{t-T}, \ \forall t \geq t_0, \\ &\int_{t-T}^{t} ||h(\chi_u(\tau, x_{t-T})) - h(\chi_u(\tau, x'_{t-T}))||^2 d\tau \geq \beta(\|x_{t-T} - x'_{t-T}\|) \end{aligned}$$

for some class $\mathcal{K}$ function β.

From the above proposed definitions of persistency and regular persistency, we recover the usual definitions already available for state affine systems (of [15] for instance):

Proposition 2. *For state affine systems, regularly persistent inputs are inputs u such that:*

$$\exists t_0, T, \alpha : \int_{t-T}^{t} \Phi_u^T(\tau, t-T) C^T C \Phi_u(\tau, t-T) d\tau \geq \alpha I > 0 \quad \forall t \geq t_0, \qquad (1.10)$$

with $\Phi_u(\tau, t)$ the transition matrix classically defined by:

$$\frac{d\Phi_u(\tau, t)}{d\tau} = A(u(\tau))\Phi_u(\tau, t), \ \Phi_u(t, t) = I.$$

This is a straight consequence of the application of definition 11 to the case of state affine systems, with $\beta(\|z\|) = \alpha\|z\|^2$.

The left-hand side quantity in (1.10) corresponds to the so-called *observability Grammian*, classically defined for LTV systems, for any $t_1 < t_2 \in I\!R$, as:

$$\Gamma(t_1, t_2) = \int_{t_1}^{t_2} \Phi^T(\tau, t_1) C^T(\tau) C(\tau) \Phi(\tau, t_1) d\tau \qquad (1.11)$$

where Φ as above denotes the transition matrix for the autonomous part of the system.

Remark 1

- Regularly persistent inputs for state affine systems are those making the system an LTV system *Uniformly Completely Observable* in the sense of Kalman [36] (since uniform complete observability for LTV systems is typically defined by (1.10);
- For general nonlinear systems, the definition is not of easy use, while for state affine or LTV systems, it is independent of initial states.

As an example of input properties, consider the following system:

$$\dot{x} = \begin{pmatrix} 0 & u \\ 0 & 0 \end{pmatrix} x; \ y = (1 \ 0)x$$

For this system, the input for instance defined by:

$u(t) = 1$ *on* $t \in [2kT, (2k+1)T[, \ k \geq 0$
$u(t) = 0$ *on* $t \in [(2k+1)T, (2k+2)T[, \ k \geq 0$

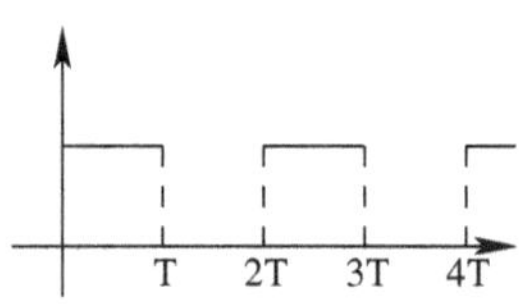

is regularly persistent, while that defined by:

$u(t) = 1$ *on* $t \in [2kT, (2k + \frac{1}{k+1})T[, \ k \geq 0$
$u(t) = 0$ *on* $t \in [(2k + \frac{1}{k+1})T, (2k+2)T[, \ k \geq 0$

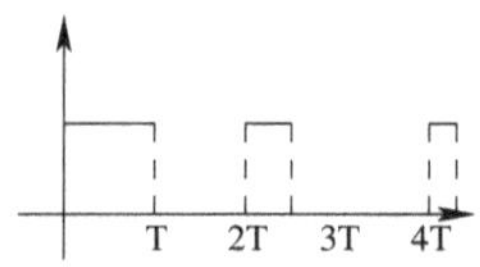

is not [15].

Notice that for the reasons previously mentioned, regular persistency appears to be the property actually needed for effective state reconstruction.

However, it can be noticed that it depends on some time T roughly required to get enough information. If one is interested by an estimation 'in short time', he will need some kind of stronger observability property, corresponding to the application of what was originally called *locally regular inputs* on the basis of state affine systems [15]. In a more general context, this property can be formulated as follows:

Definition 12. *Locally regular inputs*
An input u is a locally regular input for a system (1.1) if:

$$\exists T_0, \alpha : \forall x_{t-T}, x'_{t-T},\ \forall T \leq T_0,\ \forall t \geq T,$$
$$\int_{t-T}^{t} ||h(\chi_u(\tau, x_{t-T})) - h(\chi_u(\tau, x'_{t-T}))||^2 d\tau \geq \beta(\|x_{t-T} - x'_{t-T}\|, \frac{1}{T})$$

for some class $\mathcal{KL}$ function β.

This property characterizes in some sense observability for arbitrarily short times. Obviously when T decays to zero, the observability cannot be kept guaranteed, which explains the decaying characteristic of β. When again considering state affine systems, we can roughly recover the definition previously used in [15, 4, 11] for instance, by considering some appropriate $\beta(\|x_t - x'_t\|, \frac{1}{T})$:

Proposition 3. *For state affine systems, locally regular inputs are inputs u such that:*

$$\exists T_0, \alpha : \forall T \leq T_0,\ \forall t \geq T,$$
$$\int_{t-T}^{t} \Phi_u^T(\tau, t-T) C^T C \Phi_u(\tau, t-T) d\tau \geq \alpha \frac{1}{T} \begin{pmatrix} T & & 0 \\ & T & \\ & & \ddots \\ 0 & & T^n \end{pmatrix}^2 \tag{1.12}$$

with $\Phi_u(\tau, t)$ the transition matrix as in proposition 2.

Here β is given by the right-hand side multiplied by $\|x_{t-T} - x'_{t-T}\|$: this is in particular motivated by the form of the Grammian for the linear part of a uniformly observable system (1.8) [15]:

$$\Gamma(t-T, t) = T \begin{pmatrix} 1 & \frac{T}{2} & \frac{T^2}{6} & \cdots \\ \frac{T}{2} & \frac{T^2}{3} & \frac{T^3}{8} & \cdots \\ \frac{T^2}{6} & \frac{T^3}{8} & \frac{T^4}{20} & \\ \vdots & \vdots & & \ddots \end{pmatrix},$$

which can indeed be lower bounded as in (1.12) for α small enough.

Obviously for a linear observable system, every input is locally regular.

Notice that the characterization (1.12) actually slightly differs from that previously considered in [15, 4, 11] ($\Phi_u(\tau, t)$ was considered instead of $\Phi_u(\tau, t-T)$ in the inequality), but they become equivalent whenever Φ_u (i.e. A) is bounded.

All this will tell us on some possible observer designs for classes of systems, as discusses in next section. Notice that more specific notions of observability, which have been introduced in connection with more specific designs not presented in details here will be omitted (such as 'infinitesimal observability' or 'differential observability', related to 'high gain techniques' as in [26], or 'generic observability' used in algebraic approaches as in [19] for instance). Additionally, some final remarks can be given as follows:

Remark 2

- If a system, e.g. control affine, is not observable in the sense of rank condition, it can be decomposed into observable and non observable subsystems as follows [34]:
$$\begin{aligned}\dot{\zeta}_1 &= f_1(\zeta_1, \zeta_2) + g_1(\zeta_1, \zeta_2)u \\ \dot{\zeta}_2 &= f_2(\zeta_2) + g_2(\zeta_2)u \\ y &= h_2(\zeta_2)\end{aligned}$$
where the subsystem in ζ_2 satisfies the observability rank condition. In that case one has to work on ζ_2.
- If the considered system is not observable, but satisfies the following: $\forall u$ *such that* x_0 *and* x'_0 are indistinguishable by u :
$$\chi_u(t, x_0) - \chi_u(t, x'_0) \to 0 \, as \, t \to \infty$$
it satisfies a property of *detectability*, and in that case one may have the opportunity to design an observer in the sense of (i) and (ii).
- The analytic observability conditions which have here been presented (persistency, regular persistency, local regularity of definitions 10, 11 or 12 respectively) have been defined in terms of inputs, for controlled nonlinear systems of the form (1.1). But those definitions clearly still hold for time-varying systems (1.2). They can even be considered for uncontrolled systems (time-varying or not) since they are basically defined by output evolutions w.r.t. initial conditions. In other words, those notions could have been defined as various observability properties, parameterized by the input in the controlled case.

1.3 Nonlinear Observer Design

In view of the previously presented notions of observability, and in particular the problem of inputs which has been highlighted, it can easily be understood that observer designs will in general depend on the inputs of the system (or the analytic observability conditions previously highlighted).

However, in some cases, the observer design might be independent of the input, as in the case of uniformly observable systems for instance. Hence, following the viewpoint of [6], observer designs can be classified into 'uniform observers' and 'non uniform observers' w.r.t. inputs (or time), and we can roughly consider the following cases:

- For uniformly observable systems, one might design uniform observers;
- For non-uniformly observable systems, one might design non uniform observers.

The first ones correspond to the so-called *Luenberger observer* for LTI systems [42], while the second ones typically correspond to the case of *Kalman observers* for LTV systems [36]. Those observers will be first recalled in the next subsection, and then extended to nonlinear systems as *Luenberger-like* and *Kalman-like* designs. Subsection 1.3.2 will then discuss possible extensions of such 'basic designs'.

Notice that one might also design *non uniform* observers for uniformly observable systems (for instance using a Kalman approach), while in some cases *uniform* designs might be achieved for non uniformly observable systems (when the system satisfies some detectability property for instance, as discussed in [12]).

1.3.1 Basic Structures

Some observers are presented here for particular structures of systems. In the whole section, an observer is to be understood as a *global, exponential, tunable* observer.

Remember that the observer approach we consider is that of designing an auxiliary system intended to give an estimate $\hat{x}$ of the actual state vector x in the sense that $\hat{x}(t) - x(t) \to 0$ as $t \to \infty$. Hence the main problem turns to be an observer design so as to make the origin asymptotically stable for the corresponding observer error system. In all the presented results hereafter, this can be mostly studied by classical Lyapunov tools, as they are recalled in the appendix section 1.5.

Observer Designs for Linear Structures

The cases of LTI and LTV systems are here first considered.

Luenberger observer (for LTI systems)

Let us consider here LTI systems of the following form:

$$\begin{aligned} \dot{x}(t) &= Ax(t) + Bu(t) \\ y(t) &= Cx(t) \end{aligned} \tag{1.13}$$

For those systems we have the following classical (Luenberger) result [42]:

Theorem 3. *If system (1.13) satisfies the observability rank condition then there exists an observer of the form:*

$$\dot{\hat{x}}(t) = A\hat{x}(t) + Bu(t) - K(C\hat{x}(t) - y(t))$$

with K such that $A - KC$ is stable.

Remark 3. The rate of convergence can be arbitrarily chosen by appropriate design of K.

This can be established by showing that observability guarantees the existence of a transformation into a so-called observability canonical form, for which the design of an appropriate observer gain is straightforward (see e.g. [40]).

Kalman observer (for LTV systems)

Let us consider here LTV systems of the following form:

$$\begin{aligned} \dot{x}(t) &= A(t)x(t) + Bu(t) \\ y(t) &= C(t)x(t) \end{aligned} \tag{1.14}$$

with $A(t), C(t)$ uniformly bounded.

For those systems we have the following (Kalman-related) result [36, 16, 32, 10, 26]:

Theorem 4. *If system (1.14) is uniformly completely observable, then there exists an observer of the form:*

$$\dot{\hat{x}}(t) = A(t)\hat{x}(t) + B(t)u(t) - K(t)(C(t)\hat{x}(t) - y(t))$$

with $K(t)$ given by:

$$\begin{aligned} \dot{M}(t) &= A(t)M(t) + M(t)A^T(t) - M(t)C^T(t)W^{-1}C(t)M(t) + V + \delta M(t) \\ M(0) &= M_0 = M_0^T > 0, \ W = W^T > 0 \\ K(t) &= M(t)C^T(t)W^{-1} \end{aligned} \tag{1.15}$$

with either $\delta > 2\|A(t)\|$ for all t, or $V = V^T > 0$.

Remark 4

- The rate of convergence can be tuned by δ or V.
- For $\delta = 0$, we get the classical Kalman observer, the usual related condition for convergence being that (A, V) be *uniformly completely controllable* (dual of uniform complete observability).
- For $\delta = 0$, the observer is optimal in the sense of minimizing w.r.t. z:

$$\begin{aligned} &\int_0^t [(C(\tau)z(\tau) - y(\tau))^T W^{-1}(C(\tau)z(\tau) - y(\tau)) + v^T(\tau)V^{-1}v(\tau)]d\tau \\ &+(z_0 - \hat{x}_0)^T M_0^{-1}(z_0 - \hat{x}_0) \end{aligned}$$

  under $\dot{z}(t) = A(t)z(t) + v(t)$ $y(t) = C(t)z(t)$.

  Namely, it provides an explicit solution to the optimization-based approach mentioned in the introduction.

  It is also optimal in the sense of minimizing the mean of the square estimation error for a system affected by state white noises and measurement white noises, uncorrelated to each other, with V and W as respective variance matrices [40].

- The observer gain can also be computed as $K(t) = S^{-1}(t)C^T W^{-1}$ where S is the solution of:

$$\begin{aligned}\dot{S}(t) &= -A^T(t)S(t) - S(t)A(t) + C^T(t)W^{-1}C(t) - \delta S(t) - S(t)VS(t)\\ S(0) &= S^T(0) > 0\end{aligned}$$

  which makes it a linear equation in S whenever V is chosen equal to 0. This is also true for all subsequent *Kalman-like* designs, even if they will be expressed in terms of (1.15).

The result of theorem 4 can be established by showing that:

(i) $\exists \alpha_1, \alpha_2, t_0$ such that $\forall t \geq t_0 : \; 0 < \alpha_1 I \leq M^{-1}(t) \leq \alpha_2 I$ basically from the condition of uniform complete observability;
(ii) $V(e,t) = e^T(t)M^{-1}(t)e(t)$ where $e := \hat{x} - x$ is a Lyapunov function for the observer error equation, which is exponentially decaying with a rate of decay tunable via δ or the minimal eigenvalue of V.

This can be shown either when $V = 0$ and $\delta > 2\|A(t)\|$ [10, 32], or when $V = V^T > 0$ and $\delta = 0$ [26].

On the basis of theorem 4, an extension can be intuitively derived for *nonlinear* systems relying on its first order approximation along the estimated trajectories, and known as *Extended Kalman Filter* (see e.g. [27]):

Definition 13. *Extended Kalman Filter (EKF)*
Given a nonlinear system of the form:

$$\begin{aligned}\dot{x}(t) &= f(x(t), u(t))\\ y(t) &= h(x(t))\end{aligned}$$

the corresponding Extended Kalman Filter is given by:

$$\dot{\hat{x}}(t) = f(\hat{x}(t), u(t)) - K(t)(h(\hat{x}(t)) - y(t))$$

where $K(t)$ is given as in the Kalman observer (1.15) with:

$$A(t) := \frac{\partial f}{\partial x}(\hat{x}(t), u(t)), \quad C(t) := \frac{\partial h}{\partial x}(\hat{x}(t))$$

This yields a candidate for a systematic observer design in front of a nonlinear system, but in general the convergence is not guaranteed, except under specific structure conditions (or domain of validity). This motivates the inspection of more specific nonlinear structures.

Observer Designs for Nonlinear Structures

Some observer designs are here presented for specific structures of nonlinear systems, extending the Luenberger and Kalman observers above recalled for linear systems.

Luenberger-like design (for UO systems)

Let us first consider classes of systems for which observability does not depend on the input, namely Uniformly Observable systems.

The idea is basically to rely on a linear time-invariant part in order to design a gain as in Luenberger observers, and either compensate exactly all nonlinear elements when possible (by output injection for instance), or dominate them via the linear part.

Additive output nonlinearity

Consider here a system of the form:

$$\begin{aligned} \dot{x} &= Ax + \varphi(Cx, u) \\ y &= Cx \end{aligned} \tag{1.16}$$

Here the nonlinearity can be constructed from direct measurements and thus compensated in the observer design (as originally proposed in [38, 39] for instance):

Theorem 5. *If (A, C) is observable, system (1.16) admits an observer of the form:*

$$\dot{\hat{x}} = A\hat{x} + \varphi(y, u) - K(C\hat{x} - y)$$

with K such that $A - KC$ is stable.

Remark 5

Clearly here, the observer error is exactly linear, and thus the convergence rate can be arbitrarily tuned by appropriate choice of K as in the case of linear systems.

Additive triangular nonlinearity

Consider here a system of the form:

$$\begin{aligned} \dot{x} &= A_0 x + \varphi(x, u) \\ y &= C_0 x \\ \textit{with} \quad A_0 &= \begin{pmatrix} 0 & 1 & & 0 \\ & & \ddots & \\ & & & 1 \\ 0 & & & 0 \end{pmatrix}, \quad C_0 = (1\ 0 \cdots 0). \end{aligned} \tag{1.17}$$

Here the idea will be to use the uniform observability, and thus a structure as in (1.8), to weight a gain based on the linear part, so as to make the linear dynamics of the observer error to dominate the nonlinear one [24, 15, 26]:

Theorem 6. *If φ is globally Lipschitz w.r.t. x, uniformly w.r.t. u and such that:*

$$\frac{\partial \varphi_i}{\partial x_j}(x, u) = 0 \; for \; j \geq i+1, \quad 1 \leq i, j \leq n,$$

system (1.17) admits an observer of the form:

$$\dot{\hat{x}} = A_0\hat{x} + \varphi(\hat{x}, u) - \begin{pmatrix} \lambda & & 0 \\ & \ddots & \\ 0 & & \lambda^n \end{pmatrix} K_0(C_0\hat{x} - y)$$

with K_0 such that $A_0 - K_0C_0$ is stable, and λ large enough.

Remark 6

- This design is known as *high gain observer* since it relies on the choice of some sufficiently large tuning parameter λ;
- The larger λ is, the faster the convergence is.
- Output injection can also be used as in theorem 6.
- This design can be extended to systems of the following form [20, 25, 26]:

$$\dot{x}(t) = f(x(t), u(t)),\ y(t) = C_0x(t)$$

  where $\frac{\partial f_i}{\partial x_j} = 0$ for $j > i + 1$ and $\frac{\partial f_i}{\partial x_{i+1}} \geq \alpha_i > 0$ for all x, u;
- The design can also be extended to multi-output uniformly observable systems [17, 18];
- This design has been shown to be very useful for observer-based control.

The result of theorem 6 can be established by showing that $V(e) = e^TP(\lambda)e$ is a Lyapunov function for the observer error equation, exponentially decaying with a rate of decay being tunable via λ, where:

$$e = \hat{x} - x \quad \text{and} \quad P(\lambda) = \begin{pmatrix} \lambda & & 0 \\ & \ddots & \\ 0 & & \lambda^n \end{pmatrix}^{-1} P_0 \begin{pmatrix} \lambda & & 0 \\ & \ddots & \\ 0 & & \lambda^n \end{pmatrix}^{-1},$$

with P_0 such that:

$$P_0(A_0 - K_0C_0) + (A_0 - K_0C_0)^T = -I.$$

Kalman-like design (for non-UO systems)

In the case when observability depends on the inputs (systems which are not uniformly observable), the design will be restricted to some appropriate classes of inputs. Then the two possible cases of compensable or non compensable nonlinearities can again be considered.

<u>State affine systems</u>
Consider here a system of the form:

$$\begin{aligned} \dot{x}(t) &= A(u(t))x(t) + B(u(t)) \\ y(t) &= Cx(t) \end{aligned} \tag{1.18}$$

with $A(u(t))$ uniformly bounded.

Here the idea is that imposing the input function yields a linear time-varying system. Hence the following Kalman-like result holds [32, 15, 10]:

Theorem 7. *If u is regularly persistent for (1.18), then the system admits an observer of the form:*

$$\dot{\hat{x}}(t) = A(u(t))\hat{x}(t) + B(u(t)) - K(t)(C\hat{x}(t) - y(t))$$

with $K(t)$ given by:

$$\begin{aligned} \dot{M}(t) &= M(t)A^T(u(t)) + A(u(t))M(t) - M(t)C^TW^{-1}CM(t) + V + \delta M(t) \\ M(0) &= M^T(0) > 0,\ W = W^T > 0 \\ K(t) &= M(t)C^TW^{-1} \end{aligned}$$

with $\delta > 2\|A(u(t))\|$ or $V = V^T > 0$ as in LTV systems.

Remark 7
The convergence rate can be tuned by appropriate choice of δ or V.

This design can clearly be extended to systems which are affine in the unmeasured states, up to additive output nonlinearity, of the following form [29, 10]:

$$\begin{aligned} \dot{x}(t) &= A(u(t), Cx(t))x(t) + B(u(t), Cx(t)) \\ y(t) &= Cx(t) \end{aligned} \tag{1.19}$$

with $A(u(t), C\chi_u(t, x_0))$ bounded for any x_0.

Theorem 8. *If u is regularly persistent for (1.19), in the sense that it makes $v(t) := \begin{pmatrix} u(t) \\ C\chi_u(t, x_0) \end{pmatrix}$ regularly persistent for $\dot{x}(t) = A(v(t))x(t)$, $y(t) = Cx(t)$ for any x_0, then the system admits an observer of the form:*

$$\dot{\hat{x}}(t) = A(u(t), y(t))\hat{x}(t) + B(u(t), y(t)) - K(t)(C\hat{x}(t) - y(t))$$

with $K(t)$ given by:

$$\begin{aligned} \dot{M}(t) &= M(t)A^T(u(t), y(t)) + A(u(t), y(t))M(t) - M(t)C^TW^{-1}CM(t) \\ &\quad + V + \delta M(t) \\ M(0) &= M^T(0) > 0,\ W = W^T > 0 \\ K(t) &= M(t)C^TW^{-1} \end{aligned}$$

with $\delta > 2\|A(u(t), y(t))\|$ or $V = V^T > 0$.

State affine systems and additive triangular nonlinearity

Combining structure of system (1.18) or more generally (1.19) with that of system (1.17) leads to consider systems of the following form:

$$\begin{aligned}\dot{x} &= A_0(u,y)x + \varphi(x,u)\\ y &= C_0 x \quad \text{with}\\ A_0(u,y) &= \begin{pmatrix} 0 & a_{12}(u,y) & & 0\\ & & \ddots & \\ & & & a_{n-1n}(u,y)\\ 0 & & & 0\end{pmatrix} \; \textit{bounded}, \quad C_0 = (1\ 0 \cdots 0),\end{aligned} \tag{1.20}$$

and with φ as in theorem 6.

This means that the observer will need to rely on high gain, but for a non uniformly observable system. As a consequence, the observability property corresponding to observability for short times of proposition 3 will be here required, but parameterized by y as above:

Theorem 9. *If φ is globally Lipschitz w.r.t. x, uniformly w.r.t. u and such that:*

$$\frac{\partial \varphi_i}{\partial x_j}(x,u) = 0 \; for\, j \geq i+1, \quad 1 \leq i,j \leq n,$$

and u is locally regular for (1.17), in the sense that it makes $v(t) := \begin{pmatrix} u(t)\\ C\chi_u(t,x_0)\end{pmatrix}$ locally regular for $\dot{x}(t) = A(v(t))x(t)$, $y(t) = Cx(t)$ for any x_0, then the system admits an observer of the form:

$$\dot{\hat{x}} = A_0(u,y)\hat{x} + \varphi(\hat{x},u) - \begin{pmatrix} \lambda & & 0\\ & \ddots & \\ 0 & & \lambda^n\end{pmatrix} K_0(t)(C_0\hat{x} - y)$$

with $K_0(t)$ given by:

$$\begin{aligned}&\dot{M}(t) = \lambda(M(t)A^T(u(t),y(t)) + A(u(t),y(t))M(t) - M(t)C^T W^{-1} C M(t) + \delta M(t))\\ &M(0) = M^T(0) > 0,\ W = W^T > 0\\ &K(t) = M(t)C^T W^{-1}\end{aligned}$$

$\delta > 2\|A(u,y)\|$ and $\lambda = \frac{1}{T}$ large enough.

This can be established by showing that [4]:

(i) From local regularity assumption:

$$\exists \lambda > 0,\ \forall \lambda \geq \lambda_0,\ \forall t \geq \frac{1}{\lambda}, \quad 0 < \alpha_1 I \leq M^{-1}(t) \leq \alpha_2 I$$

for α_1, α_2 independent of λ.

(ii) $V(e,t) = e^T P(\lambda,t)e$ is a Lyapunov function for the observer error equation, exponentially decaying, with a rate of decay tunable by λ, where $e = \hat{x} - x$ and

$$P(\lambda,t) = \begin{pmatrix} \lambda & & 0\\ & \ddots & \\ 0 & & \lambda^n\end{pmatrix}^{-1} M^{-1}(t) \begin{pmatrix} \lambda & & 0\\ & \ddots & \\ 0 & & \lambda^n\end{pmatrix}^{-1}$$

Notice that this kind of design also holds for systems of the form (1.20) where the a_{ii+1}'s in A_0 are matrices instead of scalars [11].

Finally, notice again that we have presented observer designs in terms of UO systems (for which observers with constant gains have been given) and non UO systems (for which observers with varying gains have been presented). But obviously one could design an observer with a varying gain for UO systems, since in that case any input will satisfy the appropriate condition for the observer to work. Conversely, in some cases one can design an observer with a *constant* gain even if the system is not UO: this can be done provided the system satisfies some detectability property as mentioned before [12].

1.3.2 Advanced Designs

The presentation of possible observer designs in previous section has been restricted to very specific structures of systems. In this section are presented some ways to deal with nonlinear systems which do not a priori satisfy the structures previously presented.

Interconnection-Based Design

The first way to extend the class of systems for which an observer can be designed is to interconnect observers in order to design an observer for some interconnected system, when possible. If indeed a system is not under a form for which an observer is already available, but can be seen as an interconnection between several subsystems each of which would admit an observer if the states of the other subsystems were known, then a candidate observer for the interconnection of these subsystems is given by interconnecting available sub-observers (e.g. as in [13]). This is sketched by figure 1.4 below for the case of two subsystems.

As a simple example, consider the following system:

$$\begin{aligned} \dot{x}_1 &= x_2 \\ \dot{x}_2 &= u_1 \\ \dot{x}_3 &= x_4 + \varphi(x_2) \\ \dot{x}_4 &= u_2 \\ y &= \begin{pmatrix} x_1 \\ x_3 \end{pmatrix} \end{aligned} \tag{1.21}$$

Clearly here one can consider the system as the interconnection of the following two subsystems:

$$(\Sigma_1) \begin{cases} \dot{x}_1 = x_2 \\ \dot{x}_2 = u_1 \\ y_1 = x_1 \end{cases} \quad \text{and} \quad (\Sigma_2) \begin{cases} \dot{x}_3 = x_4 + \varphi(v) \\ \dot{x}_4 = u_2 \\ y = x_3 \end{cases}$$

where $v = x_2$ defines the interconnection.

It is also clear that (Σ_1) being linear and observable, it admits an observer (say O_1), as well as (Σ_2) whenever v is considered as a known input for (Σ_2) (let (say $O_2(v)$) denote the corresponding observer).

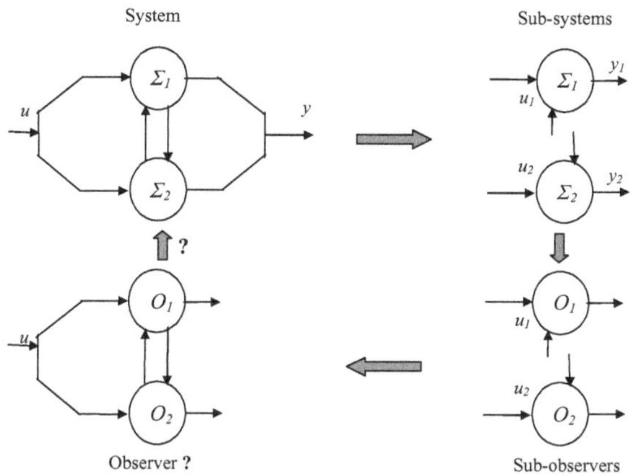

Fig. 1.4. Interconnection-based observer design

The idea is then to get an observer for the whole system from the interconnection (O_1)+$(O_2(\hat{x}_2))$ where $\hat{x}_2$ is provided by (O_1).

It can here be checked that for instance if φ is globally Lipschitz, $(O_1)+(O_2(\hat{x}_2))$ can indeed yield an observer.

Now if (Σ_2) is replaced by:

$$(\Sigma_2') \begin{cases} \dot{x}_3 = \varphi(x_2)x_4 \\ \dot{x}_4 = u_2 \\ y_2 = x_3 \end{cases}$$

it also results from previous section that an observer can be designed for (Σ_2') if x_2 is considered to be a known input, provided that this input is regularly persistent for (Σ_2'). If φ is globally Lipschitz, it can again be checked that this is enough for making it possible to get an observer for the whole system by interconnecting sub-observers (e.g. as in [10]).

This shows that under appropriate conditions separate possible designs can indeed yield some overall observer. But it does not go that well in any case. Consider for instance the following system:

$$\dot{x}_1 = -\frac{1}{2(t+1)}x_1;\ y_1 = 0 \tag{1.22}$$

$$\dot{x}_2 = -\frac{1}{4(t+1)}x_2 + x_1;\ y_2 = 0 \tag{1.23}$$

This system can be seen as an interconnection via x_1 between two subsystems respectively defined by (1.22) and (1.23). Clearly each of them admits an observer (here not tunable) as follows, as long as x_1 is assumed to be known for the second one:

$$\dot{\hat{x}}_1 = -\frac{1}{2(t+1)}\hat{x}_1;\ y_1 = 0 \tag{1.24}$$

$$\dot{\hat{x}}_2 = -\frac{1}{4(t+1)}\hat{x}_2 + x_1;\ y_2 = 0 \tag{1.25}$$

But if we inject $\hat{x}_1$ given by (1.24) into (1.25), one can check that the error equation is not stable.

This just illustrates the fact that in general, the stability of the interconnected observer is not guaranteed by that of each sub-observer, in the same way as separate designs of observer and controller do not in general result in some stable observer-based controller for nonlinear systems (no *separation principle*).

This means that the stability of interconnection of sub-observers requires a specific attention. Conditions can indeed be derived so as to guarantee a possible design by interconnection of separate subdesigns, either in the case of *cascade* interconnection as in the above examples, or even in the case of *full* interconnection [13].

Full interconnection

Let us first consider the general case of *full* interconnection, via the example of systems made of two subsystems for the sake of illustration, and described by the following representation:

$$(\Sigma)\begin{cases} \dot{x}_1 = f_1(x_1, x_2, u),\ u \subset U \subset I\!R^m;\ f_i\ \mathcal{C}^\infty \text{ function},\ i = 1,2; \\ \dot{x}_2 = f_2(x_2, x_1, u),\ x_i \in X_i \subset I\!R^{n_i},\ i = 1,2; \\ y\ = (h_1(x_1),\ h_2(x_2))^T = (y_1,\ y_2)^T,\ y_i \in I\!R^{n_i},\ i = 1,2. \end{cases} \tag{1.26}$$

Assume also that $u(.) \in \mathcal{U} \subset \mathcal{L}^\infty(I\!R^+, U)$, and set $\mathcal{X}_i := \mathcal{AC}(I\!R^+, I\!R^{n_i})$ the space of absolutely continuous function from $I\!R^+$ into $I\!R^{n_i}$. Finally, when $i \in \{1,2\}$, let $\bar{\imath}$ denote its complementary index in $\{1,2\}$.

The idea here is that system (1.26) can be seen as the interconnection of two subsystems (Σ_i) for $i = 1,2$ given by:

$$(\Sigma_i)\quad \dot{x}_i = f_i(x_i, v_{\bar{\imath}}, u),\quad y_i = h_i(x_i),\qquad (v_{\bar{\imath}}, u) \in \mathcal{X}_{\bar{\imath}} \times \mathcal{U}. \tag{1.27}$$

Assume that for each system (Σ_i), one can design an observer $(\mathcal{O}_i)$ of the following form:

$$(\mathcal{O}_i)\quad \dot{z}_i = f_i(z_i, v_{\bar{\imath}}, u) + k_i(g_i, z_i)(h_i(z_i) - y_i),\quad \dot{g}_i = G_i(z_i, v_{\bar{\imath}}, u, g_i), \tag{1.28}$$

for smooth k_i, G_i and $(z_i, g_i) \in (I\!R^{n_i} \times \mathcal{G}_i)$, $\mathcal{G}_i$ positively invariant by (1.28).

The point is to look for an observer for (1.26) under the form of the following interconnection:

$$(\mathcal{O})\begin{cases} \dot{\hat{x}}_i = f_i(\hat{x}_i, \hat{x}_{\bar{\imath}}, u) + k_i(\hat{g}_i, \hat{x}_i)(h_i(\hat{x}_i) - y_i);\ i = 1,2; \\ \dot{\hat{g}}_i = G_i(\hat{x}_i, \hat{x}_{\bar{\imath}}, u, \hat{g}_i);\ i = 1,2 \end{cases} \tag{1.29}$$

Set $e_i := z_i - x_i$, and for any $u \in \mathcal{U}, v_{\bar{\imath}} \in \mathcal{X}_i$ consider the following system (where $k_i^{v_{\bar{\imath}}}(t)$ denotes gain $k_i(g_i, z_i)$ defined in (1.28)) :

$$\mathcal{E}_i^{(u,v_{\bar{\imath}})} \begin{cases} \dot{e}_i = f_i(z_i, v_{\bar{\imath}}, u) - f_i(z_i - e_i, v_{\bar{\imath}}, u) + k_i^{v_{\bar{\imath}}}(t)(h_i(z_i) - h_i(z_i - e_i)) \\ \dot{z}_i = f_i(z_i, v_{\bar{\imath}}, u) + k_i^{v_{\bar{\imath}}}(t)(h_i(z_i) - h_i(z_i - e_i)) \\ \dot{g}_i = G_i(z_i, v_{\bar{\imath}}, u, g_i). \end{cases}$$

Then sufficient conditions for (1.29) to be an observer for (1.26) have been expressed in [13] as follows:

Theorem 10. *[13] If for $i = 1, 2$, any signal $u \in \mathcal{U}, v_{\bar{\imath}} \in AC(\mathbb{R}^+, \mathbb{R}^{n_{\bar{\imath}}})$, and any initial value $(z_i^0, g_i^0) \in \mathbb{R}^{n_i} \times \mathbb{G}_i$, $\exists V_i(t, e_i), W_i(e_i)$ positive definite functions such that:*

(i) $\forall x_i \in X_i; \forall e_i \in \mathbb{R}^{n_i}; \forall t \geq 0,$

$$\frac{\partial V_i}{\partial t}(t, e_i) + \frac{\partial V_i}{\partial e_i}(t, e_i)[f_i(x_i + e_i, v_{\bar{\imath}}(t), u(t)) - f_i(x_i, v_{\bar{\imath}}(t), u(t)) \\ + k_i^{v_{\bar{\imath}}}(t)(h_i(x_i + e_i) - h_i(x_i))] \leq -W_i(e_i)$$

(ii) $\exists \alpha_i > 0; \forall x_i \in X_i; \forall x_{\bar{\imath}} \in \mathbb{R}^{n_{\bar{\imath}}}; \forall e_i \in \mathbb{R}^{n_i}; \forall e_{\bar{\imath}} \in \mathbb{R}^{n_{\bar{\imath}}}; \forall t \geq 0,$

$$\left\| \frac{\partial V_i}{\partial e_i}(t, e_i)[f_i(x_i, x_{\bar{\imath}} + e_{\bar{\imath}}, u(t)) - f_i(x_i, x_{\bar{\imath}}, u(t))] \right\| \leq \alpha_i \sqrt{W_i(e_i)} \sqrt{W_{\bar{\imath}}(e_{\bar{\imath}})},$$

(iii) $\alpha_1 + \alpha_2 < 2,$

then (1.29) is an asymptotic observer for (1.26).

•

This can be established on the basis of Lyapunov arguments by appropriately combining V_1 and V_2. The result can be extended to more than two subsystems by using Lyapunov stability analysis of interconnected systems for instance as in [37].

As an example, this approach can yield observers for systems of the form:

$$\begin{aligned} \dot{x}_1 &= A_1 x_1 + f_1(x_1, x_2, u) \\ \dot{x}_2 &= A_2 x_2 + f_2(x_1, x_2, u) \\ y &= \begin{pmatrix} C_1 x_1 \\ C_2 x_2 \end{pmatrix} \end{aligned}$$

relying on high gain separate designs for x_1 and x_2 for instance, or even Kalman separate designs (in particular if $A_i = A_i(u)$ for some $i \in \{1, 2\}$ for instance) [6, 13].

Cascade interconnection

In the *weaker* case of *cascade* interconnection, namely when $f_1(x_1, x_2, u) = f_1(x_1, u)$ in (1.26), various results have been proposed for the stability of the interconnected system. Let us report here the weakened assumptions proposed in [13] in this context of observer design:

Theorem 11. *Assume that:*

I. System $\dot{x}_1 = f_1(x_1, u)$; $y_1 = h_1(x_1)$ admits an observer $(\mathcal{O}_1)$ as in (1.28) (without v_2), s.t. $\forall u \in \mathcal{U}$ and $\forall x_1(t)$ admissible trajectory of the system associated to u:

$$\lim_{t\to\infty} e_1(t) = 0 \text{ and } \int_0^{+\infty} \|e_1(t)\|dt < +\infty \quad (\text{with } e_1 := z_1 - x_1); \tag{1.30}$$

II. $\exists c > 0$; $\forall u \in U$; $\forall x_2 \in X_2$, $\|f_2(x_2, x_1, u) - f_2(x_2, x_1', u)\| \le c\|x_1 - x_1'\|$;

III. $\forall u \in \mathcal{U}, \forall v_1 \in \mathcal{AC}(I\!R^+, I\!R^{n_1})$, $\forall z_2^0, g_2^0$, $\exists v(t, e_2), w(e_2)$ positive definite functions s.t for every trajectory of $\mathcal{E}_2^{(u,v_1)}$ with $z_2(0) = z_2^0, g_2(0) = g_2^0$:

(i) $\forall x_2 \in X_2, e_2 \in I\!R^{n_2}, t \ge 0$,

$$\frac{\partial v}{\partial t}(t, e_2) + \frac{\partial v}{\partial e_2}(t, e_2)[f_2(x_2 + e_2, v_1(t), u(t)) - f_2(x_2, v_1(t), u(t)) + k_2^{v_1}(t)(h_2(x_2 + e_2) - h_2(x_2))] \le -w(e_2)$$

(ii) $\forall e_2 \in I\!R^{n_2}, t \ge 0$; $v(t, e_2) \ge \bar{w}(e_2)$

(iii) $\forall e_2 \in I\!R^{n_2}\backslash\mathcal{B}(0, r), t \ge 0$; $\left\|\frac{\partial v}{\partial e_2}(t, e_2(t))\right\| \le \lambda(1 + v(t, e_2(t)))$ for some constants $\lambda, r > 0$ and $\mathcal{B}(0, r) := \{e_2 : \|e_2\| \le r\}$.

Then:

$$\begin{aligned} \dot{\hat{x}}_1 &= f_1(\hat{x}_1, u) + k_1(\hat{g}_1, \hat{x}_1)(h_1(\hat{x}_1) - h_1(x_1)) \\ \dot{\hat{x}}_2 &= f_2(\hat{x}_1, \hat{x}_2, u) + k_2(\hat{g}_2, \hat{x}_2)(h_2(\hat{x}_1) - h_2(x_1)) \\ \dot{\hat{g}}_1 &= G_1(\hat{x}_1, u, \hat{g}_1); \\ \dot{\hat{g}}_2 &= G_2(\hat{x}_2, \hat{x}_1, u, \hat{g}_2). \end{aligned} \tag{1.31}$$

is an observer for (1.26) where $f_1(x_1, x_2, u) = f_1(x_1, u)$.

•

Once again this result can be established by Lyapunov analysis.

Notice that the here above proposed conditions might be modified by using more specific stability results for cascade systems.

A typical example of cascade observer design can be found in [10], where the Kalman-like design was extended to systems of the following form:

$$\begin{aligned} \dot{x}_1 &= A_1(u, y)x_1 + B_1(u, y) \\ \dot{x}_2 &= A_2(u, y, x_1)x_2 + B2(u, y, x_1) \\ y &= \begin{pmatrix} C_1x_1 \\ C_2x_2 \end{pmatrix} \end{aligned}$$

Those examples show how using available observers for systems in some particular forms, one might be able to design observers for further nonlinear systems. Next section proposes another way to do so.

Transformation-Based Design

Principle

The observer designs presented till now are still all based on particular structures of the system (either isolated or interconnected). The subsequent idea is that these designs can also give state observers for systems which can be turned into one of these forms by an appropriate transformation. The most commonapproach in that respect is to consider changes of state coordinates. Such a relationship defines some *system equivalence*:

Definition 14. *System equivalence [resp. at x_0]*
A system described by:

$$\begin{cases} \dot{x} = f(x,u) = f_u(x)\, x \in I\!R^n, u \in I\!R^m \\ y = h(x) \in I\!R^p \end{cases} \tag{1.32}$$

will be said to be **equivalent** *[resp.* **at x_0***] to the system:*

$$\begin{cases} \dot{z} = F(z,u) = F_u(z) \\ y = H(z) \end{cases} \tag{1.33}$$

if there exists a diffeomorphism $z = \Phi(x)$ defined on $I\!R^n$ [resp. some neighbourhood of x_0] such that:

$$\forall u \in I\!R^m, \quad \frac{\partial \Phi}{\partial x} f_u(x) \mid_{x=\Phi^{-1}(z)} = F_u(z) \qquad \text{and} \qquad h \circ \Phi^{-1} = H.$$

Systems (1.32) and (1.33) are then said to be equivalent by $z = \Phi(x)$.

The interest of such a property for observer design can then be illustrated by the following proposition (e.g. as in [6]):

Proposition 4. *Given two systems (Σ_1) and (Σ_2) respectively defined by:*
$(\Sigma_1)\begin{cases} \dot{x} = X(x,u) \\ y = h(x) \end{cases}$ *and* $(\Sigma_2)\begin{cases} \dot{z} = Z(z,u) \\ y = H(z) \end{cases}$
and equivalent by $z = \Phi(x)$,
If:

$$(\mathcal{O}_2)\begin{cases} \dot{\hat{z}} = Z(\hat{z},u) + k(w, H(\hat{z}) - y)) \\ \dot{w} = F(w,u,y) \end{cases}$$

is an observer for (Σ_2),
Then:

$$(\mathcal{O}_2)\begin{cases} \dot{\hat{x}} = X(\hat{x},u) + \left(\frac{\partial \Phi}{\partial x}\right)^{-1}_{|\hat{x}} k(w, h(\hat{x}) - y) \\ \dot{w} = F(w,u,y) \end{cases}$$

is an observer for (Σ_1).

From this indeed, if a system is not of an appropriate structure for an observer design in view of previous sections, but is equivalent to some other system which does have some appropriate structure, then the observer problem can be solved for the original system.

Examples

The idea of proposition 4 has motivated various works on characterizing systems which can be turned into some appropriate structures for observer design, from the linear one up to output injection [38, 14, 39] to several forms of cascade block state affine systems up to nonlinear injections from block to block as in (1.34) below for instance [30, 45, 10, 31, 8, 9, ...].

$$\left\{\begin{array}{l} \dot{z}_1 = A_1(u, y^1)z_1 + \varphi_1(u, y^1) \\ \dot{z}_2 = A_2(u, y^2, z_1)z_2 + \varphi_2(u, y^2, z_1) \\ \quad \vdots \\ \dot{z}_q = A_q(u, y^q, z_1, \dots z_{q-1})z_q + \varphi_q(u, y^q, \dots z_{q-1}) \\ y \;= \begin{pmatrix} C_1 z_1 \\ \vdots \\ C_q z_q \end{pmatrix} = \begin{pmatrix} y^1 \\ \vdots \\ y^q \end{pmatrix} \\ \qquad u \in I\!R^m, z_i \in I\!R^{n_i}, y^i \in I\!R^{\nu_i}, \end{array}\right. \tag{1.34}$$

As a simple illustrative example, let us consider here the problem of turning a nonlinear system:

$$\begin{array}{l} \dot{x} = f(x) \\ y = h(x), \;\; x \in I\!R^n \end{array}$$

into a linear observable form up to output injection as follows:

$$\begin{array}{l} \dot{x} = Ax + \varphi(Cx) \\ y = Cx \end{array}$$

Necessary and sufficient conditions for this problem to be solvable have been given in terms of differential geometry in [38].

A constructive algorithm to simultaneously check the possibility of the transformation and construct φ can alternatively be given in the spirit of [28] as follows:

1. Get the representation:

$$y^{(n)} = \Phi(y, \dot{y}, \dots y^{(n-1)})$$

   and set $z_1 := y$.
2. For $i \geq 1$, define φ_i by: $\frac{\partial \varphi_i}{\partial y} = \frac{\partial z_i^{(n-i+1)}}{\partial y^{(n-i)}}$;
   If φ_i is not only a function of y, the transformation fails and the procedure ends. Else, set: $z_{i+1} := \dot{z}_i - \varphi_i$
3. Continue until $i = n$ or the procedure aborts.

The procedure is clearly sufficient, and it can be checked that it is indeed necessary.

As a second simple example, turning some $n-$dimensional nonlinear control affine system into the appropriate structure for high gain observer design, if possible, is obtained by the following transformation [23, 24]:

$$z = \begin{pmatrix} h(x) \\ L_f h(x) \\ \vdots \\ L_f^{n-1} h(x) \end{pmatrix}$$

Finally, it can be underlined that some enlargement of the class of systems admitting an observer on the basis of the particular structures highlighted in the above presentation can also be obtained by further considering output transformations (e.g. as in [39, 28, 5]), or state extension (e.g. using immersion [22, 41, 35, 11]), for instance.

In particular, it has been shown in [5] that any control-affine system satisfying the observability rank condition can be turned into a form (1.20) for appropriate dimensions of the a_{ii+1}'s.

1.4 Conclusion

The purpose in this chapter was to give some overview on techniques of observer design for nonlinear systems. Clearly this presentation follows a particular viewpoint on the problem, and does not claim to be exhaustive. In particular the most important notions of observability (from this viewpoint) have been reviewed, and some observers have been presented according to two types of designs in that respect: uniform and non uniform ones w.r.t. input (or time). Those designs are in particular driven by specific structures of systems, and admit smooth explicit gains. Extensions of such designs to more general structures by interconnections and transformations have also been discussed. More details on some of the mentioned techniques can be found in the subsequent chapters - such as high gain designs, immersion-based results or optimization-based approaches. On the other hand, further comments on *detectability* and related designs have for instance been omitted, as well as various other technical approaches where the design is not necessarily smooth (as in sliding modes [21, 3, ...]), explicit (as in LMI-based designs [2, 44, ...]) or exact (as in many approximate approaches).

1.5 Appendix: Lyapunov Tools

The purpose of observers being asymptotic state reconstruction, an observer for a given system is to be an auxiliary system such that the error between the observer state and the system state asymptotically decays to zero, namely 0 is to be an asymptotically stable equilibrium for the error system.

From this, Lyapunov tools for stability analysis are instrumental in designing observers, and the purpose here is thus to recall the main results in that respect (as they can be found for instance in [37]).

In general the considered stability will be that of some nonautonomous system, namely a system of the form:

$$\dot{x}(t) = f(x(t), t) \tag{1.35}$$

such that $f(0,t) = 0$ for any $t \geq 0$, and where f is regular enough (at least piecewise continuous in t and locally Lipschitz in x on $D \times [0, \infty)$ where D is some state domain of $I\!R^n$ containing 0).

For such a system one can consider the following:

Definition 15. *Stability*
The equilibrium $x = 0$ of (1.35) will be said to be uniformly stable *if:*

$$\forall \epsilon > 0,\ \exists \delta > 0,\ independent\ of\ t_0 : \|x(t_0)\| < \delta \Rightarrow \|x(t)\| < \epsilon,\ \forall t \geq t_0 \geq 0.$$

The equilibrium is uniformly asymptotically stable *if it is uniformly stable and:*

$$\exists c > 0\ independent\ of\ t_0 : \forall \|x(t_0)\| < c,\ lim_{t\to\infty}\|x(t)\| = 0,\ uniformly\ in\ t_0,$$

namely:

$$\|x(t)\| \leq \beta(\|x(t_0)\|, t - t_0),\ \forall t \geq t_0 \geq 0, \forall \|x(t_0)\| < c$$

for a constant $c > 0$ independent of t_0, and a continuous function $\beta(r,s)$ vanishing at 0 and strictly increasing w.r.t. its first argument, and decreasing w.r.t. its second argument even going to zero at infinity (class $\mathcal{KL}$ function).

The equilibrium is globally uniformly asymptotically stable *if it satisfies the above inequality for any initial state $x(t_0)$.*

The equilibrium is exponentially stable *if it satisfies the above inequality with $\beta(r,s) = kre^{-\gamma s}$, $k, \gamma > 0$, and* globally exponentially stable *if this condition holds for any initial state.*

Then we can recall the following:

Theorem 12. *Let $V : D \times [0, \infty) \to I\!R$ be a C^1 function such that:*

1. $W_1(x) \leq V(x,t) \leq W_2(x)$
2. $\dfrac{\partial V}{\partial t} + \dfrac{\partial V}{\partial x} f(t,x) \leq -W_3(x)$

for $t \geq 0, x \in D$ and W_1, W_2, W_3 are continuous positive definite functions on D. Then the equilibrium $x = 0$ is uniformly asymptotically stable.

If the above conditions hold globally and in addition W_1 is radially unbounded *($W_1(x) \to \infty$ if $\|x\| \to \infty$), then $x = 0$ is* globally *uniformly asymptotically stable.*

If in fact:

$$W_i(x) \geq k_i\|x\|^c\ for\ i = 1, 3,\ \ W_2(x) \leq k_2\|x\|^c$$

for $k_1, k_2, k_3, c > 0$, in the above conditions, then then $x = 0$ is exponentially *stable.*

If those conditions hold globally, then then $x = 0$ is globally exponentially *stable.*

Notice that a function V satisfying the first inequality of item 1 above is called *proper*, and it is called *decrescent* if the second inequality holds.

References

[1] M. Alamir. Optimization-based nonlinear observers revisited. *Int. Journal of Control*, 72(13):1204–1217, 1999.
[2] M. Arcak and P. Kokotović. Nonlinear observers: a circle criterion design and robustness analysis. *Automatica*, 37(12):1923–1930, 2001.
[3] J.P. Barbot, T. Boukhobza, and M. Djemai. Sliding mode observer for triangular input form. In *Proc. 35th IEEE Conf. on Decision and Control, Kobe, Japan*, pages 1489–90, 1996.
[4] G. Besançon. Further results on high gain observers for nonlinear systems. In *Proc. 38th IEEE Conf. on Decision and Control, Phoenix, USA*, 1999.
[5] G. Besançon. On output transformations for state linearization up to output injections of nonlinear systems. *IEEE Trans. on Automatic Control*, 44(11):1975–1981, 1999.
[6] G. Besançon. A viewpoint on observability and observer design for nonlinear systems. In *New Directions in Nonlinear Observer Design - Lecture Notes in Control and Info. Sciences 244*, pages 2–21. H. Nijmeijer and T. Fossen Eds. Springer-Verlag, London, 1999.
[7] G. Besançon. Observer design for nonlinear systems. In *Advanced topics in control systems theory - Lecture Notes in Control and Info. Sciences 328*, pages 61–89. A. Loría, F. Lamnabhi-Lagarrigue and E. Panteley Eds, Springer-Verlag, London, 2006.
[8] G. Besançon and G. Bornard. State equivalence based observer design for nonlinear control systems. In *Proc. of IFAC World Congress, San Francisco, CA, USA*, pages 287–292, 1996.
[9] G. Besançon and G. Bornard. On charactering classes of observer forms for nonlinear systems. In *Proc. 4th European Control Conf., Brussels, Belgium*, 1997.
[10] G. Besançon, G. Bornard, and H. Hammouri. Observer synthesis for a class of nonlinear control systems. *European J. of Control*, 3(1):176–193, 1996.
[11] G. Besançon and A. Ţiclea. Immersion-based observer design for rank-observable nonlinear systems. *IEEE Trans. on Automatic Control*, 52(1):83–88, 2007.
[12] G. Besançon and H. Hammouri. On uniform observation of non-uniformly observable systems. *Systems & Control Letters*, 33(1):1–11, 1996.
[13] G. Besançon and H. Hammouri. On observer design for interconnected systems. *Journal of Math. Systems, Estimation and Control*, 8(3), 1998.
[14] D. Bestle and M. Zeitz. Canonical form observer design for nonlinear time-variable systems. *Int. Journal of Control*, 38(2):419–431, 1983.
[15] G. Bornard, F. Celle-Couenne, and G. Gilles. Observability and observers. In *Nonlinear Systems - T.1, Modeling and Estimation*, pages 173–216. Chapman & Hall, London., 1995.
[16] G. Bornard, N. Couenne, and F. Celle. Regularly persistent observer for bilinear systems. In *New Trends in Nonlinear Control Theory; Lecture Notes in Control and Info. Sci., 122*, pages 130–140. Springer-Verlag, Berlin, 1988.
[17] G. Bornard and H. Hammouri. A high gain observer for a class of uniformly observable systems. In *Proc. 30th IEEE Conf. on Decision and Control, Brighton, England*, pages 1494–1496, Dec 1991.
[18] G. Bornard and H. Hammouri. A graph approach to uniform observability of nonlinear multi output systems. In *40th IEEE Conf. on Decision and Control, Las Vegas, USA*, 2002.

[19] G. Conte, C. Moog, and AM. Perdon. *Nonlinear control systems: an algebraic setting.* Springer, Berlin, 1999.
[20] F. Deza, D. Bossanne, E. Busvelle, J.P. Gauthier, and D. Rakotopara. Exponential observers for nonlinear systems. *IEEE Trans. on Automatic Control*, 38(3):482–484, 1993.
[21] S. Drakunov and V. Utkin. Sliding mode observers. tutorial. In *Proc. 34th IEEE Conf. on Decision and Control, New Orleans, LA, USA*, pages 3376–8, 1995.
[22] M. Fliess and Kupka. A finiteness criterion for nonlinear input-output differential systems. *Siam Journal on Control and Optimization*, 21(5):721–728, 1983.
[23] J.P. Gauthier and G. Bornard. Observability for any $u(t)$ of a class of nonlinear systems. *IEEE Trans. on Automatic Control*, 26(4):922–926, 1981.
[24] J.P. Gauthier, H. Hammouri, and S. Othman. A simple observer for nonlinear systems - applications to bioreactors. *IEEE Trans. on Automatic Control*, 37(6):875–880, 1992.
[25] J.P. Gauthier and A.K. Kupka. Observability and observers for nonlinear systems. *Siam Journal on Control and Optimization*, 32(4):975–994, 1994.
[26] J.P. Gauthier and A.K. Kupka. *Deterministic observation theory and applications.* Cambridge Univ Press, 1997.
[27] A. Gelb. *Applied optimal estimation.* MIT Press, Cambridge, 1992.
[28] A. Glumineau, C. Moog, and F. Plestan. New algebraic-geometric conditions for the linearization by input-output injection. *IEEE Trans. on Automatic Control*, 41(4):598–603, 1996.
[29] H. Hammouri and F. Celle. Some results about nonlinear systems equivalence for the observer synthesis. In *New Trends in Systems Theory*, pages 332–339. Birkhäuser, 1991.
[30] H. Hammouri and J.P. Gauthier. Bilinearization up to output injection. *Systems & Control Letters*, 11:139–149, 1988.
[31] H. Hammouri and M. Kinnaert. A new formulation for time-varying linearization up to output injection. *Systems & Control Letters*, 28:151–157, 1996.
[32] H. Hammouri and J. De Leon Morales. Observer synthesis for state-affine systems. In *Proc. 29th IEEE Conf. on Decision and Control, Honolulu, HA, UA*, pages 784–785, 1990.
[33] R. Hermann and A.J. Krener. Nonlinear controllability and observability. *IEEE Trans. on Automatic Control*, 22(5):728–740, 1977.
[34] A. Isidori. *Nonlinear Control Systems.* Springer-Verlag, Berlin, 3rd edition, 1995.
[35] P. Jouan. Immersion of nonlinear systems into linear systems modulo output injection. *Siam Journal on Control and Optimization*, 41:1756–78, 2003.
[36] R.E. Kalman and R.S. Bucy. New results in linear filtering and prediction theory. *J. Basic Eng.*, 83:95–108, 1961.
[37] H. Khalil. *Nonlinear Systems.* Prentice Hall, 2nd edition, 1996.
[38] A. J. Krener and A. Isidori. Linearization by output injection and nonlinear observers. *Systems & Control Letters*, 3:47–52, 1983.
[39] A. J. Krener and W. Respondek. Nonlinear observers with linearizable error dynamics. *Siam Journal on Control and Optimization*, 23(2):197–216, 1985.
[40] H. Kwakernaak and R. Sivan. *Linear Optimal Control Systems.* Wiley-Interscience, New York, 1972.
[41] J. Lévine and R. marino. Nonlinear systems immersion, observers and finite dimensional filters. *Systems & Control Letters*, 7:137–142, 1986.

[42] D.G. Luenberger. Observers for multivariable systems. *IEEE Trans. on Automatic Control*, 11(2):190–197, 1966.
[43] H. Michalska and D.Q. Mayne. Moving horizon observers and observer-based control. *IEEE Trans. on Automatic Control*, 40(6):995–1006, 1995.
[44] J. Moreno. Observer design for nonlinear systems: a dissipative approach. In *IFAC Symp. Systems Structure and Control, Oaxaca, Mex.*, 2004.
[45] J. Rudolph and M. Zeitz. A block triangular nonlinear observer normal form. *Systems & Control Letters*, 23:1–8, 1994.
[46] H.J. Sussmann. Single input observability of continuous time systems. *Math. Systems Theory*, 12:371–393, 1979.
[47] G. Zimmer. State observation by on-line minimization. *Int. Journal of Control*, 60(4):595–606, 1994.

2

Uniform Observability and Observer Synthesis

Hassan Hammouri

LAGEP UMR CNRS 5007, Université Claude Bernard Lyon 1
hammouri@lagep.univ-lyon1.fr

2.1 Introduction

The single input observability is the practical observability notion that can be used for the state and parameter estimation. A system is single input observable if there exists an input which distinguishes any different initial states (see chapter 1). Such inputs are called universal inputs. For analytic systems the observability is equivalent to the single input observability (see [19]). For nonlinear systems, even if the system is single input observable, it may admit an input which renders it unobservable. However, for stationary linear systems, the single observability doesn't depend on the input and can be characterized using a Brunowsky canonical form [21]. The property that the single input observability doesn't depend on the input will be called the uniform observability. As for stationary linear systems, canonical forms can be designed in order to characterize some class of uniformly observable nonlinear systems.

In the observation context, a natural extension of stationary linear systems consists in considering linear systems up to output injection:

$$\begin{cases} \dot{x} = Ax + \varphi((u,y)) \\ y = Cx \end{cases} \tag{2.1}$$

where the state $x \in I\!R^n$, the known input $u \in I\!R^m$ and the measured output $y \in I\!R^p$.

Clearly, the observability of (C, A) is equivalent to the fact that system (2.1) is observable independently on the input.

Assume indeed that (C, A) is observable. Let u be any borelian input on some $[0, T]$ and x^0, $\overline{x}^0$ be two initial states. Assume now that the associated outputs $y(t) = y(x^0, u, t)$, $\overline{y}(t) = y(\overline{x}^0, u, t)$ are identically equal on $[0, T]$ and let us show that these initial states are the same.

Denote by $x(t)$, $\overline{x}(t)$ the associated trajectories and differentiating y, $\overline{y}$, we obtain $CAx(t) = CA\overline{x}(t)$, for every $t \in [0, T]$ (since $y(t) = \overline{y}(t)$. Differentiating $CAx(t)$ and $CA\overline{x}(t)$ and using the fact that $y(t) = \overline{y}(t)$, we deduce that $CA^2x(t) = CA^2\overline{x}(t)$. Repeating this argument until obtaining $CA^kx(t) = C\overline{A}^k\overline{x}(t)$, for $0 \leq k \leq n-1$ and using the observability of (C, A), we deduce that $x^0 = \overline{x}^0$. The same argument can be used to prove the converse.

G. Besançon (Ed.): Nonlinear Observers and Applications, LNCIS 363, pp. 35–69, 2007.
springerlink.com

An observer for systems (2.1) is a simple extension of the Luenberger observer:

$$\dot{\widehat{x}} = A\widehat{x} + \varphi((u, y)) + K(C\widehat{x} - y) \tag{2.2}$$

where K is any constant $n \times p$ constant matrix such that $A + KC$ is stable.

Based on this nice observability property and the fact that the observability is an intrinsic property (it doesn't depend on the system of coordinates), one can ask how we can transform a nonlinear system by a change of coordinates to systems of the form (2.1). This problem has been initiated by H. Krener and A. Isidori in [17] and extended to the multi-output systems in [18], [22]).

In what follows, we recall the result of [17].

Consider the single output non controlled nonlinear system:

$$\begin{cases} \dot{x} = f(x) \\ y = h(x) \end{cases} \tag{2.3}$$

The state $x(t) \in I\!R^n$, the known input $u(t) \in I\!R^m$ and the measured output $y(t) \in I\!R^p$.

In the single output case (p=1), we will recall the necessary and sufficient condition that systems (2.3) must satisfy in order to be transformed into the canonical form:

$$\begin{cases} \dot{x} = Ax + \varphi(y) \\ y = Cx \end{cases} \tag{2.4}$$

where, $A = \begin{pmatrix} 0 & 0 & & 0 \\ 1 & 0 & & 0 \\ \vdots & \ddots & & \vdots \\ 0 & \dots & 1 & 0 \end{pmatrix}$ and $C = (0, \dots, 0, 1)$

To do so, consider the family of vector fields $X_1, \dots, X_n$ defined by:

$$\begin{cases} L_{X_1}(L_f^k(h)) = 0, \quad \text{for } \ k = 0, \dots, n-1 \\ L_{X_1}(L_f^{n-1}(h)) = 1 \\ X_i = [X_{i-1}, f], \quad \text{for } \ i = 2, \dots, n \end{cases} \tag{2.5}$$

where L_{X_1} denotes the Lie derivative along the vector field X_1 and $[,]$ denotes the symbol of the Lie bracket operation.

Now, define the following transformation $\Phi = (\Phi_1, \dots, \Phi_n)$ by: $L_{X_i}(\Phi_j)(x) = \delta_i^j$, where δ_i^j is the symbol of Kronecker.

Theorem 2.1.1 *[17]*
Assuming that the system (2.3) is observable in the rank sense at some $x^0 \in I\!R^n$. A necessary and sufficient condition for which $z = \Phi(x)$ becomes a local system of coordinates around x^0 in which system (2.3) becomes of the form (2.4) is that the vector fields $X_1, \dots, X_n$ commute. Namely, $[X_i, X_j] = 0$, for every, i, j.

Proof 1. *Assuming that $X_1, \dots, X_n$ commute, then one can check that: $f = \sum_{i=1}^{n-1} \Phi_i(x) X_{i+1}$. Hence, system (2.3) becomes of the form (2.4).*

Since the relation (2.5) doesn't depend on the system of coordinates, to prove the converse, it suffices to check the commutation of vector fields $X_1, \ldots, X_n$ obtained from $f = Ax + \varphi(y)$.

In [18], [22]), the authors gave an extension of this result to the multi-output systems which can be transformed into the Brunowsky canonical form:

$$\begin{cases} \dot{x} = Ax + \varphi(y) \\ y = Cx \end{cases} \tag{2.6}$$

where, $A = \begin{pmatrix} 0 & 0 & & 0 \\ A_1 & 0 & & 0 \\ \vdots & \ddots & & \vdots \\ 0 & \ldots & A_p & 0 \end{pmatrix}$, $A_k = \begin{pmatrix} 0 & 0 & & 0 \\ 1 & 0 & & 0 \\ \vdots & \ddots & & \vdots \\ 0 & \ldots & 1 & 0 \end{pmatrix}$ is $n_k \times n_k$ matrix with $n_1 + \ldots + n_p = n$, and $C = \begin{pmatrix} C_1 & 0 & 0 \\ \vdots & \ddots & \vdots \\ 0 & \ldots & C_p \end{pmatrix}$, with $C_k = (0, \ldots, 1)$ a n_k vector.

The fact that we are only interested to nonlinear systems which can be transformed to a Brunowsky canonical form up to output injection is that every observable linear system up to output injection can be transformed by a linear change of coordinates to a Brunowsky canonical form. Clearly the class of these systems is very restricted. This is why, many authors try to obtain others classes of systems for which an observer can be designed. An extension of the above result consists to classify nonlinear systems which can be transformed to controlled state affine systems up to output injection. An important property of these systems is that a Kalman like observer can designed [10], [11] (see chapter 1). Contrarily to the extended Luenberger for linear systems up to output injection, the Kalman like observer for state affine systems converges only for inputs which render sufficiently observable the system. This comes from the fact that general observable state affine systems may admit inputs which render them unobservable. An important problem consists in asking if a nonlinear system observable for every input (uniformly observable system) admits an observer which converges for every bounded input. This problem is completely solved in the single output case (see [3], [4], [5], [6]) and many partial solutions have been obtained in the literature for the multi-output case (see for instance [1], [12]). In the following chapters, we will give these classes of canonical forms and the associated high gain observer construction.

2.2 Canonical Form and High Gain Observer : A Single Output Case

Consider nonlinear systems of the form:

$$\begin{cases} \dot{x} = f(x, u) \\ y = h(x) \end{cases} \tag{2.7}$$

where the state $x(t) \in I\!R^n$, the known input $u(t) \in I\!R^m$ and the measured output $y(t) \in I\!R^p$.

Recall that system (2.7) is uniformly observable (or observable independently on the iput), if for every input $u \in L^{\infty}([0,T], \mathbb{R}^m)$, where $T > 0$ is fixed, u is universal input. Namely, for every initial states x^0, $\overline{x}^0$, the associated outputs $y(x^O, u, t)$, $y(\overline{x}^O, u, t)$ are not identically equal on $[0, T(, x^0, \overline{x}^0, u)[$, where $T(, x^0, \overline{x}^0, u) \leq T$ is largest time such that the trajectories $x(t)$ and $\overline{x}(t)$ are will defined for every $t \in [0, T(, x^0, \overline{x}^0, u)[$.

Notice that if the linear part of systems (2.4), (2.6) is observable then system (2.4) is uniformly observable. Moreover an observer takes the form (2.5), (2.7) . This observer exponentially converges whenever the unknown trajectory $x(t)$ is defined for all $t \geq 0$. A sufficient condition which guarantees the completeness of the system (ie. the trajectories are defined on the whole $\mathbb{R}^+$) is that φ is a global Lipschitz function. Notice that the completeness is necessary for the existence of an observer which converges as $t \to \infty$.

In the following subsection, we will show that uniformly observable systems can be transformed into a canonical form. This canonical form extends the Brunowsky canonical form (2.4).

2.2.1 Observability Canonical Form for Uniformly Observable Systems

For the sake of simplicity, we consider the control affine nonlinear system:

$$\begin{cases} \dot{x} = f_0(x) + u_1 f_1 + \ldots + u_m f_m(x) \\ y = h(x) \end{cases} \tag{2.8}$$

where the state $x(t) \in \mathbb{R}^n$, the known input $u(t) \in \mathbb{R}^m$ and the measured output $y(t) \in \mathbb{R}$. The f_i's are assumed to be of class $\mathcal{C}^{\infty}$.

Given a function φ from $\mathbb{R}^n$ into $\mathbb{R}$ of class $\mathcal{C}^n$, the the Lie derivatives of φ along the vector f_0 are:

$L_{f_0}(\varphi) = \sum_{i=1}^{n} f_{0i} \dfrac{\partial \varphi}{\partial x_k}$. For $k = 1, \ldots, n$, $L_{f_0}^k(\varphi) = L_{f_0}(L_{f_0}^{k-1}(\varphi))$, with $L_{f_0}^0(\varphi) = \varphi$.

Denote by $\Phi(x) = \begin{pmatrix} \Phi_1(x) \\ \vdots \\ \Phi_n(x) \end{pmatrix}$ the transformation defined by:

$\Phi_k(x) = L_{f_0}^{k-1}(h)(x)$, for $k = 1, \ldots, n$.

Then the following theorem initially stated in [3] and reformulated in [4] with a simple new proof can be given:

Theorem 2.2.1. *If system (2.8) is uniformly observable, then there exists an open dense subset $\mathcal{M}$ of $\mathbb{R}^n$ such that for every $x^0 \in \mathcal{M}$, there exists a neighborhood V, such that the map Φ becomes a diffeomorphism from V into its range. Moreover, it transforms system (2.8) restricted to V into the following canonical form:*

$$\begin{cases} \dot{z} = Az + \psi_0(z) + \sum_{i=1}^{m} \psi_i(z) u_i \\ y = Cz \end{cases} \tag{2.9}$$

$$A = \begin{pmatrix} 0 & 1 & & 0 \\ \vdots & & \ddots & \\ & & & 1 \\ 0 & & \dots & 0 \end{pmatrix}, \ \psi_0(z) = \begin{pmatrix} 0 \\ \vdots \\ \psi_n(z) \end{pmatrix}, \ \psi_k(z) = \begin{pmatrix} \psi_{k1}(z_1) \\ \vdots \\ \psi_{kj}(z_1, \dots, z_j) \\ \vdots \\ \psi_{kn}(z) \end{pmatrix}, \ C =$$

$(1, 0, \dots, 0)$;

Conversely, if a system (2.8) can be transformed into the above canonical form using any diffeomorphism, then the system is uniformly observable on the domain of definition of the diffeomorphism.

Proof 2. *Assuming that system (2.8) is uniformly observable, then in particular $u = 0$ renders the system observable. Clearly, the property that ϕ is a local diffeomorphism is an open property. Namely the set of points of $I\!R^n$ for which $\frac{\partial \Phi}{\partial x}$ is of rank n is an open subset of $I\!R^n$. To show that this set is also a dense subset of $I\!R^n$, we will use the fact that the input $u = 0$ renders system (2.8) observable. Assume indeed that this set is not a dense one. Then there exists an open subset W of $I\!R^n$ such that for every $\xi \in W$, the rank of $\frac{\partial \Phi}{\partial x}(\xi)$ is lower than or equal to $n-1$. In particular there exists $k \leq n-1$ and differentiable functions $\alpha_1, \dots, \alpha_k$, such that $dL^j_{f_0}(h)(\xi) = \sum_{i=1}^k \alpha_i(\xi) dL^{i-1}_{f_0}(h)(\xi)$, for $1 \leq j \leq n$. Now taking any different initial states ξ^0, $\overline{\xi}^0$ such that $dL^j_{f_0}(h)(\xi^0) = dL^j_{f_0}(h)(\overline{\xi}^0)$, for $0 \leq j \leq k-1$ and consider the associated trajectories $\xi(t)$, $\overline{\xi}(t)$ of the non forced system ($u = 0$), it follows that $h(\xi(t)) = h(\overline{\xi}(t))$. Hence $u = 0$ renders system (2.8) unobservable. This contradicts the fact that the system is observable for every inputs.*

Now, let show that Φ transforms system (2.8) into the canonical form system (2.9).

Since the system is observable for every input $u_1, \dots, u_m$, then in particular, it is so for inputs of the form $(0, \dots, u_i, \dots, 0)$. Hence, it is enough to give the canonical form for the single input-single output uniformly observable system:

$$\begin{cases} \dot{x} = f_0(x) + u f_1(x) \\ y = h(x) \end{cases} \tag{2.10}$$

Let $x^0 \in I\!R^n$ such that Φ becomes a diffeomorphism from an open neighborhood of this point.

Set $z = \Phi(x)$, namely, $z_i = L^{i-1}_{f_0}(h)(x)$, $1 \leq i \leq n$. In this new system of coordinates, system (2.10) takes the form:

$$\begin{cases} \dot{z} = Az + \psi_0(z) + u\psi_1(z) \\ y = Cz \end{cases} \tag{2.11}$$

where A is the shift matrix defined above, $\psi_0(z) = \begin{pmatrix} 0 \\ \vdots \\ \psi_n(z) \end{pmatrix}$, $\psi_1(z) = \begin{pmatrix} \psi_{11}(z) \\ \vdots \\ \psi_{1j}(z) \\ \vdots \\ \psi_{1n}(z) \end{pmatrix}$,

$C = (1, 0, \ldots, 0)$.

Indeed, for $1 \le i \le n-1$,

$$\begin{cases} \dot{z}_i(t) = \dfrac{dL_{f_0}^{i-1}(h)}{dt}(x(t)) = L_{f_0}^{i}(h)(x(t)) + u(t)L_{f_1}(L_{f_0}^{i-1}(h))(x(t)) \\ = z_{i+1}(t) + u(t)\psi_{1i}(z(t)) \\ \text{where, } \quad \psi_{1i}(z) = L_{f_1}(L_{f_0}^{i-1}(h))(\Phi^{-1}(z)) \end{cases} \tag{2.12}$$

For $i = n$,

$$\begin{cases} \dot{z}_n(t) = \dfrac{dL_{f_0}^{n-1}(h)}{dt}(x(t) = L_{f_0}^{n}(h)(x(t)) + u(t)L_{f_1}(L_{f_0}^{n-1}(h))(x(t)) \\ = \psi_{0n}(z(t)) + u(t)\psi_{1n}(z(t)) \\ \text{where, } \quad \psi_{0n}(z) = L_{f_0}^{n}(\Phi^{-1}(z)) \quad \text{and} \quad \psi_{1n}(z) = L_{f_1}(L_{f_0}^{n-1}(h))(\Phi^{-1}(z)) \end{cases} \tag{2.13}$$

By construction, $z_1(t) = h(x(t))$. *Thus* $C = (1, 0, \ldots, 0)$.

Now, let us show that $\psi_{1i}(z) = \psi_{1i}(z_1, \ldots, z_i)$, *for* $1 \le i \le n$:
Assume that this is not the case, and let i_0 *be the smallest integer* $i_0 \le n-1$ *for which there exists* $j \ge i_0 + 1$ *and that* $\dfrac{\partial \psi_{1i_0}}{\partial x_j} \ne 0$. *In what follows, we will construct an input* $u_0(t)$ *which renders system (2.11) unobservable, and from the fact that the observability doesn't depend on the system of coordinates, we can conclude that* u_0 *renders system (2.8) unobservable.*

To do so, set $v(z, \overline{z}) = \dfrac{\overline{z}_{i_0+1} - z_{i_0+1}}{\psi_{1i}(z) - \psi_{1i}(\overline{z})}$. *From above, there exists* z^0, $\overline{z}^0$, *with* $z_i^0 = \overline{z}_i^0$, *for* $1 \le i \le n$; $i \ne j$ *and* $z_j^0 \ne \overline{z}_j^0$ *such that* $\psi_{1i}(z^0) - \psi_{1i}(\overline{z}^0) \ne 0$. *Thus there exist neighborhoods* V_0 *and* $\overline{V}_0$ *of* z^0 *and* $\overline{z}^0$ *respectively, such that for every* $(z, \overline{z}) \in V_0 \times \overline{V}_0$, $\psi_{1i}(z) - \psi_{1i}(\overline{z}) \ne 0$ *(in particular* $v(z, \overline{z})$ *is will defined on* $\Omega = V_0 \times \overline{V}_0$*).*

The candidate input u_0 *which renders system (2.11) unobservable will be constructed as follows.*

Consider the following system defined on Ω *by:*

$$\begin{cases} \dot{z} = Az + \psi_0(z) + v(z, \overline{z})\psi_1(z) \\ \dot{\overline{z}} = A\overline{z} + \psi_0(\overline{z}) + v(z, \overline{z})\psi_1(\overline{z}) \end{cases} \tag{2.14}$$

$(z^0, \overline{z}^0)$ *is as above, denote by* $(z^0(t), \overline{z}^0(t))$ *the solution of system (2.14) which issued from* $(z^0, \overline{z}^0)$ *at* $t = 0$. *Then on some interval (maybe small)* $[0, T]$, $(z^0(t), \overline{z}^0(t))$ *is will defined, the same is true for* $u_0(t) = v(z^0(t), \overline{z}^0(t))$. *In what follows, we will show that the outputs* $z_1^0(t)$, $\overline{z}_1^0(t)$ *associated to initial conditions* z^0 *and* $\overline{z}^0$ *are the same on* $[0, T]$, *which contradicts the fact that the system is uniformly observable.*

Set $e(t) = \overline{z}(t) - z(t)$, system (2.14) becomes equivalent to:

$$\begin{cases} \dot{z} = Az + \psi_0(z) + v(z,\overline{z})\psi_1(z) \\ \dot{e} = Ae + (\psi_0(z) - \psi_0(e-z) + v(z,e-z)(\psi_1(z) - \psi_1(e-z)) \end{cases} \tag{2.15}$$

Using the definition of v, one can remark that $e(0) = \overline{z}^0 - z^0$ is an equilibrium point of the time varying linear system:

$$\begin{aligned} \dot{e}(t) = Ae(t) &+ (\psi_0(z^0(t)) - \psi_0(e(t) - z^0(t)) \\ &+ v(z^0(t), e(t) - z^0(t))(\psi_1(z^0(t)) - \psi_1(e(t) - z^0(t))) \end{aligned} \tag{2.16}$$

Thus, for $e(0) = \overline{z}^0 - z^0$, we have $e_1(t) = 0$ on $[0,T]$, hence the outputs $z_1^0(t)$, $\overline{z}_1^0(t)$ are the same on $[0,T]$. Consequently, u_0 doesn't distinguish the different initial states z^0, $\overline{z}^0$. This is in contradiction with the fact that every input renders the system observable. Thus Φ transforms system (2.8) into the canonical form (2.9).

The converse is obvious. Indeed, since the observability doesn't depend the system of coordinates, it is enough to show that system (2.9) is uniformly observable.

The proof is straightforward. To do so, take any different initial state z^0, $\overline{z}^0$ and any $u \in L^\infty([0,T], \mathbb{R}^m))$ such that the associated trajectories $z^0(t)$, $\overline{z}^0(t)$ are well defined on $[0,T]$ and that the outputs $z_1^0(t)$, $\overline{z}_1^0(t)$ are identically equal on this interval, and let us show that $z^0 = \overline{z}^0$.

Differentiating these outputs yields that $z_2^0(t) = \overline{z}_2^0(t)$ and by repeating this processus, we get $z_k^0(t) = \overline{z}_k^0(t)$, for every k.

Let us illustrate this theorem by a simple academic example:

Example 1. *Consider the system*

$$\begin{cases} \dot{x_1} = x_1 + u \\ \dot{x_2} = -x_2 + ux_1 \\ y = x_1 + x_2 \end{cases} \tag{2.17}$$

The transformation Φ is given by:

$\Phi = \begin{pmatrix} x_1 + x_2 \\ x_1 - x_2 \end{pmatrix}$. *and then system (2.17) takes the form:*

$$\begin{cases} \dot{z_1} = z_2 + u(1 + \frac{u}{2}(z_1 + z_2)) \\ \dot{z_2} = z_1 + u(1 - \frac{u}{2}(z_1 + z_2)) \\ y = z_1 \end{cases} \tag{2.18}$$

Consequently system (2.17) cannot be put into the canonical form. Hence it is not observable for every input $u \in L^\infty([0,T], \mathbb{R})$.

One can also verify that the constant input $u = -2$ renders the system unobservable.

Based on the above canonical forms, an observer will be given in the following subsection. This observer possesses the property that its gain doesn't depend on the frequency of the signal, but only on the upper bound of the signal. This observer is called a high gain observer.

2.2.2 High Gain Observer Design

Consider a gain the observable canonical canonical form:

$$\begin{cases} \dot{z} = Az + \varphi(u, z) \\ y = Cz \\ z \in I\!R^n; \quad u \in I\!R^m \end{cases} \tag{2.19}$$

$$A = \begin{pmatrix} 0 & 1 & & 0 \\ \vdots & & \ddots & \\ & & & 1 \\ 0 & & \dots & 0 \end{pmatrix}; \quad C = (1, 0, \dots, 0) \tag{2.20}$$

$$\varphi(u, z) = \begin{pmatrix} \varphi_1(z_1, u) \\ \vdots \\ \varphi_k(z_1, \dots, z_j, u) \\ \vdots \\ \varphi_n(z, u) \end{pmatrix} \tag{2.21}$$

This canonical form extends this given in theorem 2.2.1 for systems (2.9) (here, $\varphi(u, x)$ is not necessary affine with respect to u).

In order to design our high gain observer, the following hypothesis will be required:

H) The above nonlinear function φ is a global Lipschitz function :
For all bounded subset of $I\!R^m$; $\exists c > 0, \forall z, z' \in I\!R^n$, we have $\|\varphi(z, u) - \varphi(z', u)\| \leq c\|z - z'\|$, where $\|.\|$ denotes the euclidian norm of $I\!R^n$.

Remark 2.2.1. *i) This hypothesis guarantees the completeness of the system (for every admissible control, all trajectories of the system are defined on the wall $I\!R^+$).*

ii) If the concerned trajectories of the system lie into a bounded subset Ω of $I\!R^n$, then we can prolong the nonlinear term φ to a global Lipschitz function $\widetilde{\varphi}$ outside B, so that trajectories of the new system coincide with those of the initial system (see the application of mechanical systems, at the end of this section).

Now, let $\theta > 0$ a parameter and set:

$$\Delta_\theta = \begin{pmatrix} \theta & 0 & & 0 \\ \vdots & \ddots & & \\ 0 & & \dots & \theta^n \end{pmatrix} \text{ and } K = \begin{pmatrix} k_1 \\ \vdots \\ k_n \end{pmatrix},$$

such that $A + KC = \begin{pmatrix} k_1 & 1 & & 0 \\ \vdots & & \ddots & \\ k_{n-1} & 0 & & 1 \\ k_n & 0 \dots\dots & & 0 \end{pmatrix}$ becomes a Hurwiz matrix.

Our candidate observer takes the following form:

$$\dot{\widehat{z}} = A\widehat{z} + \varphi(u, \widehat{z}) + \Delta_\theta K(C\widehat{z} - y) \tag{2.22}$$

where $y(t)$ is the measured output associated to the unknown state of system (2.19).

Theorem 2.2.2. *Under the hypothesis **H**), system (2.22) forms an exponential observer for system (2.19). More precisely, we have:*

Let U a compact subset of $\mathbb{R}^m$, then there exists a constant $\theta_0 > 0$ such that $\forall \theta > \theta_0$; $\exists \alpha > 0$; $\exists \beta > 0$; $\forall \widehat{z}(0)$, we have $\|\widehat{z}(t) - z(t)\| \leq \alpha e^{-\beta t}\|\widehat{z}(0) - z(0)\|$, where $z(t)$ is the unknown trajectory to be estimated.

Remark 2.2.2. *The convergence of the above observer can be arbitrary chosen. More precisely, β depends on the parameter θ and $\lim \beta(\theta) = +\infty$.*

Proof 3 *(of theorem 2.2.2)*
Set $e(t) = \Delta_\theta^{-1}(\widehat{z}(t) - z(t))$, to prove the theorem, we will show that $\|e(t)| \leq \lambda e^{-\gamma t}\|e(0)\|$ for some constant λ and γ which depend on θ.

Using the equation (2.19) and (2.22) and the definition of Δ_θ and e, we deduce:

$$\dot{e} = \theta(A + KC)e + \Delta_\theta^{-1}(\varphi(u, \widehat{z}) - \varphi(u, z)) \tag{2.23}$$

Set $\delta\varphi = \Delta_\theta^{-1}(\varphi(u, \widehat{z}) - \varphi(u, z))$ and denote by $\delta\varphi_i$ its ith component, we obtain:

$$|\delta\varphi_i| = |\frac{1}{\theta^i}(\varphi_i(u, \widehat{z}) - \varphi_i(u, z))| \leq c\frac{1}{\theta^i}\sqrt{\theta^2 e_1^2 + \ldots + \theta^{2i} e_i^2} \tag{2.24}$$

*where c is the Lipschitz constant (given by the above hypothesis **H**). This constant depends only on the upper bound of $\|u(t)\|$.*

Now taking $\theta \geq 1$, it follows that

$$|\delta\varphi_i| \leq c\|e\| \tag{2.25}$$

Since $A + KC$ is Hurwiz, there exists a symmetric positive definite matrix P such that:

$$(A + KC)TP + P(A + KC) = -I \tag{2.26}$$

where I is the identity matrix.

Set $V(e) = e^T Pe$, and using (2.23)-(2.26), a simple calculation gives:

$$\frac{d(V(e(t)))}{dt} = -\theta\|e(t)\|^2 + 2e^T(t)P\delta\varphi \tag{2.27}$$

Now from (2.25), it follows that:

$$\frac{d(V(e(t)))}{dt} \leq -(\theta + 2c\|P\|)\|e(t)\|^2 \tag{2.28}$$

Taking $\theta_0 > 2c\|P\| = \beta$, then for every $\theta > \theta_0$, it follows that $V(e(t)) \leq e^{-(\theta-\beta)t}V(e(0))$, which exponentially converges to 0. This ends the proof of the theorem.

Remark 2.2.3. *Notice that the triangular form given in (2.21) together with the Lipschitz condition play a capital role in the proof of the theorem.*

2.2.3 An Extension to a Simple Multi-output Canonical Form

In this subsection, we will give an extension of the above single output canonical form to a multi-output case. The considered class contains the model of mechanical systems in the case when the positions are measured.

$$\begin{cases} \dot{z} = Az + \varphi(u, z) \\ y = Cz \end{cases} \tag{2.29}$$

where $z = \begin{bmatrix} z_1 \\ z_2 \\ \cdots \\ z_q \end{bmatrix} \in R^n$; $z_k =\in R^p$ with $qp = n$, $u \in R^m$, $y = z_1$, it means that $C = (I_p, \ldots, 0)$ is $p \times n$ matrix, where I_p is $p \times p$ identity matrix.

$A = \begin{pmatrix} 0 & I_p & & 0 \\ \vdots & & \ddots & \\ & & & I_p \\ 0 & & \ldots & 0 \end{pmatrix}$ is a $n \times n$ matrix, and $\varphi = \begin{bmatrix} \varphi_1 \\ \varphi_2 \\ \cdots \\ \varphi_q \end{bmatrix}$; $\varphi_k(u, z) \in I\!R^p$.

$$\varphi_k(u, z) = \varphi_k(u, z_1, \ldots, z_k) \tag{2.30}$$

System (2.29) can be rewritten:

$$\begin{cases} \text{for } \ 1 \le k \le q-1, \\ \dot{z}_k = z_{k+1} + \varphi_k(u, z_1, \ldots, z_k) \\ \text{for } \ k = 1, \\ \dot{z}_q = \varphi_q(u, z) \end{cases} \tag{2.31}$$

Set $\Delta_\theta = \begin{pmatrix} \theta I_p & 0 & & 0 \\ \vdots & & \ddots & \\ 0 & & \ldots & \theta^p I_p \end{pmatrix}$

Using similar argument as for the single output case, an candidate observer for system (2.29) takes the form:

$$\dot{\widehat{z}} = A\widehat{z} + \varphi(u, \widehat{z}) + \Delta_\theta K(C\widehat{z} - y) \tag{2.32}$$

More precisely, we can state a similar theorem as theorem 2.2.2 above.

An application of this theorem can for instance be given by mechanical systems:

Many mechanical system have a mathematical model of the form:

$$\Gamma(\xi)\frac{d^2\xi}{dt^2} + \gamma(\xi, \dot{\xi}) + Bu = 0 \tag{2.33}$$

where $\xi \in I\!R^p$ is the position vector and $\dot{\xi}$ is the velocity vector of the system. $\Gamma(\xi)$ is the inertial matrix which is in general symmetric positive definite. $\gamma(\xi, \dot{\xi})$ is a vector which contains the Coriolis forces.

Set $z_1 = \xi$ and $\dot{\xi} = z_2$, if the position vector is measured, system (2.33) takes the form:

$$\begin{cases} \dot{z}_1 = z_2 \\ \dot{z}_2 = \varphi(u, z) \\ y = z_1 \end{cases} \tag{2.34}$$

where, $\varphi(u, z) = -\Gamma^{-1}(\xi)(\gamma(\xi, \dot{\xi}) + Bu)$.

Assuming that the state of the system belongs to a bounded set Ω. Let χ be a $\mathcal{C}^\infty$ which takes 1 on Ω and vanishes outside a bounded set containing this set. Then trajectories of system (2.35) coincide with the following:

$$\begin{cases} \dot{z}_1 = z_2 \\ \dot{z}_2 = \widetilde{\varphi}(u, z) \\ y = z_1 \end{cases} \tag{2.35}$$

where, $\widetilde{\varphi}(u, z) = \chi(z)\varphi(u, z)$.

This construction permits to render the nonlinearity a global Lipschitz one. Hence, one can estimate the state of the system by using the observer:

$$\dot{\widehat{z}} = A\widehat{z} + \widetilde{\varphi}(u, \widehat{z}) + \Delta_\theta K(C\widehat{z} - y) \tag{2.36}$$

where $A = \begin{pmatrix} 0 & I_p \\ 0 & 0 \end{pmatrix}$, $C = (I_p \;\; 0)$, $K = \begin{pmatrix} K_1 \\ K_2 \end{pmatrix}$, $A + KC = \begin{pmatrix} K_1 & I_p \\ K_2 & 0 \end{pmatrix}$ is Hurwiz.

In [5] the authors have extended the above canonical form (2.9) to single output uniformly observable systems which are not necessarily control affine:

$$\begin{cases} \dot{x} = f(u, x) \\ y = h(u, x) \end{cases} \tag{2.37}$$

where $x \in M$, $u \in U$ both M and U are analytic manifolds. f and h are analytic with respect (x, u). The extension to the multi-output case is stated in [1] and [12], see the following section.

2.3 High Gain Observer for a Multi-output Canonical Form

2.3.1 The Considered Class of Systems

We consider nonlinear systems which are equivalent by diffeomorphism to systems of the form:

$$\begin{cases} \dot{z} = Az + \varphi(u, z) \\ y = Cz \end{cases} \tag{2.38}$$

where $z = \begin{bmatrix} z_1 \\ z_2 \\ \cdots \\ z_p \end{bmatrix} \in R^n$; $z_k = \begin{bmatrix} z_{k1} \\ z_{k2} \\ \cdots \\ z_{kn_k} \end{bmatrix} \in R^{n_k}$ with $\sum_{i=1}^{p} n_i = n$, $u \in R^m$,

$y = \begin{bmatrix} y_1 \\ y_2 \\ \cdots \\ y_p \end{bmatrix} \in R^p$. Moreover:

$$A = \begin{bmatrix} A_1 & & \\ & \ddots & \\ & & A_p \end{bmatrix}, \; A_k = \begin{bmatrix} 0 & 1 & & 0 \\ \vdots & & \ddots & \\ 0 & \dots & 0 & 1 \\ 0 & \dots & 0 & 0 \end{bmatrix},$$

$$C = \begin{bmatrix} C_1 & & \\ & \ddots & \\ & & C_p \end{bmatrix}, \; C_k = [1, 0, \dots, 0] \text{ and } \varphi = \begin{bmatrix} \varphi_1 \\ \varphi_2 \\ \cdots \\ \varphi_p \end{bmatrix}; \; \varphi_k = \begin{bmatrix} \varphi_{k1} \\ \varphi_{k2} \\ \cdots \\ \varphi_{kn_k} \end{bmatrix}.$$

We assume the following :

A1) there exist two sets of integers $\{\sigma_1 \dots \sigma_p\}$, $\{\delta_1, ..., \delta_p\}$, with $\delta_k > 0$, $k = 1, \dots, p$, such that:
for $k, l = 1, \dots, p$; $i = 1, \dots, n_k$ and $j = 2, \dots, n_l$ we have :

$$\text{If } \; \frac{\partial \varphi_{ki}}{\partial z_{lj}}(u, z) \not\equiv 0 \; \text{ for some } \; j \geq 2 \; \text{ then } \; \sigma_i^k + \frac{\delta_k}{2} > \sigma_j^l - \frac{\delta_l}{2} \tag{2.39}$$

where $\sigma_i^k = \sigma_k + (i-1)\delta_k$. In particular, assumption A1) implies that :

$$\frac{\partial \varphi_{ki}}{\partial z_{kj}}(u, z) \equiv 0 \; \text{ for } \; j \geq i + 1 \; \text{ and } \; 1 \leq i \leq n_k - 1.$$

A2) the function φ is global Lipschitz with respect to z locally uniformly in u.

Before stating our main theorem, let us analyse the meaning of the condition given by assumption A1) more closely.
Set:

$$I = \Big\{ (i, k) \;\; ; 1 \leq i \leq n_k \; , 1 \leq k \leq p \quad \text{for which there exist} \\ (l, j), \; 1 \leq l \leq p, \; 2 \leq j \leq n_l, \; \text{such that } \frac{\partial \varphi_{ki}}{\partial z_{lj}}(u, z) \not\equiv 0 \Big\} \tag{2.40}$$

Let $(i, k) \in I$ and set also:

$$I(i, k) = \left\{ l; 1 \leq l \leq p \text{ for which } \; \exists j, 2 \leq j \leq n_l \text{ such that } \frac{\partial \varphi_{ki}}{\partial z_{lj}}(u, z) \not\equiv 0 \right\} \tag{2.41}$$

Notice that I and $I(i, k)$ may be empty.

Now, let $(i,k) \in I$, $l \in I(i,k)$ and set:

$$j(i,l,k) = max\left\{ j, 2 \leq j \leq n_l \text{ such that } \frac{\partial \varphi_{ki}}{\partial z_{lj}}(u,z) \not\equiv 0 \right\}$$

The above condition is equivalent to :

$$\forall (i,k) \in I; \quad \forall l \in I(i,k), \quad \sigma_i^k + \frac{\delta_k}{2} > \sigma_{j(i,l,k)}^l - \frac{\delta_l}{2} \tag{2.42}$$

or equivalently,

$$\sigma_k - \sigma_l + \frac{(2i-1)}{2}\delta_k - \frac{(2j(i,l,k)-3)}{2}\delta_l > 0. \tag{2.43}$$

Hence, assumption A1) holds if the linear problem:

$$MX > 0 \tag{2.44}$$

has a solution in $\mathbf{N}^{2p}$ where $X = \begin{bmatrix} \sigma \\ \delta \end{bmatrix}$, $\sigma = (\sigma_1, \dots, \sigma_p)^T$, $\delta = (\delta_1, \dots, \delta_p)^T$ and M is a constant matrix derived from (2.39).

Note that the linear problem (2.44) admits a solution in $\mathbf{N}^{2p}$, if and only if, it admits a solution in $(R_+^*)^{2p}$. Indeed, if (2.39) is satisfied for some positive real numbers $\sigma_1, .., \sigma_p; \delta_1, .., \delta_p$, then for every integer $N > 0$, we have:

$$N\sigma_k - N\sigma_l + \frac{(2i-1)}{2}N\delta_k - \frac{(2j(i,l,k)-3)}{2}N\delta_l > 0$$

Now, taking N sufficiently large, the set of integer parts $([N\sigma_1], \cdots, [N\sigma_p]; [N\delta_1], \cdots, [N\delta_p])$ forms a solution for linear program (2.44).

Consequently, a solution of the linear problem (2.44) can obtained in two stages: First, we calculate a solution of (2.44) in $(R_+^*)^{2p}$ by using a linear programming technic (see for instance the simplex algorithm). Next, we consider an integer N such that the set of integers $([N\sigma_1], \cdots, [N\sigma_p]; [N\delta_1], \cdots, [N\delta_p])$, becomes a solution of (2.44).

Example: consider the following system

$$\left\{ \begin{array}{l} \dot{z_{11}} = z_{12} + \varphi_{11}(u, z_{11}, z_{21}, z_{22}) \\ \dot{z_{12}} = z_{13} + \varphi_{12}(u, z_{11}, z_{12}, z_{21}, z_{22}) \\ \dot{z_{13}} = \varphi_{13}(u, z_{11}, z_{12}, z_{13}, z_{21}, z_{22}) \\ \dot{z_{21}} = z_{22} + \varphi_{21}(u, z_{11}, z_{21}) \\ \dot{z_{22}} = z_{23} + \varphi_{22}(u, z_{11}, z_{12}, z_{21}, z_{22}) \\ \dot{z_{23}} = \varphi_{23}(u, z_{11}, z_{12}, z_{13}, z_{21}, z_{22}, z_{23}) \\ y = \begin{bmatrix} z_{11} \\ z_{21} \end{bmatrix} \end{array} \right. \tag{2.45}$$

For this system, one can chose $\sigma_1 = \sigma_2 = 1$ and $\delta_1 = 3$, $\delta_2 = 2$.

2.3.2 A High Gain Observer

Consider the dynamical system :

$$\dot{\hat{z}} = A\hat{z} + \varphi(u, \underline{\hat{z}}) - S_{\Theta}^{-1}C^T(C\hat{z} - y) \tag{2.46}$$

with :
i)

$$\begin{aligned} \underline{\hat{z}}_{k1} &= y_k \quad \text{for} \quad k = 1, \ldots, p \quad (\textit{output injection}) \\ \underline{\hat{z}}_{ki} &= \hat{z}_{ki}, \quad \text{for } i \neq 1 \end{aligned}$$

ii) u and y are the known output and input of the system (2.38) respectively

iii) $S_{\Theta} = \begin{bmatrix} S_{\theta^{\delta_1}} & & \\ & \ddots & \\ & & S_{\theta^{\delta_p}} \end{bmatrix}$ is a block diagonal matrix with $S_{\theta^{\delta_k}}$, $k = 1, \ldots, p$, the unique solution of :

$$\theta^{\delta_k} S_{\theta^{\delta_k}} + A_k^T S_{\theta^{\delta_k}} + S_{\theta^{\delta_k}} A_k = C_k^T C_k. \tag{2.47}$$

We then state the following :

Theorem 2.3.1. *Assume that system (2.38) satisfies assumptions A1)-A2), then:* $\forall M > 0$; $\exists \theta_0 > 0$; $\forall \theta \geq \theta_0$; $\exists \lambda_\theta > 0$; $\exists \mu_\theta > 0$ *such that* $\|\hat{z}(t) - z(t)\|^2 \leq \lambda_\theta e^{-\mu_\theta t}\|\hat{z}(0) - z(0)\|^2$ *for every admissible control* u *s.t.* $\|u(t)\| \leq M$, $\forall t \geq 0$. *Moreover,* $\lim\limits_{\theta \to +\infty} \mu_\theta = +\infty$. *This means that system (11) is an exponential observer for system (2.38) which works for bounded inputs.*

Proof 4. *First of all, it can be shown that the explicit solution of equation (12) is given by :*

$$S_{\theta^{\delta_k}}(i, j) = \frac{(-1)^{i+j} C_{i+j-2}^{j-1}}{\theta^{\delta_k(i+j-1)}} \quad \textit{for} \quad 1 \leq i, j \leq n_k \quad \textit{where} \quad C_n^p = \frac{n!}{(n-p)!p!}.$$

Moreover, $S_{\theta^{\delta_k}}$ *is symmetric positive definite for every* $\theta > 0$ *(see [4]).*
A simple algebraic computation also gives :

$$S_{\theta^{\delta_k}} = \frac{1}{\theta^{\delta_k}} \Delta_k(\theta) S_{1k} \Delta_k(\theta) \tag{2.48}$$

where $S_{1k} = S_{\theta^{\delta_k}}|_{\theta=1}$ *and*

$$\Delta_k(\theta) = \begin{bmatrix} 1 & & & \\ & \frac{1}{\theta^{\delta_k}} & & \\ & & \ddots & \\ & & & \frac{1}{\theta^{\delta_k(n_k-1)}} \end{bmatrix}, \qquad k = 1, \ldots p.$$

Set $e(t) = \hat{z}(t) - z(t)$. *Then the error equation is given by :*

$$\dot{e} = \left(A - S_{\Theta}^{-1}C^TC\right)e + \varphi(u,\hat{\underline{z}}) - \varphi(u,z).$$

where u *is an admissible control such that* $\|u\|_\infty \leq M$, $M > 0$ *is a given constant. In particular for the k'th subsystem, we have:*

$$\dot{e_k} = \left(A_k - S_{\theta^{\delta_k}}^{-1}C_k^TC_k\right)e_k + \varphi_k(u,\hat{\underline{z}}) - \varphi_k(u,z).$$

Now set $\Lambda_k(\theta) = \begin{bmatrix} \frac{1}{\theta^{\sigma_1^k}} & & \\ & \ddots & \\ & & \frac{1}{\theta^{\sigma_{n_k}^k}} \end{bmatrix}$ *for* $k = 1, \ldots p$ *where* σ_i^k *are the integers defined in A1).*

Then, the following equalities hold :

- $\Lambda_k(\theta)A_k\Lambda_k^{-1}(\theta) = \theta^{\delta_k}A_k$
- $\Lambda_k(\theta)\Delta_k^{-1}(\theta) = \theta^{-\sigma_1^k}I_k$ *(I_k is the $n_k \times n_k$ identity matrix)*
- $C_k\Lambda_k^{-1}(\theta) = \theta^{\sigma_1^k}C_k$
- $C_k\Delta_k^{-1}(\theta) = C_k$

Set $\bar{e}_k = \Lambda_k(\theta)e_k$ *for* $k = 1, \ldots, p$. *Using the above equalities, we get :*

$$\begin{aligned}
\dot{\bar{e}}_k &= \Lambda_k(\theta)\left(A_k - S_{\theta^{\delta_k}}^{-1}C_k^TC_k\right)\Lambda_k^{-1}(\theta)\bar{e}_k + \Lambda_k(\theta)\left(\varphi_k(u,\hat{\underline{z}}) - \varphi_k(u,z)\right) \\
&= \theta^{\delta_k}\left(A_k - S_{1k}^{-1}C_k^TC_k\right)\bar{e}_k + \Lambda_k(\theta)\left(\varphi_k(u,\hat{\underline{z}}) - \varphi_k(u,z)\right)
\end{aligned}$$

Consider the function $V(\bar{e}) = \bar{e}^TS\bar{e} = \sum_{i=1}^{p} V_i(\bar{e}_i)$ *where* $V_i(\bar{e}_i) = \bar{e}_i^TS_{1i}\bar{e}_i$ *and* $S = diag\left(S_{11}, \ldots, S_{1p}\right)$.

We have :

$$\begin{aligned}
\dot{V_k} &= \dot{\bar{e}}_k^TS_{1k}\bar{e}_k + \bar{e}_k^TS_{1k}\dot{\bar{e}}_k \\
&= 2\theta^{\delta_k}\bar{e}_k^TS_{1k}\left(A_k - S_{1k}^{-1}C_k^TC_k\right)\bar{e}_k + 2\bar{e}_k^TS_{1k}\Lambda_k(\theta)\left(\varphi_k(u,\hat{\underline{z}}) - \varphi_k(u,z)\right) \\
&= -\theta^{\delta_k}\left(\bar{e}_k^TS_{1k}\bar{e}_k + \bar{e}_k^TC_k^TC_k\bar{e}_k\right) + 2\bar{e}_k^TS_{1k}\Lambda_k(\theta)\left(\varphi_k(u,\hat{\underline{z}}) - \varphi_k(u,z)\right)
\end{aligned}$$

Therefore,

$$\begin{aligned}
\dot{V_k} &\leq -\theta^{\delta_k}V_k + 2\bar{e}_k^TS_{1k}\Lambda_k(\theta)\left(\varphi_k(u,\hat{\underline{z}}) - \varphi_k(u,z)\right) \\
&\leq -\theta^{\delta_k}V_k + 2\|S_{1k}\bar{e}\|\|\Lambda_k(\theta)\left(\varphi_k(u,\hat{\underline{z}}) - \varphi_k(u,z)\right)\| \\
&\leq -\theta^{\delta_k}V_k + 2\sqrt{\lambda_{max}^k}\sqrt{V_k}\sum_{i=1}^{n_k}\frac{1}{\theta^{\sigma_i^k}}|\varphi_{ki}(u,\hat{\underline{z}}) - \varphi_{ki}(u,z)|
\end{aligned}$$

where λ_{max}^k *is the maximum eigenvalue of* S_{1k}.

Now,

$$\dot{V}_k \le -\theta^{\delta_k} V_k + 2\rho_k\sqrt{\lambda^k_{max}}\sqrt{V_k}\sum_{i=1}^{n_k}\sum_{l=1}^{p}\sum_{j=1}^{n_l}\chi_{i,j}\theta^{\sigma^l_j-\sigma^k_i}|\bar{e}_{lj}|$$

where $\rho_k = \sup\left\{\frac{\partial\varphi_{ki}}{\partial z}(u,z); z\in R^n \text{ and } \|u\|_\infty \le M\right\}$ *and* $\chi_{i,j}=0$ *if* $\frac{\partial\varphi_{ki}}{\partial z_{lj}}(u,z)\equiv 0$, $\chi_{i,j}=1$ *otherwise.*

Therefore,

$$\dot{V}_k \le -\theta^{\delta_k} V_k + 2\rho_k\sqrt{\lambda^k_{max}}\sqrt{V_k}\sum_{i=1}^{n_k}\sum_{l=1}^{p}\sum_{j=1}^{n_l}\chi_{i,j}\theta^{\sigma^l_j-\sigma^k_i}\|\bar{e}_l\|$$

$$\le -\theta^{\delta_k} V_k + 2\rho_k\sqrt{\lambda^k_{max}}\sqrt{V_k}\sum_{i=1}^{n_k}\sum_{l=1}^{p}\sum_{j=1}^{n_l}\chi_{i,j}\theta^{\sigma^l_j-\sigma^k_i}\frac{\sqrt{V_l}}{\sqrt{\lambda^l_{min}}}$$

where λ^l_{min} *is the minimum eigenvalue of* S_{1l}.

Thus,

$$\dot{V}_k \le -\left(\sqrt{\theta^{\delta_k}V_k}\right)^2 + 2\rho_k\mu_S\sqrt{\theta^{\delta_k}V_k}\sum_{i=1}^{n_k}\sum_{l=1}^{p}\sum_{j=1}^{n_l}\chi_{i,j}\theta^{\sigma^l_j-\sigma^k_i-\delta_k/2-\delta_l/2}\sqrt{\theta^{\delta_l}V_l}$$

where μ_S *is the conditioning number of* S *(i.e. the square root of the ratio of the maximum and the minimum eigenvalue of* S*).*

Now, according to A1) there exists $\epsilon_k > 0$ *such that :*

$$\sigma^l_j - \sigma^k_i - \frac{\delta_k}{2} - \frac{\delta_l}{2} \le -\epsilon_k.$$

Then, assuming $\theta \ge 1$*, we have* $\theta^{\sigma^l_j-\sigma^k_i-\delta_k/2-\delta_l/2} \le 1$.

Therefore,

$$\dot{V}_k \le -\left(\sqrt{\theta^{\delta_k}V_k}\right)^2 + 2\rho_k\mu_S\sqrt{\theta^{\delta_k}V_k}\sum_{i=1}^{n_k}\sum_{l=1}^{p}\sum_{j=1}^{n_l}\theta^{-\epsilon_k}\sqrt{\theta^{\delta_l}V_l}$$

$$\le -\left(\sqrt{\theta^{\delta_k}V_k}\right)^2 + 2n_k\rho_k\mu_S\theta^{-\epsilon}\sqrt{\theta^{\delta_k}V_k}\sum_{l=1}^{p}\sum_{j=1}^{n_l}\sqrt{\theta^{\delta_l}V_l}$$

where ϵ *is the minimum of the* ϵ_k*'s.*

Now set $V^\star_k = \theta^{\delta_k}V_k$ *for* $k = 1, \dots, p$ *and* $V^\star = \sum_{k=1}^{p} V^\star_k$. *Notice that* $\theta^{\delta_{min}}V \le V^\star \le \theta^{\delta_{max}}V$, *where* δ_{max}, δ_{min} *is the maximum respectively the minimum of the* δ_k*'s.*

Then,

$$\dot{V}_k \le -V^\star_k + 2n_k\rho_k\mu_S\theta^{-\epsilon}\sqrt{V^\star_k}\sum_{l=1}^{p}\sum_{j=1}^{n_l}\sqrt{V^\star_l}$$

$$\le -V^\star_k + 2n_k n\rho_k\mu_S\theta^{-\epsilon}\sqrt{V^\star_k}\sqrt{V^\star}$$

$$\le -V^\star_k + 2n_k n\rho_k\mu_S\theta^{-\epsilon}V^\star$$

Hence,

$$\begin{aligned}\dot{V} &\leq -V^\star + 2n^2 \rho \mu_S \theta^{-\epsilon} V^\star \\ &\leq -\left(1 - 2n^2 \rho \mu_S \theta^{-\epsilon}\right) V^\star\end{aligned}$$

where $\rho = \max\{\rho_k, 1 \leq k \leq p\}$.

Finally,

$$\dot{V} \leq -\theta^{\delta_{min}} \left(1 - 2n^2 \rho \mu_S \theta^{-\epsilon}\right) V$$

Now, choosing θ_0 *such that* $1 - 2n^2 \rho \mu_S \theta_0^{-\epsilon} > 0$, *we obtain :*

$$\forall \theta \geq \theta_0; \quad \bar{e}(t)^T S \bar{e}(t) \leq \exp(-\mu_\theta t) \bar{e}(0)^T S \bar{e}(0)$$

where $\mu_\theta = \theta^{\delta_{min}} \left(1 - 2n^2 \rho \mu_S \theta^{-\epsilon}\right)$.

Otherwise,

$\|\bar{e}(t)\|^2 \leq \mu_S^2 \exp(-\mu_\theta t) \|\bar{e}(0)\|^2$ *and consequently* $\|e(t)\|^2 \leq \lambda_\theta \exp(-\mu_\theta t) \|e(0)\|^2$ *where* $\lambda_\theta = \mu_S^2 \dfrac{c_1^2(\theta)}{c_0^2(\theta)}$ *with* $c_1(\theta) = \max\left\{\dfrac{1}{\theta^{\sigma_k}}, 1 \leq k \leq p\right\}$ *and* $c_2 = \min\left\{\dfrac{1}{\theta^{\sigma_{n_k}^k}}, 1 \leq k \leq p\right\}$. *This ends the proof of Theorem 2.3.1.*

We have discussed observers for multi-output nonlinear systems which are observable for every input. Under adequate structure conditions, it is possible to deal with arbitrary nonlinearities. The lack of canonical observability forms for multi-output systems is a difficulty, and the condition given is coordinate-dependent. Another difficulty is due to the computation of the coordinate changes that are required first, before applying the results presented. The observer structure presented in this section extends those given in the above section. Nevertheless, the canonical form discussed in this section doesn't cover all the uniformly observable system. In the following section, we will give another structure which allows to broaden the class of uniformly observable systems proposed above.

2.4 Uniformly Observable Structure and Observer Synthesis

2.4.1 Some Observability Concepts and Related Results

Different notions of observability have been presented in [13]. More recent concepts of observability can be found in [6]. The purpose of this section is to establish the definitions of some classical as well as relatively new concepts of observability. We therefore recall some theoretical results and implications associated to these concepts.

Consider the MIMO nonlinear system:

$$\begin{cases} \dot{x} = f(u, x) \\ y = h(x) \end{cases} \tag{2.49}$$

$x(t) \in M$, a n-dimensional manifold; $u(t) \in U$, a borelian set of $\mathbb{R}^m$; u and y are the known input and output of (2.49) respectively.

In this section, system (2.49) is assumed to be smooth. This means that there exists an open set $\tilde{U}$ containing U such that:

$$f : \tilde{U} \times M \longrightarrow TM \quad \text{and} \quad h : M \longrightarrow \mathbb{R}^p$$

are of class $\mathcal{C}^\infty$.

For every fixed $u \in U$, $f_u : M \longrightarrow TM$ denotes the vector field defined by $f_u(x) = f(u,x)$ and the map $h = (h_1, \ldots, h_p)$ is an almost everywhere local submersion. It means that, $Rank(\frac{\partial h}{\partial x}(x)) = p$ for almost every x.

• Some well-known observability notions

Let $u \in L^\infty([0,T], U)$, $x \in M$ an initial state and $x^u(\cdot)$ the trajectory associated to the initial state x and to the input u. This trajectory is well-defined on the maximal interval $[0, T(x,u)[\subset [0,T]$. When $T(u,x) < T$, $T(u,x)$ is called the positive escape time. In such a case, $T(u,x)$ has the following property: for every sequence $(t_n)_{n \geq 0}$ s.t. $\lim_{n \to +\infty} t_n = T(u,x)$, the set $\{x^u(t_n), n \geq 0\}$ has no accumulation point.

System (2.49) is said to be **observable** if for every pair of different initial states $x, \bar{x}$, there exists an input $u \in L^\infty([0,T], U)$ s.t. $h(x^u(\cdot))$ is not identically equal to $h(\bar{x}^u(\cdot))$ on $[0, T(x,\bar{x},u)[$ where $T(x,\bar{x},u) = min\{T(u,x), T(u,\bar{x})\}$. We say that such an input **distinguishes** the considered initial states x, $\bar{x}$ on $[0,T]$.

An input which distinguishes every pair of different initial states on $[0,T]$ is called a **universal** input on $[0,T]$. A non universal input is called a **singular** input on $[0,T]$. Notice that unlike linear systems, observable nonlinear systems may admit singular inputs. Obviously, a system which admits a universal input is observable. The converse is also true in the analytical case (it means U, M, f and h are analytic). The proof of this result is given in [19].

Now, denote by $\mathcal{O}$ the smallest vector space containing $h_1, \ldots, h_p$ and closed under the Lie derivatives L_{f_u}, $u \in U$ (i.e. $\forall u \in U; \forall \tau \in \mathcal{O}, L_{f_u}(\tau) \in \mathcal{O}$). This vector space is the classical observation space. Let $\tilde{\mathcal{O}}$ be the co-distribution spanned by $\{d\tau, \tau \in \mathcal{O}\}$, system (1) is said to be **rank observable at** $x \in M$ if $dim\tilde{\mathcal{O}}(x) = n$. It is said to be **rank observable** if $\forall x \in M, dim\tilde{\mathcal{O}}(x) = n$. This rank observability condition is related to the concept of local weak observability notion (see for instance [13] for more details and precise definitions).

• Uniform observability concepts

The following definitions and results are useful for the characterization of systems of the form (2.49) which are observable independently on the input.

Definition 2.4.1. *Let E be any borelian subset of U, system (2.49) is said to be:*

(i) E-**uniformly observable** *iff for every $T > 0$ and every $u \in L^\infty([0,T],E)$, u is a universal input on $[0,T]$.*
(ii) **locally** E-**uniformly observable** *iff every $x \in M$ admits an open neighborhood V_x s.t. system (2.49) restricted to V_x is E-uniformly observable.*
(iii) **locally** E-**uniformly observable almost everywhere** *iff there exists an open dense subset M' of M s.t. the restriction of system (2.49) to M' is locally E-uniformly observable.*

For single output control affine systems if the system is locally $\mathbb{R}^m$-uniformly observable, then locally almost everywhere it can be steered by the local change of coordinates to a canonical form (2.9) (see theorem 2.2.1 of subsection 2.2.1 of section 2.2.

In the single output case, this canonical form as well as its associated high gain observer have been extended in [5] to single output analytic systems of the form:

$$\begin{cases} \dot{x} = f(u,x) \\ y = h(u,x) \end{cases} \tag{2.50}$$

To do so, the authors used the uniform infinitesimal observability concept:

Consider the tangent map $Tf_u : TM \longrightarrow T(TM)$ associated to $f_u : M \longrightarrow TM$, for every $u \in U$. The family of vector fields $(Tf_u)_{u \in U}$, defines, in a unique sense, a lifted system on TM :

$$\dot{\xi} = T_M f_u(\xi)$$

Finally the lifted system associated to system (2.50) is given by:

$$\begin{cases} \dot{\xi} = T_M f_u(\xi) \\ \tilde{y} = d_M h(\xi, u) \end{cases} \tag{2.51}$$

where $d_M h(\cdot, u)$ is the classical differential map from $TM \longrightarrow \mathbb{R}$.

When $M = \mathbb{R}^n$,TM can be identified with $\mathbb{R}^n \times \mathbb{R}^n$ and $\xi = (x,z)$. Thus system (2.51) takes the form:

$$\begin{cases} \dot{x} = f(u,x) \\ \dot{z} = \dfrac{\partial f}{\partial x}(u,x).z \\ \tilde{y} = \dfrac{\partial h}{\partial x}(u,x).z \end{cases} \tag{2.52}$$

Definition 2.4.2 *[5]. Let $u \in L^\infty([0,T],U)$ and $x \in M$*

i) System (2.50) is said to be **infinitesimally observable** *at (u,x) if the linear map:*

$$T_xM \longrightarrow L^\infty\left([0,T(u,x)[, I\!R\right) \quad \left(\xi \longrightarrow d_M h\left(u(\cdot),\xi^u(\cdot)\right)\right)$$

is one to one.

ii) System (2.50) is called **uniformly infinitesimally observable** *iff for every $T > 0$; for every $(u,x) \in L^\infty([0,T],U) \times M$, system (2.50) is infinitesimally observable at (u,x).*

The following result is stated in [5] (Theorem 3.1):

Theorem 2.4.1 *[5]. Assume that the single output system (2.50) is analytic and uniformly infinitesimally observable and that either one of the following conditions holds:*

(i) *U is a compact connected analytic manifold.*
(ii) *$U = I\!R^m$ and f, h are polynomial in u.*

Then, there exists a subanalytic (resp. semi-analytic in the case of (ii)) subset M' of codimension 1 in M such that system (2.50) is locally everywhere diffeomorphic to the triangular canonical form:

$$\begin{cases} \dot{z}^1 = F^1(u,z^1,z^2) \\ \dot{z}^2 = F^2(u,z^1,z^2,z^3) \\ \quad \cdots \\ \dot{z}^i = F^i(u,z^1,\ldots,z^{i+1}) \\ \quad \cdots \\ \dot{z}^n = F^n(u,z^1,\ldots,z^n) \\ y = H(u,z^1) \end{cases} \tag{2.53}$$

with

$$\frac{\partial H}{\partial z^1}(u,z) \neq 0 \;\; \textit{and} \;\; \frac{\partial F^i}{\partial z^{i+1}}(u,z) \neq 0; \; \forall (u,z) \in U\times V, \; \textit{and}\, i = 1,\ldots,n-1 \tag{2.54}$$

where V is the domain in which the local transformation takes its values.

As in subsection 2.2 an observer for single output systems (2.53) can be obtained. The structure of the observer takes the following form:

$$\dot{\hat{z}} = F(u,\hat{z}) + K(C\hat{z} - y) \tag{2.55}$$

In the following subsections 2.4.3 and 2.4.4, we will extend the above observer design (2.55) to a class of MIMO nonlinear systems which generalizes systems (2.53). The class of nonlinear systems which can be steered by a change of coordinates to such a canonical form will be characterized in subsection 2.4.5.

2.4.2 Preliminary

The canonical form that we consider has the following triangular structure:

$$\begin{cases} \dot{z} = F(u,z) \\ y = Cz \end{cases} \tag{2.56}$$

where $F(u,z) = \begin{pmatrix} F^1(u,z) \\ \vdots \\ F^q(u,z) \end{pmatrix}$, $z = \begin{pmatrix} z^1 \\ \vdots \\ z^q \end{pmatrix}$; $u \in U$ a compact submanifold of $I\!R^m$; $z^i \in I\!R^{n_i}$; $n_1 \geq n_2 \geq \ldots \geq n_q$; $n_1 + \ldots + n_q = n$. Each function $F^i(u,z)$, $i = 1, \ldots, q-1$ satisfies the following structure:

$$F^i(u,z) = F^i(u, z^1, \ldots, z^{i+1}), \; z^i \in I\!R^{n_i} \tag{2.57}$$

with the following rank condition:

$$Rank(\frac{\partial F^i}{\partial z^{i+1}}(u,z)) = n_{i+1} \quad \forall z \in I\!R^n; \forall u \in U \tag{2.58}$$

In section 4, we will show that condition (2.58) characterizes a subclass of locally U-uniformly observable systems.

Definition 2.4.3. *A constant gain exponential observer for system (2.56) is a dynamical system of the form:*

$$\dot{\hat{z}} = F(u, \hat{z}) + K(C\hat{z} - y) \tag{2.59}$$

where K is a constant matrix such that:

$$\|\hat{z}(t) - z(t)\| \leq \lambda e^{-\mu t} \|\hat{z}(0) - z(0)\|$$

where $\lambda > 0$ and $\mu > 0$ are constants which do not depend on the input $u \in L_\infty(I\!R^+, U)$ nor on $\hat{z}(0), z(0)$.

In the following subsections, we will give two observer constructions. First, we give a sufficient condition allowing to design a constant gain exponential observer for system (2.56). Next, we propose an observer construction for general systems of the form (2.56)-(2.57)-(2.58).

2.4.3 Constant Gain Exponential Observer

Consider again system (2.56) where the inputs $u(t)$ take their values in some Borelian and bounded subset of $I\!R^m$. As in many works related to high gain observer synthesis, we need the following assumption :

H1) Global Lipschitz condition:

$\exists c > 0; \forall u \in U; \forall z, z' \in I\!R^n$, $\|F(u,z) - F(u,z')\| \leq c\|z - z'\|$.

Notice that such assumption can be omitted in the case where the state of the system lies into a bounded set (this remark is formulated in many papers concerning the high gain observers, see for instance [4]).

Now, let $p_1 \geq p_2$ be two positive integers and denote by $\mathbb{M}(p_1, p_2; I\!R)$ the space of $p_1 \times p_2$ real matrices. Let $N \in \mathbb{M}(p_1, p_2; I\!R)$ with $rank(N) = p_2$ and consider the convex cone of $\mathbb{M}(p_1, p_2; I\!R)$ given by $\mathcal{C}(p_1, p_2; \alpha; N) = \{M \in \mathbb{M}(p_1, p_2; I\!R);\ s.t.\ M^T N + N^T M < \alpha I_{p_2}\}$ where α is a constant real number and I_{p_2} is the $p_2 \times p_2$ identity matrix.

Theorem 2.4.2. *Assume that assumption H_1) holds. Then, a sufficient condition for the existence of a constant gain exponential observer for system (2.56)-(2.57)-(2.58) is:*

$$\left\{\begin{array}{l} \text{For every } k, 1 \leq k \leq q-1, \text{ there exists a } n_k \times n_{k+1} \text{ constant matrix } S_{k,k+1} \\ \text{such that:} \dfrac{\partial F^k}{\partial z^{k+1}}(u,z) \in \mathcal{C}(n_k, n_{k+1}; -1; S_{k,k+1});\ \text{for every } (u,z) \in U \times I\!R^n \end{array}\right. \tag{2.60}$$

$n_1 \geq n_2 \geq \ldots \geq n_q$, n_i is the dimension of z^i-space.

Remark 2.4.1. *In the single output case, condition (2.54) is equivalent to condition (2.60) of Theorem 2.4.2.*

The proof of the theorem requires the following proposition:

Proposition 2.4.1. *Assume that **H1)** and (2.60) hold. Then, there exists a $n \times n$ S.P.D. matrix P satisfying the following condition:*

There exist $\rho > 0$; $\eta > 0$ such that for every $(u,z) \in U \times I\!R^n$, we have :

$$PA(u,z) + A^T(u,z)P - \rho C^T C \leq -\eta I \tag{2.61}$$

where $A(u,z) = \begin{pmatrix} 0 & A_1(u,z) & 0 & \ldots & 0 \\ \vdots & 0 & A_2(u,z) & 0 \ \ldots & \vdots \\ \vdots & \ldots & \ddots & \ddots \ \ddots & \vdots \\ \vdots & \ldots & \ldots & \ddots \ \ddots & 0 \\ \vdots & \ldots & \ldots & \ddots \ \ddots & A_{q-1}(u,z) \\ 0 & \ldots & \ldots & \ldots\ \ldots & 0 \end{pmatrix}$ *with* $A_k(u,z) = \dfrac{\partial F^k}{\partial z^{k+1}}(u,z)$

Proof of Proposition 2.4.1. Set $\Gamma_k = \{\dfrac{\partial F^k}{\partial z^{k+1}}(u,z);\ (u,z) \in U \times I\!R^n\}$. From condition (2.60), we can choose matrices $S_{k,k+1}$,$1 \leq k \leq q-1$, such that:

$$\forall M_k \in \Gamma_k,\ S^T_{k,k+1} M_k + M_k^T S_{k,k+1} < -I_{n_{k+1}}$$

Now consider the following symmetric bloc tridiagonal matrix:

$$P = \begin{pmatrix} P_{11} & P_{12} & 0 & \dots & \dots & 0 \\ P_{12}^T & P_{22} & P_{23} & 0 & \dots & 0 \\ 0 & \ddots & \ddots & \ddots & \ddots & \vdots \\ \vdots & \ddots & \ddots & \ddots & \ddots & 0 \\ 0 & \dots & 0 & P_{q-2,q-1}^T & P_{q-1,q-1} & P_{q-1,q} \\ 0 & \dots & 0 & 0 & P_{q-1,q}^T & P_{q,q} \end{pmatrix} \tag{2.62}$$

where $P_{k,k+1} = \rho_{k+1} S_{k,k+1}$ and P_{kk} is a $n_k \times n_k$ S.D.P. matrix, ρ_{k+1} and P_{kk} will be specified below.

In the sequel, if M is a $k \times l$ matrix, we denote by $\|M\|$ the subordinate $\| \|_2$-norm: $\|M\| = \sup\limits_{\|\xi\|=1} \|M\xi\|$, where $\|\xi\|$ and $\|M\xi\|$ are the L_2-norms.

Let $\sigma_k = \lambda_{max}(P_{kk})$ and $\widehat{\sigma_k} = \lambda_{min}(P_{kk})$ be the respective largest and smallest eigenvalues of P_{kk}. From hypothesis **H1)**, we know that Γ_k is a bounded subset of $M(n_k, n_{k+1}, I\!R)$. Set $m_k = sup\{\|M_k\|;\ M_k \in \Gamma_k\}$ and choose P such that:

(i) $4\rho_{k+1}^2 \|S_{k,k+1}\|^2 < \widehat{\sigma}_k \widehat{\sigma}_{k+1}$, for $1 \le k \le q-1$
(ii) $4\sigma_k^2 m_k^2 < \rho_k \rho_{k+1}$, for $1 \le k \le q-1$
(iii) $4\rho_{k+1}^2 m_{k+1}^2 \|S_{k,k+1}\|^2 < \rho_k \rho_{k+2}$, for $1 \le k \le q-2$

To obtain such a matrix, it is enough to choose $P_{kk} = \sigma_k I_k$ ($\sigma_k = \widehat{\sigma}_k$), and numbers ρ_k such that $\rho_k << \sigma_k << \rho_{k+1} << \sigma_{k+1}$, where the notation $a << b$ means that $\frac{a}{b}$ is sufficiently small.

Before proving inequality (2.61) of Proposition 2.4.1, let us show that P is a S.D.P. matrix. Indeed, let $x \in I\!R^n$, $x \neq 0$, a simple calculation shows that:

$$\begin{aligned} x^T P x &= \sum_1^q ({x^k}^T P_{kk} x^k) + 2 \sum_1^{q-1} ({x^k}^T P_{k,k+1} x^{k+1}) \\ &= \frac{1}{2} {x^1}^T P_{11} x^1 + \frac{1}{2} {x^q}^T P_{qq} x^q + \\ &\quad \frac{1}{2} \sum_1^{q-1} ({x^k}^T P_{kk} x^k + 4 {x^k}^T P_{k,k+1}^T x^{k+1} + {x^{k+1}}^T P_{k+1,k+1} x^{k+1}) \\ &\ge \frac{\widehat{\sigma}_1}{2} \|x^1\|^2 + \frac{\widehat{\sigma}_q}{2} \|x^q\|^2 + \\ &\quad \frac{1}{2} \sum_1^{q-1} (\widehat{\sigma}_k \|x^k\|^2 - 4\rho_{k+1} \|S_{k,k+1}\| \|x^k\| \|x^{k+1}\| + \widehat{\sigma}_{k+1} \|x^{k+1}\|^2) \end{aligned}$$

Using condition (i) above, we get $x^T P x > 0$.

Now let us show inequality (2.61) of Proposition 2.4.1. Set $A(u,z) = A$, a simple computation gives:

$$
\begin{aligned}
& x^T(PA + A^TP - \rho_1 C^TC)x \\
= & -\rho_1\|x^1\|^2 + 2{x^1}^T P_{11}A_1x^2 + 2{x^1}^T P_{12}A_2x^3 \\
+ & {x^2}^T(P_{12}^T A_1 + A_1^T P_{12})x^2 + 2{x^2}^T P_{22}A_2x^3 + 2{x^2}^T P_{23}A_3x^4 \\
+ & \dots \\
+ & {x^{q-2}}^T(P_{q-3,q-2}^T A_{q-3} + A_{q-3}^T P_{q-3,q-2})x^{q-2} \\
+ & 2{x^{q-2}}^T P_{q-2,q-2}A_{q-2}x^{q-1} + 2{x^{q-2}}^T P_{q-2,q-1}A_{q-1}x^q \\
+ & {x^{q-1}}^T(P_{q-2,q-1}^T A_{q-2} + A_{q-2}^T P_{q-2,q-1})x^{q-1} \\
+ & 2{x^{q-1}}^T P_{q-1,q-1}A_{q-1}x^q + {x^q}^T(P_{q-1,q}^T A_{q-1} + A_{q-1}^T P_{q-1,q})x^q \\
= & \frac{1}{2}\left\{-\rho_1\|x^1\|^2 + 4{x^1}^T P_{11}A_1x^2 + {x^2}^T(P_{12}^T A_1 + A_1^T P_{12})x^2\right\} \\
+ & \frac{1}{2}\left\{-\rho_1\|x^1\|^2 + 4{x^1}^T P_{12}A_2x^3 + {x^3}^T(P_{23^T}A_2 + A_2^T P_{23})x^3\right\} \\
+ & \frac{1}{2}\left\{{x^2}^T(P_{12}^T A_1 + A_1^T P_{12})x^2 + 4{x^2}^T P_{22}A_2x^3 + {x^3}^T(P_{23}^T A_2 + A_2^T P_{23})x^3\right\} \\
+ & \dots
\end{aligned}
\tag{2.63}
$$

$$
\begin{aligned}
+ & \dots \\
+ & \frac{1}{2}\left\{{x^{q-2}}^T(P_{q-3,q-2}^T A_{q-3} + A_{q-3}^T P_{q-3,q-2})x^{q-2} + 4{x^{q-2}}^T P_{q-2,q-2}A_{q-2}x^{q-1}\right. \\
+ & \quad \left.{x^{q-1}}^T(P_{q-2,q-1}^T A_{q-2} + A_{q-2}^T P_{q-2,q-1})x^{q-1}\right\} \\
+ & \frac{1}{2}\left\{{x^{q-2}}^T(P_{q-3,q-2}^T A_{q-3} + A_{q-3}^T P_{q-3,q-2})x^{q-2}\right. \\
+ & \left. 4{x^{q-2}}^T P_{q-2,q-1}A_{q-1}x^q + {x^q}^T(P_{q-1,q}^T A_{q-1} + A_{q-1}^T P_{q-1,q})x^q\right\} \\
+ & \frac{1}{2}\left\{{x^{q-1}}^T(P_{q-2,q-1}^T A_{q-2} + A_{q-2}^T P_{q-2,q-1})x^{q-1}\right. \\
+ & \left. 4{x^{q-1}}^T P_{q-1,q-1}A_{q-1}x^q + {x^q}^T(P_{q-1,q}^T A_{q-1} + A_{q-1}^T P_{q-1,q})x^q\right\} \\
\le & \frac{1}{2}\sum_1^{q-1}\{-\rho_k\|x^k\|^2 + 4\|x^k\|\|x^{k+1}\|\sigma_k m_k - \rho_{k+1}\|x^{k+1}\|^2\} \\
+ & \frac{1}{2}\sum_1^{q-2}\{-\rho_k\|x^k\|^2 + 4\|x^k\|\|x^{k+2}\|\rho_{k+1}m_{k+1}\|S_{k,k+1}\| - \rho_{k+2}\|x^{k+2}\|^2\}
\end{aligned}
\tag{2.64}
$$

Using conditions (ii) and (iii) above, we obtain:

$x^T(PA + A^TP - \rho_1 C^TC)x \le -\eta I$, where $\eta > 0$ is a constant.

This ends the proof of the proposition.

Proof of Theorem 2.4.2. We assume that hypothesis **H1)** and condition (2.60) of Theorem 2.4.2 hold, and we will construct a constant matrix K, such that the following system:

$$\dot{\hat{z}} = F(u, \hat{z}) - \Delta_\theta K(C\hat{z} - y) \tag{2.65}$$

is a constant gain exponential observer, where $\Delta_\theta = \begin{pmatrix} \theta I_{n_1} & 0 & \dots & 0 \\ 0 & \theta^2 I_{n_2} & 0 & \vdots \\ \vdots & & \ddots & 0 \\ 0 & \dots & 0 & \theta^q I_{n_q} \end{pmatrix}$, I_{n_k} is the $n_k \times n_k$ identity matrix, $k = 1, \dots, q$ and $\theta > 0$ is a constant real number.

Let P be the S.D.P. matrix given by (2.62) and set $K = P^{-1}C^T$. We will show that for θ sufficiently large, $\hat{z}(t) - z(t)$ exponentially converges to 0.

As in [4, 5] and many other references related to high gain observer synthesis, consider the change of coordinates $\hat{\overline{z}} = \Delta_\theta^{-1}\hat{z}$, $\overline{z} = \Delta_\theta^{-1}z$, and set $\varepsilon = \hat{\overline{z}} - \overline{z}$. Let us show that $\varepsilon(t)$ exponentially converges to 0, for θ sufficiently large.

Set $\delta F = \begin{pmatrix} \delta F^1 \\ \vdots \\ \delta F^q \end{pmatrix}$, where

$\delta F^i = F^i(u, \hat{z}^1, \dots, \hat{z}^{i-1}, z^{i+1}) - F^i(u, z^1, \dots, z^{i-1}, z^{i+1})$ for $1 \leq i \leq q-1$ and $\delta F^q = F^q(u, \hat{z}) - F^q(u, z)$.

A simple calculation gives :

$$\dot{\varepsilon}(t) = \theta \left(A(t) - \rho P^{-1}C^TC\right)\varepsilon + \Delta_\theta^{-1}\delta F \tag{2.66}$$

$$A(t) = \begin{pmatrix} 0 & \frac{\partial F^1}{\partial z^2}(u, \hat{z}^1, \xi^2) & 0 & & \dots & 0 \\ \vdots & 0 & \frac{\partial F^2}{\partial z^3}(u, \hat{z}^1, \hat{z}^2, \xi^3) & 0 & \dots & \vdots \\ \vdots & \dots & \ddots & & \ddots & \vdots \\ \vdots & \dots & \dots & & \ddots & 0 \\ \vdots & \dots & \dots & & \ddots & \frac{\partial F^{q-1}}{\partial z^q}(u, \hat{z}^1, \dots, \hat{z}^{q-1}, \xi^q) \\ 0 & \dots & \dots & & \dots\dots & 0 \end{pmatrix}$$

with $\xi^i = \hat{z}^{i+1} + \omega_i(\hat{z}^{i+1} - z^{i+1})$ and ω_i is a diagonal matrix whose elements are in $[0,\ 1]$.

To show the exponential convergence to zero of $\varepsilon(t)$, it is enough to show:

$$\frac{d}{dt}(\varepsilon^T(t)P\varepsilon(t)) \leq -\alpha\varepsilon^T(t)P\varepsilon(t) \tag{2.67}$$

for some constant $\alpha > 0$.

Set $V(t) = \varepsilon^T(t)P\varepsilon(t)$, we obtain:

$$\dot{V}(t) = -\theta\varepsilon^T(t)(A^T(t)P - PA(t) - \rho C^TC)\varepsilon(t) + 2\varepsilon(t)P\Delta_\theta^{-1}\delta F \tag{2.68}$$

Using the Lipschitz condition (assumption **H1)**) and the triangular structure (2.57), it is not difficult to see that there exists a constant $\beta > 0$, such that for every $\theta \geq 1$, we have:

$$\|\Delta_\theta^{-1}\delta F\| \leq \beta\|\varepsilon\| \tag{2.69}$$

where β is a constant which only depends on the Lipschitz constant of F.

Combining (2.61), (2.68) and (2.69), we deduce:

$$\begin{aligned}\dot{V}(t) &\leq (-\theta\eta + 2\beta\|P\|)\|\varepsilon(t)\|^2 \\ &\leq \frac{1}{\lambda_{min}(P)}(-\theta\eta + 2\beta\|P\|)V(t)\end{aligned} \tag{2.70}$$

To end the proof of the theorem, it is enough to take $\theta > max(1, 2\frac{\beta}{\eta}\|P\|)$.

In this subsection, we have shown that the design of a constant high gain observer requires condition (2.60). However, this condition is not always satisfied by general systems of the form (2.56), (2.57), (2.58).

Consider indeed the following system:

$$\begin{cases} \dot{z^1} = u_1 z^3 \\ \dot{z^2} = u_2 z^3 \\ \dot{z^3} = 0 \\ y = Cz = \begin{pmatrix} z^1 \\ z^2 \end{pmatrix} \end{cases} \tag{2.71}$$

where $u = (u_1, u_2)$ belongs to the unit circle $U = \{u \;\; s.t. \;\; \|u\| = 1\}$.

It is obvious to see that system (2.71) is of the form (2.56), (2.57), (2.58) and satisfies hypothesis **H1)**.

Now , assume that system (2.71) admits a constant gain exponential observer:

$$\dot{\hat{z}} = A(u)\hat{z} + K(C\hat{z} - y) \tag{2.72}$$

where, $A(u) = \begin{pmatrix} 0 & 0 & u_1 \\ 0 & 0 & u_2 \\ 0 & 0 & 0 \end{pmatrix}$, $K = \begin{pmatrix} k_{11} & k_{12} \\ k_{21} & k_{22} \\ k_{31} & k_{32} \end{pmatrix}$ is a constant matrix and $C = \begin{pmatrix} 1 & 0 & 0 \\ 0 & 1 & 0 \end{pmatrix}$.

Thus, for every $u \in L^\infty(\mathcal{R}^+, U)$, the error equation:

$$\dot{e} = (A(u) + KC)e \tag{2.73}$$

is exponentially stable at the origin.

In particular, the error equations associated to inputs $u(t) = (1, 0)$ and $u(t) = (-1, 0)$ are exponentially stable. This implies that:

$\begin{pmatrix} k_{11} & k_{12} & 1 \\ k_{21} & k_{22} & 0 \\ k_{31} & k_{32} & 0 \end{pmatrix}$ and $\begin{pmatrix} k_{11} & k_{12} & -1 \\ k_{21} & k_{22} & 0 \\ k_{31} & k_{32} & 0 \end{pmatrix}$ are both Hurwitz matrices.

A simple calculation shows that this yields to the following contradiction: $k_{21}k_{32} - k_{31}k_{22} < 0$ and $k_{21}k_{32} - k_{31}k_{22} > 0$.

The next section gives a method allowing to design an exponential observer for systems of the form (2.56), (2.57), (2.58).

2.4.4 Extension to a More General Structure

Consider again systems of the form (2.56), (2.57), (2.58). We know that for every $(u,\ z^1,\ \ldots,\ ,z^k)$ and for $1 \leq k \leq q-1$, the functions $z^{k+1} \mapsto F^k(u,\ z^1,\ \ldots,\ ,z^k, z^{k+1})$ are locally one to one. In the sequel, we will assume the following:

H2) For $1 \leq k \leq q-1$, $F^k(u,\ z^1,\ \ldots,\ ,z^k,\ .)$ is one to one from $I\!R^{n_{k+1}}$ into $I\!R^{n_k}$

Notice that in the single output case, condition (i) (resp. (ii)) with condition (2.54) of Theorem 2.4.1 imply H2).

Before giving our candidate observer, we need some notations and assumptions.

Consider the following functions:

$$\begin{aligned} &\Phi^1(u,z^1) = z^1 \\ &\Phi^k(u,z^1,\ldots,z^k) = \frac{\partial \Phi^{k-1}}{\partial z^{k-1}}(u,z^1,\ldots,z^{k-1})F^{k-1}(u,z^1,\ldots,z^k);\ 2 \leq k \leq q \end{aligned} \tag{2.74}$$

From the triangular structure (2.57), it is easy to see that

$$\Phi^k(u,z^1,\ldots,z^k) = \frac{\partial F^1}{\partial z^2}(u,z^1,z^2)\ \ldots\ \frac{\partial F^{k-2}}{\partial z^{k-1}}(u,\ z^1,..,z^{k-1})F^{k-1}(u,z^1,..,z^k)$$

Using assumption **H2)** and the rank condition (2.58), it easy to show that :

- for every $(u,z^1,\ldots,z^{k-1})$, $\Phi^k(u,z^1,\ldots,z^{k-1},.)$ is one to one from $I\!R^{n_k}$ into $I\!R^{n_{k-1}}$
- for every $(u,z^1,\ldots,z^{k-1})$, $\zeta^k = \Phi^k(u,z^1,\ldots,z^{k-1},z^k)$ implies that $z^k = \varphi^k(u,z^1,..,z^{k-1},\zeta^k)$ where $\varphi^k(u,z^1,..,z^{k-1},\zeta^k)$ is a function which smoothly depends on $(u,z^1,\ldots,z^{k-1})$.

In the sequel, we will assume that φ^k admits a smooth extension $\widetilde{\varphi}^k$ i.e.:

- $\widetilde{\varphi}^k$ is a smooth function w.r.t. $(u,\zeta^1,\ldots,\zeta^{k-1},\zeta^k)$
- moreover, if $\zeta^i = \Phi^i(u,z^1,..,z^i)$ for $1 \leq i \leq k$, then $z^k = \widetilde{\varphi}^k(u,z^1,..,z^{k-1},\ \zeta^k)$.

Our candidate observer for system (2.56), (2.57), (2.58) takes the following form:

$$\dot{\hat{z}} = F(u,\hat{z}) - \Lambda(u,\hat{z})\Delta_\theta \widetilde{K}(C\hat{z} - y) \tag{2.75}$$

where F is given in (2.56); $\Lambda(u,\hat{z}) = \left[\left(\frac{\partial \Phi}{\partial z}(u,\hat{z})\right)^T \frac{\partial \Phi}{\partial z}(u,\hat{z})\right]^{-1} \left(\frac{\partial \Phi}{\partial z}(u,\hat{z})\right)^T$

with

$$\Phi = \begin{pmatrix} \Phi^1 \\ \Phi^2 \\ \vdots \\ \Phi^q \end{pmatrix}, \ \Delta_\theta = \begin{pmatrix} \theta I_{n_1} & 0 & \dots & 0 \\ 0 & \theta^2 I_{n_1} & 0 & \vdots \\ \vdots & & \ddots & 0 \\ 0 & \dots & 0 & \theta^q I_{n_1} \end{pmatrix}, \ I_{n_1} \text{ is the } n_1 \times n_1 \text{ identity matrix.}$$

$\widetilde{K}$ is a $qn_1 \times n_1$ constant matrix such that $\widetilde{A} - \widetilde{K}\widetilde{C}$ is Hurwitz, where $\widetilde{A}$ and $\widetilde{C}$ are respectively $qn_1 \times qn_1$ and $n_1 \times qn_1$ matrices defined by:

$$\widetilde{A} = \begin{pmatrix} 0 & I_{n_1} & 0 & \dots & & 0 \\ \vdots & 0 & I_{n_1} & 0 & \dots & \vdots \\ \vdots & \dots & \ddots & \ddots & \ddots & \vdots \\ \vdots & \dots & \dots & \ddots & \ddots & 0 \\ \vdots & \dots & \dots & \dots & \ddots & I_{n_1} \\ 0 & \dots & \dots & \dots & \dots & 0 \end{pmatrix} \tag{2.76}$$

$$\widetilde{C} = \left(I_{n_1}\ 0 \dots 0\right) \tag{2.77}$$

In order to prove the convergence of the above observer, we need some notations and assumptions.

Consider the following functions defined on $U \times I\!R^m \times I\!R^n$:

$$G^1(u,v,\zeta) = G^1(\zeta^1) = 0$$

$$G^k(u,v,\zeta) = G^k(u,v,\zeta^1,\dots,\zeta^k) \tag{2.78}$$

$$= \frac{\partial \Phi^k}{\partial u}(u,\zeta^1,\widetilde{\varphi}^2(u,\zeta^1,\zeta^2),\dots,\widetilde{\varphi}^k(u,\zeta^1,\ \dots,\ \zeta^k))v \tag{2.79}$$

$$+\sum_1^{k-1} \frac{\partial \Phi^k}{\partial z^i}\left(u,\zeta^1,\widetilde{\varphi}^2(u,\zeta^1,\zeta^2),\dots,\widetilde{\varphi}^k(u,\zeta^1,\dots,\zeta^k)\right)\zeta^{i+1};\quad 2 \le k \le q$$

As in the previous section, let us assume the following:

H3) (i) $\exists \alpha > 0$; $\forall u \in U$; $\forall z, z'$, $\|\Phi^k(u,z^1,\dots,z^k) - \Phi^k(u,z'^1,\dots,z'^k)\| \ge \alpha\|z - z'\|$, for $1 \le k \le q$

(ii) $\forall \rho > 0$; $\exists \beta > 0$; $\forall u \in U$; $\forall v \in I\!R^m$, $\|v\| \le \rho$; $\forall \zeta, \zeta'$; $\|G^k(u,v,\zeta) - G^k(u,v,\zeta')\| \le \beta\|\zeta - \zeta'\|$, for $1 \le k \le q$

Notice that condition (i) of (H3) implies that for every $u \in U$, the embedding map $z \mapsto \Phi(u,z)$ preserves the uniform topology.

In the sequel, we denote by $\mathcal{U}$ the set of bounded absolutely continuous functions $u(.)$ from $I\!R^+$ into U, with bounded derivatives (ie. $\dot{u} \in L^{+\infty}(I\!R^+)$).

Now, we can state our main results:

Theorem 2.4.3. *Assume that system (2.56)-(2.57)-(2.58) satisfies hypotheses **H2)** and **H3)**, then:*

For every $u \in \mathcal{U}$, $\exists\theta_0 > 0$; $\forall\theta > \theta_0$; $\exists\lambda > 0$, $\exists\sigma > 0$,

$$\|z(t) - \hat{z}(t)\| \leq \lambda e^{-\sigma t}\|z(0) - \hat{z}(0)\|, \text{ for every } t \geq 0.$$

Moreover, σ may be chosen large by taking θ sufficiently large.

If we omit hypothesis **H3)**, then we can state:

Corollary 2.4.1. *Consider system (2.56)-(2.57)-(2.58), and assume that **H2)** is satisfied. Let $u \in \mathcal{U}$ such that every trajectory associated to u and issued from a given compact subset K_1, lies into a compact subset K_2. Then, an exponential observer of the form (2.75) can be designed in order to estimate such bounded trajectories.*

Proof of Theorem 2.4.3. Let $u \in \mathcal{U}$ and consider the following systems:

$$\begin{cases} \dot{\zeta} = \widetilde{A}\zeta + G(u, \dot{u}, \zeta) \\ y = \widetilde{C}\zeta \end{cases} \tag{2.80}$$

$$\dot{\hat{\zeta}} = \widetilde{A}\hat{\zeta} + G(u, \dot{u}, \hat{\zeta}) - \Delta_\theta \widetilde{K}(\widetilde{C}\hat{\zeta} - y) \tag{2.81}$$

where, $G = \begin{pmatrix} G^1 \\ G^2 \\ \vdots \\ G^q \end{pmatrix}$; the G^k's are defined in (2.79) and $\widetilde{A}$, $\widetilde{C}$ are given by (2.76) and (2.77), and $\widetilde{K}$ is such that $\widetilde{A} - \widetilde{K}\widetilde{C}$ is Hurwitz.

We can easily check that if $z(t)$ (resp. $\hat{z}(t)$) is a trajectory of system (2.56) (resp. of system (2.75)) associated to an input $u \in \mathcal{U}$, then $\Phi(u(t), z(t))$ (resp. $\Phi(u(t), \hat{z}(t))$) is also a trajectory of system (2.80) (resp. of system (2.81)). According to hypothesis **H3-(i)**, if $\|\hat{\zeta}(t) - \zeta(t)\|$ exponentially converges to zero, then so does $\|\hat{z}(t) - z(t)\|$. Hence, it is enough to show that system (2.81) is an exponential observer for (2.80).

To do so, we shall proceed as in [4] and [5]. Set $\varepsilon(t) = \Delta_\theta^{-1}(\hat{\zeta}(t) - \zeta(t))$, we obtain:

$$\dot{\varepsilon} = \theta(\widetilde{A} - \widetilde{K}\widetilde{C})\varepsilon + \Delta_\theta^{-1}\delta G \tag{2.82}$$

where $\delta G = G(u, \dot{u}, \hat{\zeta}) - G(u, \dot{u}, \zeta)$.

Since $\widetilde{A} - \widetilde{K}\widetilde{C}$ is Hurwitz, there exists a S.P.D. matrix P such that $P(\widetilde{A} - \widetilde{K}\widetilde{C}) + (\widetilde{A} - \widetilde{K}\widetilde{C})^T P = -I$
where I is the identity matrix. To end the proof of the exponential convergence, it suffices to show the following:

$$\frac{d(\varepsilon^T P\varepsilon)}{dt}(t) \leq -\mu\|\varepsilon(t)\|^2 \tag{2.83}$$

for θ sufficiently large and for some constant $\mu > 0$.

A simple calculation yields:

$$\frac{d(\varepsilon^T P\varepsilon)}{dt}(t) = -\theta\|\varepsilon\|^2 + 2\varepsilon^T P\Delta_\theta^{-1}\delta G \tag{2.84}$$

As in the proof of Theorem 2.4.2, using the triangular structure of $G(u,\dot{u},\zeta)$ w.r.t. ζ and hypothesis **H3)-(ii)**, and taking $\theta \geq 1$, it follows that

$$\|\Delta_\theta^{-1}\delta G\| \leq \tilde{\beta}\|\varepsilon\| \tag{2.85}$$

where $\tilde{\beta}$ is a constant which does not depend on θ. Combining (2.84) and (2.85), we obtain:

$$\frac{d(\varepsilon^T P\varepsilon)}{dt}(t) \leq (-\theta + 2\tilde{\beta})\|\varepsilon\|^2$$

Now, let us choose $\theta_0 > max\{2\tilde{\beta},\ 1\}$, it follows that for $\theta > \theta_0$, we have:

$$\|\varepsilon(t)\| \leq \lambda_1 e^{-\lambda_2 t}\|\varepsilon(0)\| \tag{2.86}$$

where $\lambda_1 > 0$, $\lambda_2 > 0$ are constants, and $\lambda_2 = \lambda_2(\theta) \to +\infty$ as $\theta \to +\infty$.

Proof of Corollary 2.4.1

Let $u \in \mathcal{U}$ and consider the functions Φ^k defined in (2.74). Let Ω_k be a compact set containing all $\{(\Phi^1(u(t),z(t)),\ldots,\Phi^k(u(t),z(t)))\}$, where $z(t)$ is any trajectory of system (2.56), associated to the input u and issued from K_1. Let Ξ_k be any $\mathcal{C}^1$-function which takes value 1 on Ω_k and vanishes outside a bounded open set containing Ω_k. By construction of the system of coordinates $(\zeta^1,\ldots,\zeta^q)$ (see above), Ξ^k is a function only of $(\zeta^1,\ldots,\zeta^k)$ and having a compact support.

Now set $\widetilde{G}(u,\dot{u},\zeta) = \begin{pmatrix} \Xi^1.G^1(u,\dot{u},\zeta) \\ \vdots \\ \Xi^q.G^q(u,\dot{u},\zeta) \end{pmatrix}$ (the $G^k(u,\dot{u},\zeta)$'s are given by (2.79)), then for every trajectory $z(t)$ of system (2.56) associated to u and issued from K_1, $\Phi(u(t)(t),z(t))$ is also a trajectory of the following system:

$$\begin{cases} \dot{\zeta} = \widetilde{A}\zeta + \widetilde{G}(u,\dot{u},\zeta) \\ y = \widetilde{C}\zeta \end{cases} \tag{2.87}$$

Thus, it is enough to construct an exponential observer for system (2.87).

Now since $\widetilde{G}$ has a triangular structure similar to that of G and since it is a global Lipschitz function w.r.t. ζ, we can proceed in a similar way as above to show that the following system:

$$\dot{\hat{\zeta}} = \widetilde{A}\hat{\zeta} + \widetilde{G}(u,\dot{u},\hat{\zeta}) - \Delta_\theta^{-1}\widetilde{K}(\widetilde{C}\zeta - y) \tag{2.88}$$

forms an exponential observer for system (2.87).

2.4.5 Uniform Observability Structure

Many observability concepts are stated in section 2. In this section, we will characterize systems (1) which can be steered by a change of coordinates into the form (2.56)-(2.57)-(2.58).

Consider nonlinear systems of the form (2.49). Notice that, in general, observability (resp. rank observability) of system (2.49) does not imply observability (resp. rank observability) of the associated autonomous system :

$$(\Sigma_u) \qquad \begin{cases} \dot{x} = f_u(x) \\ y = h(x) \end{cases} \tag{2.89}$$

where u is a fixed constant control and $f_u(x) = f(u, x)$.

The uniform observability structure that we will define in particular possesses the property that if system (2.49) is rank observable, then for every fixed $u \in U$, (Σ_u) defined by (2.89) is also rank observable.

To do so, let $u \in U$ and consider the following codistributions :

• $\mathcal{E}_1^u$ is spanned by $\{dh_1, ..., dh_p\}$ (notice that $\mathcal{E}_1^u$ does not depend on u since $h_i = h_i(x)$).

• For $k \geq 1$, let $\mathcal{E}_{k+1}^u$ be the codistribution spanned by $\mathcal{E}_k^u$ and $\left\{dL_{f_u}^k(h_1), \ldots, dL_{f_u}^k(h_p)\right\}$

Clearly, we have $\mathcal{E}_1^u \subset \ldots \subset \mathcal{E}_{n-1}^u \subset \mathcal{E}_n^u \subset \ldots$

Definition 2.4.4. *System (2.49) is said to have a* U**-uniform observable structure (**U**-u.o.s)** *if and only if:*

(i) $\forall u, u' \in U; \forall x \in M$, $\mathcal{E}_k^u(x) = \mathcal{E}_k^{u'}(x)$.
(ii) For each i, *the codistribution* $\mathcal{E}_i^u$ *is of constant dimension* ν_i^u *(dim* $\mathcal{E}_i^u(x) = \nu_i^u, \forall x \in M; \forall u \in U$*).*

In a similar way, let $(\mathcal{E}_i)_{i \geq 1}$ be the family of codistributions defined by:

• $\mathcal{E}_1 = span\{dh_1, ..., dh_p\}$
• $\mathcal{E}_{i+1} = \mathcal{E}_i + span\left\{dL_{f_{u_i}} \ldots L_{f_{u_1}}(h_j); u_1, \ldots, u_i \in U, j = 1, \ldots, p\right\}$.

Remark 2.4.2
a) From (i), we can deduce that for every $u \in U$ *and every* $i \geq 1$, $\mathcal{E}_i = \mathcal{E}_i^u$.
b) From (ii), if system (2.49) is rank observable at some x *and has a* U*-u.o.s., then it is rank observable at each point of* M.

Taking account of Remark 2.4.2, we shall denote indifferently $\mathcal{E}_k^u$ by $\mathcal{E}_k$, ν_k^u by ν_k and we shall denote by q the smallest integer s.t. $\mathcal{E}_q = \mathcal{E}_{q+1}$.

In what follows, we assume that U is such that every compact subset of U is also a compact subset of $\mathbb{R}^m$. This property holds in particular if U is a closed or an open subset of $\mathbb{R}^m$. For the sake of simplicity, we will also assume that h is a local submersion.

We now state the main result of this section :

Theorem 2.4.4. *Assume that system (1) is rank observable at some point of M and has a U-u.o.s. Then, for every compact subset U' of U; system (2.49) is locally U'-uniformly observable (see Definition 2.4.4).*

Remark 2.4.3. *Notice that if we omit the compactness hypothesis of U', the theorem is no longer true.*

Indeed, consider the following example:

$$\begin{cases} \dot{x}_1 = \cos(1+u^2)x_3 \\ \dot{x}_2 = \sin(1+u^2)x_3 \\ \dot{x}_3 = 0 \\ y = (x_1, x_2) \end{cases} \tag{2.90}$$

with $U = \mathbb{R}$ and $M = \mathbb{R}^3$. Clearly, $\mathcal{E}_1^u$ and $\mathcal{E}_2^u$ are respectively spanned by $\{dx_1, dx_2\}$ and $\{dx_1, dx_2, dx_3\}$. Thus (2.90) is rank observable and it has a $\mathbb{R}$-u.o.s.

Now, taking any $x^0 = (x_1^0, x_2^0, x_3^0) \in \mathbb{R}^3$ and any neighborhood $V_{x^0}^{\varepsilon} =]x_1^0 - \varepsilon, x_1^0 + \varepsilon[\times]x_2^0 - \varepsilon, x_2^0 + \varepsilon[\times]x_3^0 - \varepsilon, x_3^0 + \varepsilon[$. Consider any constant control u such that $\dfrac{2k\pi}{1+u^2} < \varepsilon$ and take two initial states $x, \bar{x}$ with $x_1 = \bar{x}_1, x_2 = \bar{x}_2$ and $x_3 - \bar{x}_3 = \dfrac{2k\pi}{1+u^2}$. Let $x(\cdot), \bar{x}(\cdot)$ be the trajectories corresponding to u and respectively issued from x and $\bar{x}$. Obviously, $x_1(t) = \bar{x}_1(t)$ and $x_2(t) = \bar{x}_2(t)$ for every $t \geq 0$. Thus, such u is not universal on any $[0, T], T > 0$. Hence system (2.90) restricted to $V_{x^0}^{\varepsilon}$ is not locally U'-uniformly observable for any unbounded interval of U' of $\mathbb{R}$.

The proof of Theorem 2.4.4 requires the following lemma:

Lemma 2.4.1. *Assume that system (2.49) is rank observable at some point and has a U-u.o.s. Then :*

i) for every $x \in M$, there exist a neighbourhood V and a diffeomorphim:

$\Phi : V \longrightarrow W$ which transforms system (2.49) restricted to V into the following form:

$$\begin{cases} \dot{z}^1 = F^1(u, z^1, z^2) \\ \dot{z}^2 = F^2(u, z^1, z^2, z^3) \\ \quad \dots \\ \dot{z}^q = F^q(u, z) \\ y = z^1 \end{cases} \tag{2.91}$$

where

$$z^i \in \mathbb{R}^{n_i}, \quad z = \begin{bmatrix} z^1 \\ \vdots \\ z^q \end{bmatrix} \in W, \quad u \in U$$

Moreover, we have

$$p = n_1 \geq n_2 \geq \ldots \geq n_q \tag{2.92}$$

ii) $\forall u \in U; \forall z \in W; \forall i,\ 1 \leq i \leq q-1,$ *we have:*

$$rank\left(\frac{\partial F^i}{\partial z^{i+1}}(u,z)\right) = n_{i+1} \tag{2.93}$$

where $n_{i+1} = dim\ \mathcal{E}_{i+1} - dim\ \mathcal{E}_i$

Proof of Lemma 2.4.1. since system (2.49) has a U-u.o.s., we have: $\forall u \in U, \mathcal{E}_1^u = \mathcal{E}_1 \subset \mathcal{E}_2^u = \mathcal{E}_2 \subset \ldots \subset \mathcal{E}_q^u = \mathcal{E}_q$ and for $i \geq q+1$, $\mathcal{E}_i^u = \mathcal{E}_i = \mathcal{E}_q^u$.

Now, let u° be a fixed element of U; from the definition of the $\mathcal{E}_i^{u^\circ}$'s, we know that $(dh_1, \ldots, dh_p)$ forms a basis of $\mathcal{E}_1 = \mathcal{E}_1^{u^\circ}$ (since $h : M \longrightarrow \mathbb{R}^p$ is assumed to be an almost everywhere local submersion). For $i = 2, \ldots, q$, and after reordering adequately $(h_1, \ldots, h_p)$, a basis of $\mathcal{E}_i = \mathcal{E}_i^{u^\circ}$ is given by

$$\mathcal{B}_i = \left(dh_1, .., dh_p, dL_{f_{u^\circ}}(h_1), .., dL_{f_{u^\circ}}(h_{n_2}), .., dL_{f_{u^\circ}}^{i-1}(h_1), .., dL_{f_{u^\circ}}^{i-1}(h_{n_i-1})\right)$$

Set $n_1 = p$ and $n_i = dim\mathcal{E}_i - dim\mathcal{E}_{i-1}$, and using the construction of the $\mathcal{B}_i$, we obtain $n_1 \geq n_2 \geq \ldots \geq n_q$.

Now, using the fact that system (2.49) is rank observable at some point, from Remark (2.4.2), it becomes rank observable at any point of M and hence, the dimension of $\mathcal{E}_q = \mathcal{E}_q^{u^\circ}$ is equal to n. Moreover, from Definition 2.4.4-(ii), it follows that $\Phi = \left(h_1, \ldots, h_{n_1}, \ldots, L_{f_{u^\circ}}^{n_q-1}(h_1), \ldots, dL_{f_{u^\circ}}^{n_q-1}(h_{n_q})\right)$ becomes a local diffeomorphism around each point of M. Now, a simple calculation shows that for every $x \in M$; there exists a neighbourhood V of x such that system (2.49) restricted to V can be transformed by Φ into a system of the form (2.91). Indeed, since $\mathcal{E}_k = \mathcal{E}_k^{u^\circ}$, $\forall u$, we have that $L_{f_u}(L_{f_{u^\circ}}^{i-1}(h_j))$ depends only on $h_1, \ldots, h_{n_1}, \ldots, L_{f_{u^\circ}}(h_1), \ldots, L_{f_{u^\circ}}^{i}(h_{n_{i+1}})$.

To end the proof of the lemma, it remains to prove (2.93).

Denote by $\widetilde{\mathcal{E}}_i$, $\widetilde{\mathcal{E}}_i^u$ the codistributions associated to system (2.91) defined in a similar manner as the $\mathcal{E}_i$'s and $\mathcal{E}_i^u$'s. Notice that $\mathcal{E}_i$ (resp. $\mathcal{E}_i^u$) is the pull-back of $\widetilde{\mathcal{E}}_i$ (resp. $\widetilde{\mathcal{E}}_i^u$) $\left(\mathcal{E}_i = \Phi_\star\widetilde{\mathcal{E}}_i\ (resp.\ \mathcal{E}_i^u = \Phi_\star\widetilde{\mathcal{E}}_i^u)\right)$. Since Φ is a diffeomorphism, the properties (i) and (ii) of Definition 2.4.4 are then preserved for the $\widetilde{\mathcal{E}}_i^u$'s. But $\widetilde{\mathcal{E}}_i^u$ is spanned by $\left(dz_1^1, \ldots, dz_{n_1}^1, \ldots, \frac{\partial F_1^{i-1}}{\partial z^i}dz^i, \ldots, \frac{\partial F_{n_i-1}^{i-1}}{\partial z^i}dz^i\right)$, where

$$\frac{\partial F_j^{i-1}}{\partial z^i} = \left(\frac{\partial F_j^{i-1}}{\partial z_1^i}, \ldots, \frac{\partial F_j^{i-1}}{\partial z_{n_i}^i}\right) \text{ and } dz^i = \begin{pmatrix} dz_1^i \\ \vdots \\ dz_{n_i}^i \end{pmatrix}$$

Since $\widetilde{\mathcal{E}}_i^u$ is of a constant dimension $\sum_{j=1}^{i} n_j$, it follows that :

$$rank\frac{\partial F^{i-1}}{\partial z^i}(u,z) = n_i,\ \forall(u,z) \in U \times W \quad \text{where} \quad W = \Phi(V)$$

We can now prove Theorem 2.4.4.

Proof of Theorem 2.4.4. let U' be a compact subset of U. Let us show that for every $x \in M$; there exists a neighborhood V_x of x such that the restriction of system (1) to V_x is U'-uniformly observable. Using Lemma 2.4.1 and the fact that the observability is an intrinsic property (it does not depend on the system of coordinates), it is enough to show that the restriction of system (2.91) to $W = \Phi(V_x)$ is U' uniformly observable.

To do so, we need the following notations:

Set $\nu_i = n_1 + \ldots + n_i$ the dimension of $\mathcal{E}_i$, and denote by π_i (resp. $\underline{\pi}_i$) the canonical projection from $\mathbb{R}^{\nu_i}$ to $\mathbb{R}^{n_i}$ defined by $(z^1, \ldots, z^i) \mapsto z^i$ (resp. from $\mathbb{R}^{\nu_i+1}$ to $\mathbb{R}^{\nu_i}$: $(z^1, \ldots, z^{i+1}) \mapsto (z^1, \ldots, z^i)$).

Set $W_i = \pi_i(W)$, $\underline{W}_i = \underline{\pi}_i(W)$, $\underline{z}^i = (z^1, \ldots, z^i)$ and denote by $F_{u,\underline{z}^i}$ the map from W_{i+1} into $\mathbb{R}^{n_i}$ defined by $F_{u,\underline{z}^i}(z^{i+1}) = F^i(u, \underline{z}^i, z^{i+1})$, where the F^i's are defined in (2.91).

To prove Theorem 2.4.4, one just have to show that there exists a neighborhood $W = \Phi(V_x)$ of $\Phi(x)$ (maybe small) such that for $1 \leq i \leq q-1$; for every $u \in U'$ and for every $\underline{z}^i \in \underline{W}_i$, $F^i_{u,\underline{z}^i}$ is one to one. Assume indeed that the $F^i_{u,\underline{z}^i}$'s are one to one and let us show that system (2.91) restricted to W is U'-uniformly observable.

Let $u^\circ(\cdot) \in L^\infty([0,T], U')$ be any admissible input, we will show that $u^\circ(\cdot)$ is a universal input on $[0,T]$. Otherwise said, let z, $\bar{z}$ be two initial states such that the corresponding outputs $z^1(t)$, $\bar{z}^1(t)$ are identically equal on $[0,T]$ and let us show that $z = \bar{z}$.

Since $z^1(t) = \bar{z}^1(t)$, $\forall t \in [0,T]$, differentiating this equality, we get:

$$\begin{aligned} F^1(u^\circ(t), z^1(t), z^2(t)) &= F^1(u^\circ(t), \bar{z}^1(t), \bar{z}^2(t)) \\ &= F^1(u^\circ(t), z^1(t), \bar{z}^2(t)) \end{aligned}$$

thus,

$$F^1_{u^\circ(t),z^1(t)}(z^2(t)) = F^1_{u^\circ(t),\bar{z}^1(t)}(\bar{z}^2(t))$$

hence,

$$z^2(t) = \bar{z}^2(t) \quad (\text{since } F^1_{u,z^1} \text{ is one to one}).$$

Differentiating this equality and proceeding in a similar way, we get $z^3(t) = \bar{z}^3(t)$. Repeating this procedure and using the same arguments for $i = 3, \ldots,$ we get $z = \bar{z}$.

Now, let us show the injectivity of the $F^i_{u,\underline{z}^i}$'s. Assume that for every neighbourhood W of $\Phi(x)$, there exist $i, 1 \leq i \leq q-1$; $u \in U'$and $\underline{z}^i \in \underline{W}_i$ such that the restriction of $F^i_{u,\underline{z}^i}$ is not injective. Thus, one can find sequences $(\varepsilon_k)_{k\geq 0}$ $\left(\varepsilon_k > 0, \lim_{k\to+\infty} \varepsilon_k = 0\right)$, $(u_k)_{k\geq 0}$, $u_k \in U'$ and $(z_{\varepsilon_k})_{k\geq 0} \in B(\Phi(x), \varepsilon_k) = \{z / \parallel z - \Phi(x) \parallel < \varepsilon_k\}$ such that $F^{i_0}_{u_k, \underline{z}^{i_0}_{\varepsilon_k}}$ is not one to one for some fixed $i_0 \in \{1, \ldots, q-1\}$. It means that, $\forall k$; $\exists z_{\varepsilon_k}^{i_0+1}$, $\bar{z}_{\varepsilon_k}^{i_0+1} \in \pi_k(B(\Phi(x), \varepsilon_k))$, $z_{\varepsilon_k}^{i_0+1} \neq \bar{z}_{\varepsilon_k}^{i_0+1}$ and such that $F^{i_0}_{u_k, \underline{z}_{\varepsilon_k}^{i_0}}(z_{\varepsilon_k}^{i_0+1}) = F^{i_0}_{u_k, \underline{z}_{\varepsilon_k}^{i_0}}(\bar{z}_{\varepsilon_k}^{i_0+1})$.

Applying the Mean Value Theorem, we get:

$$\left[\frac{\partial F^{i_0}}{\partial z^{i+1}}\left(z_{\varepsilon_{k+1}}^{i_0+1} + \Theta_{i_0}(z_{\varepsilon_k}^{i_0+1} - \bar{z}_{\varepsilon_k}^{i_0+1})\right)\right].(z_{\varepsilon_k}^{i_0+1} - \bar{z}_{\varepsilon_k}^{i_0+1}) = 0$$

where Θ_{i_0} is the $n_{i_0+1} \times n_{i_0+1}$ diagonal matrix $diag(\theta_1, \ldots, \theta_{n_{i_0+1}})$ for some $\theta_j \in [0,1]$, $1 \leq j \leq n_{i_0+1}$.

Since $z_{\varepsilon_k}^{i_0+1} \neq \bar{z}_{\varepsilon_k}^{i_0+1}$, set $\zeta_{\varepsilon_k}^{i_0+1} = \dfrac{z_{\varepsilon_k}^{i_0+1} - \bar{z}_{\varepsilon_k}^{i_0+1}}{\| z_{\varepsilon_k}^{i_0+1} - \bar{z}_{\varepsilon_k}^{i_0+1} \|}$, we obtain :

$$\left[\frac{\partial F^{i_0}}{\partial z^{i+1}}\left(z_{\varepsilon_{k+1}}^{i_0+1} + \Theta_i(z_{\varepsilon_{k+1}}^{i_0+1} - \bar{z}_{\varepsilon_{k+1}}^{i_0+1})\right)\right].\zeta_{\varepsilon_k}^{i_0+1} = 0 \quad \text{and} \quad \| \zeta_{\varepsilon_k}^{i_0+1} \| = 1.$$

Since $(u_k)_{k\geq 1}$, $(\zeta_{\varepsilon_k}^{i_0+1})_{k\geq 1}$ are bounded sequences, then one can extract subsequences $(u_{k_l})_{l\geq 1}$, $(\zeta_{\varepsilon_{k_l}}^{i_0+1})_{l\geq 1}$ such that $\lim\limits_{k_l \to +\infty} u_{k_l} = u$ and $\lim\limits_{k_l \to +\infty} \zeta_{\varepsilon_{k_l}}^{i_0+1} = \zeta^{i_0+1}$ with $u \in U'$ and $\|\zeta^{i_0+1}\| = 1$.

Now, using the continuity of the map : $(u, z^1, .., z^{i_0+1}) \to \dfrac{\partial F^{i_0}}{\partial z^{i+1}}(u, z^1, .., z^{i_0+1})$ and the fact that $\lim\limits_{k \to +\infty}\left(z_{\varepsilon_{k_l}}^{i_0+1} + \Theta_{i_0}(z_{\varepsilon_{k_l}}^{i_0+1} - \bar{z}_{\varepsilon_{k_l}}^{i_0+1})\right) = z^{i_0+1}$ and $\lim\limits_{k \to +\infty} \underline{z}_{\varepsilon_{k_l}}^{i_0} = \underline{z}^{i_0}$ we obtain :

$$\frac{\partial F^{i_0}}{\partial z^{i+1}}\left(u, z^1, \ldots, z^{i_0+1}\right)\zeta^{i_0+1} = 0 \;\; with \;\; \|\zeta^{i_0+1}\| = 1 \tag{2.94}$$

But, from Lemma 2.4.1, the $n_{i_0} \times n_{i_0+1}$ matrix $\dfrac{\partial F^{i_0}}{\partial z^{i+1}}(u, z)$ is of rank n_{i_0+1} and $n_{i_0} \geq n_{i_0+1}$, thus $ker \dfrac{\partial F^{i_0}}{\partial z^{i+1}}(u, z) = \{0\}$. This is in contradiction with (2.94).

References

[1] G. Bornard and H. Hammouri (1991) "A high gain observer for a class of uniformly observable systems"Proc. 30th IEEE Conference on Decision and Control, Brighton, vol.122,pp.176-192.

[2] F. Deza and E. Busvelle and J.P. Gauthier (1992)"Exponentially converging observers for distillation columns and internal stability of the dynamic output feedback" Chemical Engineering Science,vol.47, N.15-16, pp.3935-3941.

[3] J.P. Gauthier and G. Bornard (1981) "Observability for any u(t) of a class of nonlinear systems" IEEE Trans. on Aut. Control,vol.26, pp.922-926.

[4] J.P. Gauthier and H. Hammouri and S. Othman (1992) "A simple observer for nonlinear systems - Application to bioreactors" IEEE Trans. on Aut. Control, vol.37, pp.875-880.

[5] J.P. Gauthier and I.A.K. Kupka (1994)" Observability and observers for nonlinear systems" SIAM J. Control. Optim., vol. 32, pp. 975-994.

[6] J.P. Gauthier and I.A.K. Kupka (2001) "Deterministic Observation Theory and Applications" Cambridge University Press.

[7] H. Hammouri, M. Kinnaert and P. Kabore (2000) "A differential geometric approach to fault detection and isolation for state affine systems" 39th IEEE Conference on Decision and Control.

[10] H. Hammouri and J. P. Gauthier (1988) "Bilinearization up to output injection." Systems & Control Letters, N.11, pp.139-149.

[11] H. Hammouri and J.P. Gauthier (1992) "Global time-varying linearization up to output injection" SIAM J. Control. Optim., vol.30, pp.1295-1310.

[10] H. Hammouri and J. P. Gauthier (1988)"Bilinearization up to output injection" Systems & Control Letters, N.11, pp.139-149.

[11] H. Hammouri and J.P. Gauthier (1992) "Global time-varying linearization up to output injection"SIAM J. Control. Optim., vol.30, pp.1295-1310.

[12] Hammouri, H. and Farza, M.(2003) "Nonlinear observer for locally uniformly observable systems" ESAIM Control, Optimisation and Calculus of Variations SAIM Control, Optimisation and Calculus of Variations, vol.9, pp.353-370.

[13] R. Hermann and A.J. Krener (1977)" Nonlinear Controllability and Observability" IEEE Trans. on Aut. Control, vol.22, pp728-740.

[14] R. E.Kalman and R.S. Bucy (1961)"New results in linear filtering and prediction theory" J. of Basic Engineering, pp.95-108.

[15] N. Kazantzis and C. Kravaris (1998)"Time scaling for observer design with linearization erro dynamics" Systems and Control Letters, vol.34, pp.241-247.

[16] A.J. Krener (1986)"Conditoned invariant and locally conditined invariant distributions" Systems and control letters, N.8, pp. 69-74.

[17] A.J. Krener and A. Isidori (1983)"Linearization by output injection and nonlinear observers" Syst. and Control Letters, vol.3, pp.47-52.

[18] A.J. Krener and W. Respondek (1985)"Nonlinear observers with linealizable error dynamics" SIAM J. Control. Optim., vol.23, pp.197-216.

[19] H.J. Sussman (1979)"Single-Input observability of continuous-time systems" Math. System Theory, vol.12, pp.371-393.

[20] D. Williamson (1977)"Observability of bilinear systems, with application to biological= control" Automatica, vol.32, pp.143-254.

[21] W.M. Wonham (1985) "Linear Multivariable Control - A geometric approach" Springer Verlag, New York, 1985.

[22] X.H. Xia and W.B. Gao (1989)"Nonlinear observer design by observer error linearization" SIAM J. Control and Opt., vol.27, pp.199-216.

3

Adaptive-Gain Observers and Applications

Nicolas Boizot[1] and Eric Busvelle[2]

[1] LASSY, Universit du Luxembourg, 6 rue Coudenhove Kalergi, L-1359 Luxembourg
nicolas.boizot@uni.lu
[2] LE2I, IUT d'Auxerre, Route des plaines de l'Yonne, 89000 Auxerre, France
busvelle@u-bourgogne.fr

3.1 Introduction

We distinguish two kinds of observers for nonlinear systems which are used by scientists and engineers: empirical observers and converging observers.

The first class of observers are based on some approximation of the nonlinear system or approximation of a theoretical best estimation. The most common example is of course the extended Kalman filter. Although, for linear systems, the Kalman filter is a converging observer and an optimal observer for some quadratic cost function, the nonlinear version is based on a linearization of the nonlinear system in a neighborhood of its estimation. Hence, the extended Kalman filter is a good – almost optimal – local observer but it is not a globally converging observer. Intuitively, if the *a priori* estimation is far from the actual state value, the linearization around the estimate has no sense (Section 3.2.2).

There is a lot of empirical observers, based on neural networks, genetic algorithms, fuzzy logic, and so on. These observers are also based on an approximation of the process.

Another type of observers are based on the approximation of the exact solution. Setting indeed the problem as a stochastic problem, the optimal solution is given by the Duncan-Mortensen-Zakaï (DMZ) equation. The solution of this nonlinear stochastic partial differential equation is the law of the state knowing observations. Hence, the conditional expectation of the state knowing observations can be expressed using the solution of the DMZ equation. However, this PDE equation is very complicated. There exist some algorithms in order to calculate an approximation of the solution, and therefore to obtain an approximate observer. For instance, some Monte-Carlo methods can be used in order to calculate the conditional density of probability of the conditional law. In this case, these methods are called particle filtering methods. They consist in the simulation (by Monte-Carlo methods) of several processes, which allows the calculation of the law of the state. The observation appears in the DMZ equation as a killing process. Although this approach has some theoretical justifications (it converges when a finite parameter – the number of particles – goes to infinity), observers based on this approach are always approximate observers.

G. Besançon (Ed.): Nonlinear Observers and Applications, LNCIS 363, pp. 71–113, 2007.
springerlink.com

Although these empirical observers are not proved to converge, they are used by many engineers for many processes, including some critical processes. During normal operation, these observers are often very reliable and give very good practical results.

The second class of observers are theoretically converging observers. In the present chapter, we mainly discuss about high-gain observers. Nevertheless, there exist also some other classes of converging observers. Most of them only deal with a small class of nonlinear systems. Most of them also have some bad performances in the presence of noise.

Here, we will not speak about sliding observers, algebraic observers, or finite dimensional filters, but we will focus on high-gain observers, and their performances compared to extended Kalman filter.

Our purpose is to present a uniform framework where nonlinear filtering, empirical observers and exponentially converging observers are compared. We mainly discuss about their similarities, and we propose an observer based on empirical observers (as those used by engineers), which is an exponentially converging observer.

Despite the lack of theoretical justification, the extended Kalman filter (EKF) is one of the most famous algorithm used to estimate unknown state variables from measurements in dynamical nonlinear systems. It is also used to estimate unknown constant or slowly varying parameters in linear systems and sometimes to perform failure detection. In this last case, it is necessary to quantify the efficiency of the EKF with time. This task is usually based on the innovation process, which is the integrated difference between actual measurements and predicted measurements. The innovation process can be monitored, and a large value of the innovation can be used to send an alarm or to switch from an old model to a new one. It can also be used to estimate the noise entering into the process or to estimate the measurement noise.

The empirical EKF is even used for critical processes. Therefore, in order to increase the performance and the reliability of the EKF several engineers and researchers already tried to develop an adaptive version. Using innovation and state estimation, it seems possible to estimate parameters that characterize the state of the process. These parameters can then be used to adapt the gain matrix by online automatic tuning of some of the covariance matrices used in the computation of the gain matrix. These kinds of adaptive EKF are empirical but seem to have nice behavior compared to the EKF.

Because of the difficulty to ensure robustness when adaptive quantity is continuously updated, some authors used an adaptive algorithm based on switching between several models. For instance, in [33], authors have developed an application on a highly critical process (from a robustness point of view). They proposed to switch between two covariances matrix Q_1 and Q_2 depending on the state of the process.

There exist many papers dealing with adaptive observers and adaptive extended Kalman filtering especially in the GPS and DGPS community, see [22, 12, 26]. In [12] for instance, authors present an adaptive extended Kalman filter using innovation in order to adapt Q and R matrices, exactly in the same spirit as in the present chapter, except that they do not give any theoretical proof. Nevertheless, the need for this kind of observer is clearly established.

In those papers, adaptation of the filter is done using empirical rules (genetic algorithms [35], neural networks [44], statistics [33]...), and no proof is given. But in all cases, efficiency of the adaptive observer is highlighted. Let us remark that for neural networks based extended Kalman filters (N-EKF), the system is split into a linear part and a nonlinear part, and the extended Kalman filter is applied to the nonlinear part, which is approximated by neurons. The weights of neurons can be calculated using EKF, making the algorithm adaptive. In this case, some proofs can be established, but only if the neural network can approximate the system.

An intuitive theoretical justification of adaptive gain is based on the high gain observer theory. It has been shown from a long time ([17]) that high gain observers have very nice theoretical properties. The first one is that they required to study the observability property of the model. This study prevents from developing an observer for a non-observable system. But high gain observers are also exponential observers: one can prove the convergence of the high gain observer. In our opinion, the convergence property is a minimum requirement for an observer which is used on some critical processes, and sometimes as a diagnostic tool. Therefore, it is a good idea to adapt the gain of observers in the following way:

- use an EKF when the estimation is close to the true state, because EKF is a good (optimal) local observer (as already stated) and
- use a high-gain observer when large perturbations occur, because these observers are nonlinear converging observers.

In [14, 15, 20], the high-gain extended Kalman filter (HG-EKF) has been introduced. Compared with the Luenberger observer, HG-EKF is also an exponentially converging observer, but with the property that it is more efficient in the presence of noise. Indeed, the high sensitivity of high-gain observers is a well known drawback: the high gain ensures convergence but also increases noise effects. In [8], a new algorithm, based on classical and high-gain EKF, has been developed. This algorithm is based on a theoretical result, which states that a time-dependent HG-EKF, which is asymptotically equivalent to a classical EKF, may be an exponentially converging observer, if the transition from HG-EKF to EKF is slow enough. But this result is based on a time-dependent observer and, in order to make its convergence property persistent, it is necessary to use several observers and to switch from one to another, depending on the innovation process. Although it is an efficient observer, as shown in the reference above, but also in [9, 10], it is rather complicated and CPU intensive. Moreover, even if the final algorithm can be considered as an adaptive high-gain extended Kalman filter (AG–EKF), its implementation is far from classical observers as used by engineers.

In this chapter, we will present a time-independent adaptive-gain extended Kalman filter. The adaptation of θ will depend on the innovation process.

As usual for the HG-EKF, the parameter θ appears in the Riccati equation of the Kalman filter, and more precisely in the matrix Q, denoted by Q_θ. But in this new case, the high-gain parameter appears also in the matrix R (denoted by R_θ), as in [12] (for a practical application). It is the first difference with the result of [8]. The second difference is that θ may increase if the innovation is high and decrease if the innovation is low. This idea is the basis of practical applications: it is also the cornerstone of the proof of the theorem.

Before considering extended Kalman filtering, we will present in the next section some results concerning nonlinear filtering. A nonlinear filter is similar to a nonlinear observer, in the sense that it is supposed to estimate the state of a system given some measurements. But nonlinear filtering deals with stochastic equations. In the deterministic case, one has in mind that the model approximates the system, that some un-modeled and unmeasured perturbations can enter continuously into the system, and that measurements are corrupted by noise. Therefore, an observer should be robust to these perturbations. In the filtering problem, these perturbations are taken into account in the synthesis of the algorithm. Hence, the stochastic approach seems to be more adapted to the problem, which is better defined (and the stochastic problem is completely solved by the DMZ equation).

As we will see however, both approaches yields similar tools. In fact, the main difference between the two theories is the observability property:

- In the stochastic case, the system has not to be observable. A nonlinear filter can be developed even for unobservable systems since it gives only the conditional law of the state knowing observations. Typically, an observable system gives rise to a unimodal law.
- In the deterministic case, an observer has no sense for a non observable system (except perhaps if the system is globally asymptotically stable in which case the model itself is a – slow – observer).

The "nonlinear filtering" section may be read even by a reader which is not specialist in probability. It can also be omitted by a reader which is not interested by the filtering/observation comparison.

3.2 Nonlinear Filtering

3.2.1 Duncan-Mortensen-Zakaï Equation

We study the observer problem in a stochastic setting. Let us consider the following stochastic system

$$\begin{cases} dX(t) = f(X(t), u)\, dt + Q^{\frac{1}{2}} dW(t) \\ dY(t) = h(X(t), u)\, dt + R^{\frac{1}{2}} dV(t) \end{cases} \tag{3.1}$$

where

- $X(t) \in \mathbb{R}^n$, $X(0)$ being a random variable, $Y(t) \in \mathbb{R}^p$, and u is a $\mathbb{R}^d$–valued measurable function,

- $W(t)$ and $V(t)$ are two independent Wiener processes (also independent from $X(0)$).

In this chapter, we will omit to specify the time variable whenever no confusion is possible, writing X instead of $X(t)$.

Therefore,

$$E\left[\left(Q^{\frac{1}{2}}W(t)\right)\left(Q^{\frac{1}{2}}W(t)\right)'\right] = Q.t$$

(where M' denotes the transpose of a matrix M) so Q is the covariance matrix of the state noise, and R is the covariance matrix of the measurement noise (the notation $Q^{\frac{1}{2}}$ represents the Cholesky decomposition of Q, also called square root of Q).

In this section, we denote by $X(t)$ a process or random variable and $x(t)$ its realization, that is $x(t) = X(t)(\omega)$.

$X(0)$ is supposed to be an $L^2(\mathbb{R}^n)$ random variable independent from W and V. For simplicity, we will assume that this random variable admits a density function, denoted by $p(0,x) = \frac{dP(\{X(0)\leq x\})}{dx}$.

Considering equations in the Ito sense, if f is a Lipschitz function w.r.t. x with a Lipschitz constant independent of u, then system (3.1) admits a unique solution.

In this stochastic context, the observer problem is an estimation problem: we want to calculate the best estimation of $X(t)$ knowing measurements Y from 0 to t, denoted by the σ–algebra $\mathcal{F}_t^Y$. Hence, we want to calculate the conditional expectation $E\left[X(t) \mid \mathcal{F}_t^Y\right]$, or more generally $E\left[\phi(X(t)) \mid \mathcal{F}_t^Y\right]$ for any test function ϕ. Finally, this is equivalent to calculate the conditional law of $X(t)$ knowing $\mathcal{F}_t^Y$.

We assume that this law admits a density denoted by $p(t,x)$, *i.e.* the conditional law is absolutely continuous with respect to Lebesgue measure (this restrictive assumption is not necessary but it simplifies some formulas, especially the DMZ equation). Then, $p(t,x)$ is the solution of the well known Duncan–Mortensen–Zakaï (DMZ) equation. We will not explain this equation here: it is a stochastic partial differential equation, which has to be regularized before to be used, and which is difficult to use for practical problems, especially if n is large (see [37] for a clear statement of the DMZ equation).

The DMZ equation has been used in several ways:

- First, this equation may be simplified in some very special cases. One of them is the linear case, where the solution of the DMZ equation is the Kalman filtering equation. There also exist some nonlinear cases where the DMZ equation gives a computable solution, for instance for systems which are linearizable up to a change of coordinates, or an immersion. In these cases, it is of course a very good approach to build an optimal observer.
- Second, despite its complexity, the (regularized version) of the DMZ equation can be approximately solved, for instance using Monte-Carlo methods. In this context, Monte–Carlo methods are called particle methods. The main idea is

to approximate the initial law of $X(0)$, given by its density $p(0,x)$, by a set of "particles", *i.e.* a set of independent random variables $X_i(0)$ such that

$$p(0,x) \simeq \sum_{i=1}^{N} \delta_{X_i(0)}$$

where δ_x denotes the Dirac measure at x. The notation $\simeq$ will be precisely defined in Theorem 2.

The principle of a particle method is then to approach the probability law of $X(t)$ knowing $\mathcal{F}_t^Y$ by a (weighted) sum of Dirac measures at points $X_i(t)$. When applied to filtering, this just consists in approaching the law of the current state knowing observations by means of a particular weighted sum of Dirac distributions. This kind of method is well adapted to the case in which the dimension of the state is large, because in this case one usually uses the Monte-Carlo method to compute the conditional expectation

$$E\left[\phi(X(t)) \mid \mathcal{F}_t^Y\right] = \int \phi(x)p(t,x)dx$$

and this method requires a sample of the law $p(t,x)$ which is given by $X_i(t)$, $i = 1,\ldots,N$.

To characterize a particle method, it is sufficient to give some rules such as

- how to calculate weights of particles (e.g. Dirac measures)
- how to move particles $X_i(t)$ in the state space

Let us give an example of a particle filtering. As we will see in next section, this algorithm have some similarities with the observer construction (Section 3.3.3), although it has been obtained by a totally different way.

We will study the nonlinear filtering problem with linear discrete-time observation, that is to say, the second equation in (3.1) is replaced by

$$Y_k = CX(t_k) + R^{\frac{1}{2}}V(k) \tag{3.2}$$

where $(t_k)_{k\in\mathbb{N}}$ is the sample time and $(V(k))_{k\in\mathbb{N}}$ is an independent (w.r.t. W and $X(0)$) Gaussian white noise. The limitation to a linear observation function is not necessary but is a simplification when one wants to implement this algorithm. The choice of discrete-time observation simplifies the mathematical background necessary to define the DMZ equation. Indeed, in this case, the conditional density $p(t,x)$ is given by the discrete version of the DMZ equation:

$$p(t_k,x) = \frac{1}{f_{Y_k}^{Y^{k-1}=y^{k-1}}(y_k)} f_{Y_k}^{X(t)=x}(y_k) \int_{\underline{X}} f_{X(t_k)}^{X(t_{k-1})=\xi}(x)p(t_{k-1},\xi)d\xi \tag{3.3}$$

with the following notations

- $f_{Y_k}^{X(t)=x}(y_k)$ represents the conditional density of Y_k knowing $X(t) = x$;
- $f_{X(t)}^{X(s)=\xi}(x)$ represents the conditional density of $X(t)$ knowing $X(s) = \xi$;
- $f_{Y_k}^{Y^{k-1}=y^{k-1}}(y_k)$ represents the conditional density of Y_k knowing Y between time 0 and time t_{k-1} is equal to $(y_0, \ldots, y_{k-1})$ so that for instance,

$$p(t,x) = f_{X(t)}^{Y^k=y^k}(x)$$

Equation (3.3) is nothing else than the Bayes formula applied to the problem.

Remark 1. We point out that the DMZ equation (3.3) gives an exhaustive information on $X(t)$ knowing all informations available at time t. Hence it gives the best possible estimate and, if the system is observable (Definition 1), it is a very good observer.

As usual with equations describing evolution of a density of probability, the un–normalized version of the DMZ is more tractable: (3.3) is equivalent to

$$q(t_k, x) = f_{Y_k}^{X(t)=x}(y_k) \int_{\underline{X}} f_{X(t_k)}^{X(t_{k-1})=\xi}(x) q(t_{k-1}, \xi) d\xi \tag{3.4}$$

with

$$p(t_k, x) = \frac{q(t_k, x)}{\int_{\underline{X}} q(t_k, \xi) d\xi}$$

There are several ways to solve the un–normalized DMZ equation using particle methods. The first way is to recognize the composition/rejection theorem in this formula ([27]), and therefore to consider this equation as a simulation formula, which is the basis of a Monte-Carlo method. The algorithm consists in simulating the process (by "particles" Z_i) and killing some of them thanks to measurements (the "bad" particles). At a time $t_k < t \leq t_{k+1}$, the number of particles which are still alive is a random variable $N(k)$. If this random number is large enough, the conditional density is approximated by

$$p(t,x) \simeq \sum_{i=1}^{N(k)} \delta_{Z_i(t)}(x)$$

This approach can not be applied exactly as explained here, since $N(k)$ is a decreasing integer which goes almost surely to 0 (each measurement kill particles). In order to obtain a more efficient algorithm, one usually consider a weighted sum of Dirac measures.

Let us introduce coefficients $a_i(t) \in [0,1]$. These numbers represent the degree of confidence in each particle, and replace binary coefficients 1 (the particle is alive) or 0 (the particle is dead). As for the DMZ equation itself, we consider an un–normalized set of coefficients $b_i(t) \in \mathbb{R}^+$ such that

$$a_i(t) = \frac{b_i(t)}{\sum_{j=1}^{N} b_i(t)}$$

We consider an algorithm $\mathcal{P}$ which describes the trajectory of particles $z_i(t)$ and weight coefficients $b_i(t)$. The law truncated at n particles given by $\mathcal{P}$ is denoted by $P_n(t)$ ($dP_n(t) = p_n(t,x)\,dx$) and defined by

$$P_n(t) = \frac{\sum_{i=1}^{n} b_i(t)\,\delta_{z_i(t)}}{\sum_{i-1}^{n} b_i(t)} = \sum_{i-1}^{n} a_i(t)\,\delta_{z_i(t)}$$

Algorithm 1. *Initialization*

$z_i(0)$ is the realization of a random variable with respect to the initial law $p(0)$;

$b_i(0) = 1;$

Loop

$z_i(t_k)$ is a Gaussian variable with respect to $f_{X(t_k)}^{X(t_{k-1})=z_i(t_{k-1}),Y_k=y_k}$;

$b_i(t_k)$ is defined by

$$b_i(t_k) = b_i(t_{k-1})\; f_{Y_k}^{X(t_{k-1})=z_i(t_{k-1})}(y_k)$$

Let us remark that this algorithm is easy to implement on a computer, in particular on a parallel computer.

Theorem 2. *Let us consider the system*

$$\begin{cases} dX = f(X,u)\,dt + Q^{\frac{1}{2}}dw(t) \\ Y_k = CX(t_k) + R^{\frac{1}{2}}V(k) \end{cases}$$

and $P(t)$ being the conditional law of $X(t)$ knowing $\mathcal{F}_t^Y$. If $P_n(t)$ represents the law given by the algorithm $\mathcal{P}$ with n particles, then we have

$$P_n(t) \to P(t) \;\; \text{as } n \to \infty \text{ weakly almost surely}$$

Remark 2. This theorem is true at t fixed. It is never true for any t. In order to obtain an asymptotic result (as in observer theory), it is necessary to add some correlations between particles. This is particularly simple here (see [39]).

In order to illustrate this theorem, we consider a continuous stirred tank reactor (CSTR). The dimensionless form of the model is:

$$dX_1 = \left(-X_1 + D_A(1-X_1)\exp\{\frac{X_2}{1+X_2/\gamma}\}\right)dt + dW_1$$

$$dX_2 = \left(-X_2(1+\beta) + H_aD_a(1-X_1)\exp\{\frac{X_2}{1+X_2/\gamma}\} + \beta u\right)dt + dW_2$$

where W_1 and W_2 are two independent Wiener processes. X_1 is the reactant concentration and X_2 is the temperature into the tank. We suppose that X_2 is measured in discrete time and that we want to control X_1 using the control variable u. The system can also be written in the following generic form

$$X(t_{k+1}) = X(t_k) + \int_{t_k}^{t_{k+1}} f(X(s))ds + \int_{t_k}^{t_{k+1}} BdW(s)$$
$$Y_k = CX(t_k) + V_k$$

with $C = \begin{pmatrix} 0 & 1 \end{pmatrix}$. We suppose that W is a two-dimensional Wiener process and that V_k is a Gaussian process independent of W and with covariance R. We propose the following discretization scheme for the continuous-time equation

$$X(t_{k+1}) = \Phi(t_k, t_{k+1}, X(t_k)) + \frac{\partial \Phi(t_k, t_{k+1}, X(t_k))}{\partial x} B\sqrt{t_{k+1} - t_k} W_k$$

where $\Phi(s, t, x)$ is the solution of

$$\begin{cases} \frac{dx(t)}{dt} = f(x(t)) \\ x(s) = x \end{cases}$$

at time t.

The right-hand part of this scheme is the first order development of

$$\Phi(t_k, t_{k+1}, X(t_k) + \int_{t_k}^{t_{k+1}} BdW(s))$$

which naturally comes from the diffusion equation. A classical theorem of probability, see for instance [21], shows that this scheme converges in law to the solution of the diffusion equation when the step of the discretization goes to zero.

Our main goal is to estimate the reactant concentration X_1 and its confidence intervals, in order to control as well as possible the CSTR.

If we solve the equations, we can see that for each particle $z(t)$ at time t and for each weight $b(t)$, we have, thanks to the algorithm of the theorem

- **Correction** at time t_k

$$\begin{cases} z(t_k) = z(t_k^-) + P(t_k^-)C^T(CP(t_k^-)C^T + R)^{-1}(y_k - Cz(t_k^-)) \\ \qquad + \left(P(t_k^-) - P(t_k^-)C^T(CP(t_k^-)C^T + R)^{-1}CP(t_k^-)\right)\bar{w}_k \\ P(t_k) = BB^T(t_{k+1} - t_k) \\ b(t_k) = b(t_{k-1}) \frac{\exp\left(-\frac{1}{2}(y_k - Cz(t_k^-))^T(CP_{t_k^-}C^T + R)^{-1}(y_k - Cz(t_k^-))\right)}{\sqrt{2\pi . det(CP(t_k^-)C^T + R)}} \end{cases} \tag{3.5}$$

  where $\bar{w}_k$ is a Gaussian white noise.
- **prediction** between t_k and t_{k+1}

$$\begin{cases} \frac{dz}{dt} = f(z(t)) \\ \frac{dP}{dt} = f^*(\xi(t))P(t) + P(t)f^*(\xi(t))' \\ \frac{db}{dt} = 0 \end{cases} \tag{3.6}$$

3.2.2 Extended Kalman filter

The previous algorithm is CPU-time consuming and rather complicated to implement, especially in the linear case. Indeed, for a linear system, there exist a very simple and famous solution. Let us consider the following linear system:

$$\begin{cases} dX = (A(t)X + B(t)u)\,dt + Q^{\frac{1}{2}}dW(t) \\ dY = C(t)Xdt + R^{\frac{1}{2}}dV(t) \end{cases} \tag{3.7}$$

with $X(0)$ a random variable with Gaussian law $\mathcal{N}(m_0, P_0)$, the DMZ equation reduces itself to the well-known Kalman filter. More precisely, solving the DMZ equation yields the following result: the conditional law of $X(t)$ knowing $y(s)$ from 0 to t $(\mathcal{F}_t^Y)$ is the Gaussian law $\mathcal{N}(z(t), P(t))$ where, for an output trajectory $y(t)$, $z(t)$ and $P(t)$ are the solutions of the finite-dimensional system of ordinary differential equations:

$$\begin{cases} dz = (A(t)z + B(t)u)\,dt + PC(t)'R^{-1}(dy - C(t)z\,dt) \\ \frac{dP}{dt} = A(t)P + PA(t)' + Q - PC(t)'R^{-1}C(t)P \end{cases} \tag{3.8}$$

with $z(0) = m_0$ and $P(0) = P_0$. Therefore, $z(t) = E\left[X(t) \mid \mathcal{F}_t^Y\right](\omega)$ is the best estimation of $X(t)$ knowing measurements up to time t. When applied to a deterministic observable linear system, Q and R being considered as tuning parameters, the Kalman filter is called the Kalman observer. The observable property is not crucial in the stochastic case since the conditional law is defined even for non observable systems. But the observability property implies that the covariance matrix of the conditional expectation of $X(t)$ knowing $Y(s)$, $0 \le s \le t$ is bounded.

In the deterministic case, this property is crucial. Recall also that, for linear systems, observability does not depends on inputs.

The Kalman filter/observer algorithm has been used for a long time by engineers for linear systems. For nonlinear systems, engineers introduced and successfully used the extended Kalman filter (EKF), either in its stochastic or its deterministic form. The EKF is just the standard Kalman filter for linear time-dependent systems, applied to the linearized system along the estimate trajectory. The EKF is the heart of our approach.

Let us consider a nonlinear system

$$\begin{cases} dX = f(X,u)\,dt + Q^{\frac{1}{2}}dW(t) \\ dY = h(X)\,dt + R^{\frac{1}{2}}dV(t) \end{cases} \tag{3.9}$$

where f and h are smooth Lipschitz functions, the linear Kalman filter does not apply anymore, and the exact solution should be obtained by solving the DMZ equation. But if one wants an approximate solution, it is very common to consider the first order approximation of the previous system. The right way to do this is to consider an *a priori* solution $\hat{x}(t)$ of the deterministic system associated to (3.9) and to use the Kalman filter to estimate the first order difference $\delta x(t) = x(t) - \hat{x}(t)$ between the *a priori* solution and the estimated solution.

This approach yields the following first order Kalman filter, for a given output trajectory:

$$\left\{\begin{array}{ll}\frac{d(\delta x)}{dt} & = f^*(\hat{x},u)\delta x + Ph^*\left(\hat{x},u\right)' R^{-1}(y(t) - h\left(\hat{x},u\right)) \\ \frac{dP}{dt} & = f^*\left(\hat{x},u\right)P + Pf^*\left(\hat{x},u\right)' + Q \\ & \quad -Ph^*\left(\hat{x},u\right)' R^{-1}h^*\left(\hat{x},u\right)P\end{array}\right. \tag{3.10}$$

where f^* and h^* are the Jacobian of f and h w.r.t. x respectively. But this approach has a major weakness: the choice of the *a priori* solution $\hat{x}(t)$ is not obvious if there is no precise *a priori* information on the initial state. This is usually the case, especially in the deterministic case, since the only missing information on the system is precisely the initial state. Moreover, if one makes a bad choice of $\hat{x}(t)$, the first order equation has no significant meaning since the actual state is far from the initial guess. At the opposite, if $\delta x(0)$ is small (that is the *a priori* solution is close to the actual solution, at least at time 0), then $\hat{x}(t) + \delta x$ will be a good approximation of the optimal filter, when state and measurement noises are small ([38]).

To overcome this difficulty, engineers have an attractive idea: to replace the *a priori* solution by the estimated solution at current time. The main advantage of this approach is that the estimated solution is supposed to be close to the actual solution, hence the first order approximation should be small and hence the linear approximation should be a good approximation. This remark yields the extended Kalman filter:

$$\left\{\begin{array}{ll}\frac{dz}{dt} & = f(z,u) + Ph^*\left(\hat{x},u\right)' R^{-1}(y(t) - h\left(z,u\right)) \\ \frac{dP}{dt} & = f^*\left(z,u\right)P + Pf^*\left(z,u\right)' + Q \\ & \quad -Ph^*\left(z,u\right)' R^{-1}h^*\left(z,u\right)P\end{array}\right. \tag{3.11}$$

where z is the estimated state. Here again, if P_0, Q and R are small, this filter is close to the optimal filter (see all works of Picard, [38] for instance).

In a deterministic context, the extended Kalman filter is a converging local observer (see [4, 8]), that is if $z(0) \simeq x(0)$ then $z(t) - x(t) \longrightarrow 0$ as $t \longrightarrow +\infty$ (exponentially). Nevertheless, the extended Kalman filter has no global converging properties. Indeed, it is well known that, if the initial guess $z(0)$ is far from $x(0)$, the extended Kalman filter may not converge. Moreover, the mathematical study of (3.11) is difficult because it has no clear mathematical meaning: it is not a first order approximation of a nonlinear object around a given trajectory. In other words, the behavior of (3.11) is not intrinsic and depends on a choice of coordinates. Hopefully, this mathematical difficulty will give us a way to chose a good system of coordinates and to prove some convergence results, thanks to this crucial choice of coordinates.

To conclude, the EKF is very efficient in a lot of practical problems. It is used as a filter or as an observer in many various systems. From a theoretical point of view, it is not an optimal filter (it differs from the DMZ equation). Nevertheless, when the system has some observability properties, it has very nice local properties: in the stochastic case, it is a good filter when noises are small (see [38]) and in the deterministic case, it is a local observer ([4, 8]).

3.2.3 Continuous-Discrete Stochastic Systems

Before considering deterministic systems and observers, let us recall a result concerning discrete measurements. Continuous-discrete time are very common in practise: the nonlinear differential equation describes a mechanical, physical or chemical process. Therefore, it is a continuous time system. But measurements are usually sampled at times t_k. Therefore, the system can be written

$$\begin{cases} dX(t) = f(X(t), u(t))dt + dW(t) \\ y_k = h(X(t_k)) + V(k) \end{cases} \tag{3.12}$$

where h is a differentiable function from the state space to $\mathbb{R}^p$.

For this system, the EKF has two set of equations: the correction step which is applied at each measurement time and the prediction step which is used to predict the system according to the model.

Correction step

$$\begin{cases} Z\left(t_k^+\right) = Z(t_k) + G(k)\left(y_k - h\left(Z(t_k)\right)\right) \\ G(k) = P(t_k)\, h^*\left(Z(t_k)\right)'\left(h^*\left(Z(t_k)\right)P(t_k)\, h^*\left(Z(t_k)\right)' + R\right)^{-1} \\ P\left(t_k^+\right) = \left(I - G(k)\, h^*\left(Z(t_k)\right)\right)P(t_k) \end{cases} \tag{3.13}$$

Prediction step

$$\begin{cases} \frac{dZ}{dt} = f(Z, u) \\ \frac{dP}{dt} = f^*(Z, u)\, P + P f^*(Z, u)' + Q \end{cases} \tag{3.14}$$

These equations present some similarities with equations (3.5,3.6). As we will see in the end of Section 3.3.4, if the system is observable, then equations (3.13,3.14) may give an observer. In the non observable case, one should use (3.5,3.6).

Although this kind of model is closer to the practical case, it is less used than continuous-time systems. The main reason is a practical one: the sampled time is usually chosen small enough w.r.t. time constants of the process. Therefore, the continuous EKF can be applied. Sometimes (for very fast processes or for slow measurement devices), the sampled time is a constraint and can not be neglected. In this case, continuous-discrete EKF should be applied.

3.3 Nonlinear Observers

3.3.1 Canonical Form of Observability

From now on, we study deterministic nonlinear systems of the general form

$$\Sigma \begin{cases} \frac{dx}{dt} = f(x, u) \\ y = h(x, u) \end{cases} \tag{3.15}$$

on a smooth n–dimensional manifold X, $y \in \mathbb{R}^p$, $u \in U$, subset of $\mathbb{R}^d$. We want to develop an observer. Our approach is closely related to observation theory, as explained in the book from Gauthier and Kupka [20], which is itself a summary of papers [16, 17, 18, 19, 32].

This theory leads to the consideration of systems under the normal form (3.21), or similar multi-output normal forms. Here, by "observability", we mean "observability for every fixed input function $u(t)$". For details, see [20].

In this introduction part, we summarize the main observability results of the observation theory developed in [20].

First of all, the state–output mapping $PX_{\Sigma,u}$ is the function $x(0) \longrightarrow (y(t))_{t\geq 0}$. In this definition (and the following ones), we do not speak about explosion times, in order to simplify the notations.

Definition 1. *The system (3.15) is said uniformly* ***observable****, or just* ***observable****, w.r.t. a certain class $\mathcal{C}$ of inputs ($L^\infty(U)$ in most cases) if, for each $u(.) \in \mathcal{C}$, the state output mapping $PX_{\Sigma,u}$ is injective.*

This first definition is the natural definition of observability. Nevertheless, injectivity is not a very tractable property, since it is not stable (even for standard mappings between finite dimensional spaces -example: $x \to x^3, \mathbb{R} \to \mathbb{R}$). Therefore, in order to state results, we need a few other definitions. The uniform infinitesimal observability makes the observable property stable.

Let us define the lift of Σ on TX, also called the first variation of Σ. Let us consider $T_X f : TX \times U \longrightarrow TTX$(the tangent bundle of TX) the tangent mapping of $f : X \times U \longrightarrow TX$ and $d_X h : TX \times U \longrightarrow \mathbb{R}^p$ the Jacobian of $h : X \times U \longrightarrow \mathbb{R}^p$. Then

$$T\Sigma \begin{cases} \frac{d\xi}{dt} = T_X f(\xi, u) = T_X f_u(\xi) \\ \eta \;= d_X h(\xi, u) = d_X h_u(\xi) \end{cases} \tag{3.16}$$

The state–output mapping of $T\Sigma$ is denoted by $PTX_{\Sigma,u}$. It is also the first order approximation of $PX_{\Sigma,u}$ denoted by $TPX_{\Sigma,u}$.

Definition 2. *System Σ is said uniformly infinitesimally observable if, for each $u(.) \in L^\infty(U)$, each $x_0 \in X$, all the tangent mappings $TPX_{\Sigma,u}|x_0$ are injective.*

Remark 3. This definition of observability is stable in the sense of discretization: if a system is uniformly infinitesimally observable, its continuous–discrete version (3.12) remains uniformly infinitesimally observable for a sampling time small enough. It is not the case for a system which is only observable (see [2]).

The two following definitions are another way to define observability in a stable way. Note that these definitions are important for practical purpose, since they give a way to prove observability for nonlinear systems.

Definition 3. *System Σ is said differentially observable (of order k) if for all $j^k\hat{u}$, the extension to k-jets mapping* [1] *$\Phi_k : x_0 \to j^k\hat{y};\ \ X \to \mathbb{R}^{km}$ is injective.*

[1] k-jets $j^k u$, of smooth functions u at $t = 0$ are defined as

$$j^k u = (u(0), u'(0), ..., u^{(k-1)}(0)).$$

Then, for a smooth function u and for each $x_0 \in X$, the k-jet $j^k y = (y(0), y'(0), ..., y^{(k-1)}(0))$ is well define: this is the k-jets state-output mapping Φ_k.

Definition 4. *System Σ is said strongly differentially observable (of order k) if for all $j^k u$, the extension to k-jets mapping $\Phi_{k,j^k u} : x_0 \to j^k y;\ \ X \to \mathbb{R}^{km}$ is an injective immersion*[2]

Clearly, strong differential observability implies differential observability, which implies observability for the C^∞ class, (and L^∞-observability).

It is also a consequence of the theory that for analytic systems, uniform infinitesimal observability implies observability of the restrictions of (3.15) to small open subsets of X, the union of which is dense in X.

The main result concerning observability of systems 3.15 is that, depending on the number of outputs w.r.t. the number of inputs, the property may be generic or not generic. More precisely, we distinguish two cases:

1. More measurements than control inputs ($p > d$): in that case, observability is a generic property, and generically, a system can be put globally under a normal form similar to (3.21), but the dimension of the state in the normal form is bigger than the dimension of the state of the original system: it is at most double plus one. Also, the control in the normal form contains a certain number of derivatives of the control of the initial system. But this is more or less unimportant for observation problems, where the control, and hence its derivatives, are known.

   Hence, if $p > d$, and for sufficiently smooth inputs, generic systems are very good from the point of view of observability.
2. Less or same number of measurements than control inputs ($p \leq d$): in that case observability is a non generic property. It is even a property of infinite codimension. This high degeneracy leads to the fact that, in the case of control affine systems, all observable systems can be put locally under normal forms similar to (3.21) (with $a_i = 1$, $i = 1, ..., n$).

   In the analytic case $p = 1$, $d \geq 1$, we can be more precise. If (3.15) is uniformly infinitesimally observable, then locally almost everywhere on X, the system (3.15) can be put in the form

$$\left\{ \begin{aligned} y &= h(x_1, u) \\ \tfrac{dx_1}{dt} &= f_1(x_1, x_2, u) \\ \tfrac{dx_2}{dt} &= f_2(x_1, x_2, x_3, u) \\ &\vdots \\ \tfrac{dx_{n-1}}{dt} &= f_{n-1}(x_1, x_2, .., x_n, u) \\ \tfrac{dx_n}{dt} &= f_n(x_1, x_2, ..., x_n, u) \end{aligned} \right. \tag{3.17}$$

   where

$$\frac{\partial h}{\partial x_1} \text{ and } \frac{\partial f_i}{\partial x_{i+1}}, i = 1, .., n-1 \tag{3.18}$$

   do not vanishe on $V_x \times U$.

[2] Immersion means that all the tangent mappings $T_{x_0}\Phi_{k,j^k\hat{u}}$ to this map, have full rank n at each point.

In the control affine case, where (3.15) can be written:

$$\dot{x} = f(x) + \sum_{i=1}^{d} g_i(x)u_i \qquad (3.19)$$
$$y = h(x)$$

then the canonical form of observability is

$$\begin{cases} y = x_1 \\ \frac{dx_1}{dt} = x_2 + \sum_{i=1}^{p} g_{1,i}(x_1)u_i \\ \frac{dx_2}{dt} = x_3 + \sum_{i=1}^{p} g_{2,i}(x_1, x_2)u_i \\ \vdots \\ \frac{dx_{n-1}}{dt} = x_n + \sum_{i=1}^{p} g_{n-1,i}(x_1, x_2, .., x_{n-1})u_i \\ \frac{dx_n}{dt} = \psi(x) + \sum_{i=1}^{p} g_{n,i}(x_1, x_2, .., x_{n-1}, x_n)u_i \end{cases} \qquad (3.20)$$

These two results are very important since they allow us to restrict our study to systems of the form (3.17) and (3.20) (and also because of course, these results are based on a constructive diffeomorphism).

3.3.2 High-Gain Extended Kalman Filter

We describe observers for nonlinear systems in canonical form of observability (3.20 and 3.21 below), on $\mathbb{R}^n$. The control space $\mathcal{U}_{\text{adm}}$, is supposed to be a closed subset of $\mathbb{R}^d$. In this section, the observation is assumed to be single-valued: it is a u–dependent linear form on $\mathbb{R}^n$. This hypothesis is not necessary and our observers constructions also applies for multi–output systems. From an observability point of view, the multi–output case is a little bit more complicated since canonical forms of observability are less natural. But from the observer point of view, except in section 3.3.4, the problem is exactly the same, since we simply apply some kind of EKF.

We consider systems of the form

$$\begin{cases} \frac{dx}{dt} = A(u)x + b(x, u) \\ y = C(u)x \end{cases} \qquad (3.21)$$

where $A(u)$, $C(u)$ are matrices:

$$A(u) = \begin{pmatrix} 0 & a_2(u) & 0 & \cdots & 0 \\ & & a_3(u) & \ddots & \vdots \\ \vdots & & \ddots & \ddots & 0 \\ & & & & a_n(u) \\ 0 & & \cdots & & 0 \end{pmatrix} \qquad (3.22)$$

$$C(u) = (a_1(u), 0, ..., 0) \qquad (3.23)$$

and where $a_i(.)$, $i = 1, ..., n$, are positive smooth functions, bounded from above and below:

$$0 < a_m \leq a_i(u) \leq a_M$$

Also, $b(x, u)$ is a smooth, u–dependent vector field, depending triangularly on x and compactly supported:

$$b(x,u) = \begin{pmatrix} b(x_1,u) \\ b(x_1,x_2,u) \\ \vdots \\ b(x_1,\ldots,x_n,u) \end{pmatrix} \tag{3.24}$$

These assumptions look very strong, but as we have already seen, under either genericity hypotheses or observability hypotheses, for the purpose of synthesis of observers, it is sufficient to restrict to these systems, under the normal form (3.21) (or similar multi-output normal forms), and meeting these assumptions. In fact, this form generalizes the canonical form of observability (3.20) for control affine systems. We call (3.21) (together with (3.22–3.23)) the generalized canonical form of observability. There are several reasons to study (3.21) rather than (3.20):

- It is sometimes easier to put the system into this form, using intuitive transformations, rather than a more restrictive normal form, the last transformation being based on Lie derivatives. This point will be illustrated in the application sections;
- Since we want to apply an EKF which uses the model to filter noises, and a high–gain approach to kill the nonlinear part of the system, it is better to leave the largest part of the nonlinear system in A rather to put it in b. This technical point will be developed later;
- Last but not least, our observer construction still work for these systems.

However, this form does not include the canonical form of observability for systems (3.15) when the control is not affine. For those systems, there exist a change of coordinates that put the equivalent system (3.17) into a system of the generalized canonical form of observability (3.21) [10, 23]. For this, we just need to suppose that u admits a time derivative almost everywhere.

Consider a system (3.17) on $\mathbb{R}^n$, and set:

$$z = \Phi_u(x) = (h(x,u), L_f h(x,u), ..., L_f^{n-1} h(x,u)). \tag{3.25}$$

Let $K \subset \mathbb{R}^n$ be any fixed open relatively compact subset. We deal with semi-trajectories of Σ that remain in K, only. It follows from (3.18) that, for all $u \in U$, Φ_u is an injective immersion (this is easily checked by induction on the components of Φ_u). Therefore, Φ_u is a u–dependent diffeomorphism from K onto its image. Consider the image of the system (3.17) restricted to K by the time dependent diffeomorphism Φ_u. It is of the form:

$$\begin{cases} \frac{d\xi}{dt} = A\xi + g(\xi, u, \frac{du}{dt}) \\ y = \xi_1 \end{cases} \tag{3.26}$$

where A is the antishift matrix, and where g is smooth and depends in a triangular way of ξ.

Even if some technical difficulties remains in the general theoretical case (see [10] for a precise result), it is clear that the new system is of the form (3.21) except that we use explicitly $\frac{du}{dt}$, considered as a new input.

Thanks to this result, our observers (Sections 3.3.2, 3.3.3 and 3.3.4) applies to general uniformly infinitesimally observable systems.

Let us come back to the system (3.21) and its properties. The assumption $0 < a_m \leq a_i(u) \leq a_M$ is not more restrictive than $a_i(u) \neq 0$[3] It just implies observability of systems in the normal form (3.21), by the following reasoning:

1. If the output $y(t)$ is known, the input being also known, the fact that $a_1(u)$ is nonzero implies that we can compute $x_1(t)$ from $y(t)$,
2. The fact that $a_2(u) \neq 0$ implies that we can compute $x_2(t)$ from the knowledge of $x_1(t)$,
3. By induction, we can reconstruct the whole state $x(t)$ from the knowledge of $y(t)$.

The compact support of b can be trivially achieved, by multiplying by a cut-off function, compactly supported, leaving the original vector field b unchanged on an arbitrarily large compact subset of R^n. Let us mention that this restriction to compact sets (unavoidable in a general observation theory), has not so important consequences: for instance, the high gain observers can be used in general for **global** dynamic output stabilization (again, see [20]).

The following results have been proved in [13, 14, 20].

We consider the equations of the extended Kalman filter (3.11), in which the covariance matrix Q depends on a real parameter θ, $\theta \geq 1$, in the following way:

$$Q_\theta = \theta \Delta^{-1} Q \Delta^{-1}$$

where

$$\Delta = \begin{pmatrix} 1 & 0 & 0 & \cdots & 0 \\ 0 & \frac{1}{\theta} & 0 & & \vdots \\ 0 & 0 & \frac{1}{\theta^2} & \ddots & \vdots \\ \vdots & & \ddots & \ddots & 0 \\ 0 & \cdots & \cdots & 0 & \frac{1}{\theta^{n-1}} \end{pmatrix}$$

The EKF becomes the high-gain extended Kalman filter (HG-EKF):

$$\begin{cases} \frac{dz}{dt} = A(u)z + b(z,u) + PC'R^{-1}(y(t) - Cz) \\ \frac{dP}{dt} = (A(u) + b^*(z,u))P + P(A(u) + b^*(z,u))' \\ \qquad +Q_\theta - PC'R^{-1}CP \end{cases} \tag{3.27}$$

[3] Modulo a trivial change of variables, and the fact that the a_i being smooth, restricting to a compact subset of the set of values of control implies that we can find the a_m and a_M.

If $\theta = 1$, the HG-EKF is equivalent to the EKF. If θ is large, Q_θ is a large symmetric definite positive (s.d.p.) matrix and since it appears in the Riccati equation in a positive way, P will become large (in the s.d.p. sense). Therefore, the gain of the observer, namely $PC'R^{-1}$, will be large. This is why the observer (3.27) is called high-gain extended Kalman filter.

This observer has some very nice properties. From a practical point of view, since it is based on extended Kalman filtering approach, it is well designed for filtering noise using the model. Moreover, the HG-EKF is applied to a system written in the canonical form of observability. As a matter of fact, it clearly improves the convergence of the observer, both in simulation and in practical situations. Moreover, the parameter θ has a clear meaning and can be used to tune efficiently the observer: if the observer is too slow, θ should be increased, and if the noise is not enough filtered, θ should be decreased.

This last point has also been validated from a theoretical point of view: the estimation error has arbitrarily large exponential decay, depending on θ. This holds whatever the initial error is, (that is, this is a global result). The theorem is the following:

Theorem 3. *For θ large enough and for all $T > 0$, the HG-EKF (3.27) satisfies for $t > \frac{T}{\theta}$*

$$\|z(t) - x(t)\| \leq \theta^{n-1} k(T) \left\| z\left(\frac{T}{\theta}\right) - x\left(\frac{T}{\theta}\right) \right\| \mathrm{e}^{-(\theta\omega(T) - \mu(T))\left(t - \frac{T}{\theta}\right)}$$

for some positive continuous functions $k(T)$, $\omega(T)$ and $\mu(T)$.

Remark 4. In a stochastic setting, the HG-EKG is a nonlinear filter with bounded variance ([13]).

3.3.3 High-Gain and Non High-Gain Extended Kalman Filter

The EKF is a local converging observer, and has very good properties w.r.t. noise. It is close to the Kalman filter, which is an optimal solution to estimate the unknown state.

The HG-EKF is a globally converging observer. Moreover, it converges exponentially as fast as wanted, depending on the choice of the parameter θ.

The EKF cannot be used to estimate the state from a poor *a priori* estimation, or when large unmodelized perturbations occurs. The HG-EKF is designed to do this. This is the basis of the observer construction proposed in this section. More precisely, let us recall that:

1. if one sets θ to 1 in system (3.27) then one obtains the classical extended Kalman filter, which is a local optimal observer (in the sense explained above)
2. if θ is large enough then one obtains a high-gain observer, which is a global exponential observer.

The first application of this remark was presented in [8]: we just added the equation

$$\frac{d\theta}{dt} = \lambda\left(1 - \theta\right) \tag{3.28}$$

to the system (3.27). If $\theta\left(0\right) = \theta_0$ is large enough (and the parameter λ small enough) then we obtain an observer which is a high-gain observer for small time and which converges asymptotically to a classical extended Kalman filter. Hence we can expect its convergence since the observer should converge exponentially to the state (high-gain observer property) and then stays in a neighborhood of the state (since extended Kalman filter is a local observer). Indeed this result has been proved in [8]. More precisely, the observer can be written (where Q_θ has be defined in the previous section):

$$\begin{cases} \frac{dz}{dt} = A(u)z + b(z,u) + PC'R^{-1}(y\left(t\right) - Cz) \\ \frac{dP}{dt} = (A(u) + b(z,u))P + P(A\left(u\right) + b^*\left(z,u\right))' \\ \qquad +Q_\theta - PC'R^{-1}CP \\ \frac{d\theta}{d\tau} = \lambda(1-\theta) \end{cases} \tag{3.29}$$

and the theorem says that the asymptotic behavior of the observer is the one of the extended Kalman filter, the "short term behavior" is the one of the HG-EKF. More precisely, let us denote by $\varepsilon\left(t\right) = z\left(t\right) - x\left(t\right)$:

Theorem 4. *For all $0 \leq \lambda \leq \lambda_0$, ($\lambda_0$ small enough), for all $\theta\left(0\right) = \theta_0$ large enough, depending on λ, for all $S\left(0\right) = S_0 \geq c\ Id$, for all $K \subset \mathbb{R}^n$, K a compact subset, for all z_0 such that $\varepsilon\left(0\right) = z_0 - x\left(0\right) \in K$, the following estimation holds, for all $\tau \geq 0$:*

$$\begin{aligned} ||\varepsilon(\tau)||^2 &\leq R\left(\lambda, c\right) e^{-a\ \tau}||\varepsilon_0||^2 \Lambda(\theta_0,\tau,\lambda), \\ \Lambda(\theta_0,\tau,\lambda), &= {\theta_0}^{2(n-1)+\frac{a}{\lambda}} e^{-\frac{a}{\lambda}\theta_0(1-e^{-\lambda\tau})}, \end{aligned} \tag{3.30}$$

Moreover the short term estimate

$$||\varepsilon(\tau)||^2 \leq \theta(\tau)^{2(n-1)} R\left(\lambda_0, c\right) e^{-(a_1\theta(T)-a_2)\tau}||\varepsilon(0)||^2. \tag{3.31}$$

holds for all $0 \leq \tau \leq T$ and for all θ_0 large enough. $R\left(\lambda, c\right)$ is a decreasing function of c, and a, a_1 and a_2 are three positive constants.

Remark 5. (3.31) means that, provided that λ is smaller than a certain constant λ_0, and θ_0 is large in front of λ, the estimation error goes exponentially to zero, and can be made arbitrarily small in arbitrary short time. Moreover, in (3.30), the function $\Lambda(\theta_0,\tau,\lambda)$ being a decreasing function of τ, for all $\tau > 0$, $\lambda > 0$, $\Lambda(\theta_0,\tau,\lambda)$ can be made arbitrarily small, increasing θ_0, hence the observer is an exponential observer. Therefore, the observer is an exponential observer but the asymptotic rate of convergence does not depend on $\theta(t)$ (because $\theta\left(t\right) \simeq 1$), hence this observer does not converge as fast as we want after a given time τ.

The main drawback of this observer, as presented here, is that it converges exponentially for any initial condition only in the beginning, in order to estimate the initial state of the system: if a large perturbation occurs after time τ, this observer will have the same behavior as an EKF (since $\theta\left(t\right) \simeq 1$ for t larger that τ).

In order to construct a persistent observer, we should take into account this property and construct a time-dependent observer. The simplest way is to use several observers of the form (3.29), each one initialized at different times, and using some delays between each initialization. Thus we obtain several estimations of the state, given by each one of the observers: the final estimation is the one corresponding to the observer that minimizes the innovation process. The whole construction is clearly explained in [8, 9] and we will recall the algorithm:

We consider a one parameter family $\{O_\tau, \tau \geq 0\}$ of observers of type (3.29), indexed by the time, each of them starting from S_0, θ_0, at the current time τ. In fact, in practice, it will be sufficient to consider, at time τ, a slipping window of time, $[\tau - T, \tau[$, and a finite set of observers $\{O_{t_i}, \ \tau - T \leq t_i \leq \tau\}$, with $t_i = \tau - i\frac{T}{N}$, $i = 1, ..., N$.

As usual, we call the term $I(\tau) = \hat{y}(\tau) - y(\tau)$, (the difference at time τ between the estimate output and the real output), the "innovation". Here, for each observer O_{t_i},we have an innovation $I_{t_i}(\tau)$.

Our suggestion is to take as the estimate of the state, the estimation given by the observer O_{t_i} that minimizes the absolute value of the innovation.

This is a very natural choice, according to probability theory (Section 3.2). The innovation process will also have an important role in Section 3.3.4, but we will consider its integral over small past time, which is another possible choice here.

Let us analyze what will be the effect of this procedure in a deterministic setting: after the transient part and if no un-modeled perturbation occurs, the best estimation is given by the oldest observer. Indeed, the oldest observer has converged and moreover, it is close to a classical EKF and therefore, it is more robust to measurement noise. But if a large perturbation occurs, making a jump on the state, the oldest (EKF) observer will no more converge. The youngest observer, which is a HG-EKF, will converge since it is in transient time (it's life time is less than τ). After an (arbitrary) short transient, the youngest observer will then give the best estimate and hence the smallest innovation.

This analysis is validated by our experience and we can even use these remarks to detect jumps, which correspond to abnormal operations or sensor failures.

Another remark is that this approach may be compared to a particle filtering method where the *a posteriori* estimation of the state is the maximum likelihood one. There exist several differences between these two algorithms and in fact, their use depends as usual on the observability study. If the system is not observable, a filtering approach should be used. If the system is observable, an observer can be used.

3.3.4 Adaptive Gain Extended Kalman Filter

Here, we present a much simple observer. Instead of equation (3.28), we introduce the equation

$$\frac{d\theta}{dt} = F(\theta, \mathcal{I}) \tag{3.32}$$

where

$$\mathcal{I}=\int_{t-T}^{t}\left\|y(s)-\bar{y}_{t-T}(s)\right\|^{2}ds=\left\|y-\bar{y}_{t-T}\right\|_{L^{2}(t-T,t)}^{2} \tag{3.33}$$

is the innovation from time $t-T$ to current time t. More precisely, in (3.33), y represents the output, but $\bar{y}_{t-T}$ represents the prediction of the output from the state estimation at time $t-T$ (given by the observer, $z\left(t-T\right)$). Hence $\bar{y}_{t-T}(s)$ is the solution at time s of

$$\begin{cases} \frac{d\xi}{d\tau}=A(u)\xi\left(\tau\right)+b(\xi\left(\tau\right),u) \\ \xi\left(t-T\right)=Z\left(t-T\right) \\ \bar{y}_{t-T}\left(\tau\right)=C\left(u\right)\xi\left(\tau\right) \end{cases}$$

T is a tuning parameter, representing the length of the window used to calculate the innovation. In the following theorem, the function F will be chosen in the form

$$F\left(\theta,\mathcal{I}\right)=\lambda\left(1-\theta\right)+K\left(\theta_{\max}-\theta\right)\mathcal{I} \tag{3.34}$$

In fact, F can be chosen in a more general form. We will give a version of F that is better adapted in the presence of noise in the application part of this chapter (Section 3.5). Intuitively, the role of the function F is:

- to let θ decrease if the innovation is small, because in this case the observer has already converged and a Kalman-like observer will be sufficient to correctly estimate the state
- to let θ increase if the innovation is too large, because in this case, the observer gives a bad estimation of the state and θ has to be large enough in order to ensure convergence, thanks to the exponential property of high-gain observers.

Finally, the adaptive-gain extended Kalman filter can be written

$$\begin{cases} \dfrac{dZ}{dt}=A(u)Z+b(Z,u)+S^{-1}C'R_{\theta}^{-1}(CZ-y(t)) \\ \dfrac{dS}{dt}=-(A(u)+b(Z,u))'S-S(A\left(u\right)+b^{*}\left(Z,u\right)) \\ \qquad +C'R_{\theta}^{-1}C-SQ_{\theta}S \\ \dfrac{d\theta}{dt}=\lambda\left(1-\theta\right)+K\left(\theta_{\max}-\theta\right)\mathcal{I} \end{cases} \tag{3.35}$$

We define Q_{θ} and R_{θ} from Q and R thanks to the matrix

$$\Delta=\begin{pmatrix} 1 & 0 & 0 & \cdots & 0 \\ 0 & \frac{1}{\theta} & 0 & & \vdots \\ 0 & 0 & \frac{1}{\theta^{2}} & \ddots & \vdots \\ \vdots & & \ddots & \ddots & 0 \\ 0 & \cdots & \cdots & 0 & \frac{1}{\theta^{n-1}} \end{pmatrix}$$

by $Q_{\theta}=\theta\Delta^{-1}Q\Delta^{-1}$ and $R_{\theta}=\theta^{-1}R$. Let us remark that this change of coordinates is different from the previous one (high-gain extended Kalman filters of Section 3.3.2 and Section 3.3.3).

Our main result is the following:

Theorem 5. *Let us consider a system in the canonical form of observability. We consider the adaptive-gain extended Kalman filter (3.35). Let us suppose that λ, K and $\theta_{\max}$ (in (3.34)) are three constant parameters such that λ is small enough, K is large enough, and θ_0 is large enough. Then, (3.35) is an exponentially converging observer.*

The proof is based on the following crucial lemma:

Lemma 1. *Let x_1^0, $x_2^0 \in \mathbb{R}^n$. Let us consider the outputs $y_1(t)$ and $y_2(t)$ with initial conditions respectively x_1^0 and x_2^0. The following condition (called persistant observability) holds:*

$$\forall T > 0 \quad \forall u \in L_b^1(\mathcal{U}_{\text{adm}}) \quad \exists \lambda_T > 0$$

$$\left\| x_1^0 - x_2^0 \right\| \leq \frac{1}{\lambda_T} \int_0^T \left\| y_1(\tau) - y_2(\tau) \right\| d\tau$$

The main difference between the previous observer is the fact that now, the matrix R depends on θ, which was not necessary when θ was only a decreasing parameter. The behavior of this adaptive–gain extended Kalman filter is illustrated on a DC–motor, in Section 3.5.

We point out that this AG-EKF is a very promising tool: it is a small modification of already existing adaptive–gain EKF proposed by engineers to improve the performance of EKF during abnormal operations. We propose the same approach in a theoretical framework, ensuring the exponential convergence of the algorithm.

3.3.5 Observer for Continuous–Discrete Systems

As already explain in Section 3.2.2, practical problems may often be written in continuous-discrete form (3.12). There also exist some observability results concerning these systems. Let us suppose, for simplicity, that the sampling time is constant, *i.e.* $t_k = k\,\Delta t$.

A generalized canonical form of observability for these systems is the natural extension of the generalized canonical form of observability (3.21)

$$\begin{cases} \frac{dx}{dt} = A(u)x + b(x,u) \\ y_k = C(u)\, x(t_k) \end{cases} \tag{3.36}$$

were A, b and C are defined as in (3.22), (3.24) and (3.23) and satisfies the same hypothesis. In the affine control case (3.20), with a discrete observation, the change of coordinates is the same as in the continuous case. In fact, (3.21) and (3.36) are exactly equivalent with $y_k = y(t_k)$.

The HG–EKF for continuous–discrete systems has the (not surprising) form:

Correction step

$$\begin{cases} z\left(t_k^+\right) = z\left(t_k\right) + G\left(k\right)\left(y_k - C\left(u\right) z\left(t_k\right)\right) \\ G\left(k\right) = P\left(t_k\right) C\left(u\right)'\left(C\left(u\right) P\left(t_k\right) C\left(u\right)' + \frac{1}{\Delta t} R\right)^{-1} \\ P\left(t_k^+\right) = \left(I - G\left(k\right) C\left(u\right)\right) P\left(t_k\right) \end{cases} \tag{3.37}$$

Prediction step

$$\begin{cases} \frac{dz}{dt} = A(u)x + b(x,u) \\ \frac{dP}{dt} = (A(u) + b^*(z,u))P + P(A\left(u\right) + b^*\left(z,u\right))' + Q_\theta \end{cases} \tag{3.38}$$

Then we have:

Theorem 6. *([14]) Under same assumptions as in continuous case and for Δt small enough, there is an interval $[\theta_0, \theta_1]$ such that for any $\theta \in [\theta_0, \theta_1]$, the continuous–discrete high–gain extended Kalman filter (3.37–3.38) is an exponential observer.*

Genericity and observability have also been studied for continuous–discrete systems. One can expect that same results hold when sampling time is small enough. Roughly speaking, it is more or less true. There exist continuous–discrete versions of theorems from Section 3.3.1 in the continuous–discrete case ([1, 2]).

3.3.6 A "weak" Separation Principle

In this section, we just want to give an important application concerning high-gain observers and particularly the high-gain extended Kalman filter.

Usually, observers are used in order to control nonlinear systems with a state-feedback control law. This control law $u(x)$ is calculated in order to achieve a good performance and, at least, to ensure the stability of the closed loop system. An observer is developed in order to estimate the state (which is not completely measured, in most applications) and the control law applied to the process is $u(z)$ (where z is the estimation of x given by the observer)[4].

Therefore, the closed loop system consist in a control law and an observer, and both are developed independently.

In the linear-quadratic case, the "separation principle" stated that, if an optimal state-feedback control law is applied with an optimal observer, the result is optimal. It is a very strong "superposition" result which is false for nonlinear systems.

Nevertheless, we can expect to prove a weaker version of the linear separation principle.

Let us consider again our system 3.21. Let us suppose that there exist a positively invariant compact subset of $\mathbb{R}^n$ for any control law $u(t)$.

[4] If a filter has been developped, then one should apply the more accurate control law $u(t) = E\left[u(X(t)) \mid \mathcal{F}_t^Y\right]$ which is usually different from $u(z)$ where $z = E\left[X(t) \mid \mathcal{F}_t^Y\right]$.

Theorem 7. *If $u(x)$ is a state feedback which make the system 3.21 globally asymptotically stable, then the system*

$$\begin{cases} \frac{dx}{dt} = A(u(z))\,x + b(x, u(z)) \\ \frac{dz}{d\tau} = A(u)z + b(z,u) - S(t)^{-1}C'R^{-1}(Cz - y(t)) \\ \frac{dS}{d\tau} = -(A(u) + b^*(z,u))'S - S(A(u) + b^*(z,u)) \\ \qquad +C'R^{-1}C - SQ_\theta S \end{cases}$$

is globally asymptotically stable for θ large enough.

Hence, this theorem show that the state-feedback control law can be replaced by an observer based control law and that the stability is preserved.

Remark 6. It has to be pointed out that this result is not true for the adaptive-gain extended Kalman filter (with these hypothesis) because it is necessary to have an exponentially converging observer with an arbitrary fast convergence.

3.4 Identifiability and Identification

3.4.1 Definitions

The problem of identification is a generalization of the observation problem: very often, practical control systems depend on some functions, (with physical meaning), that are not well known, and that have to be determined on the basis of experiments. Systems under consideration have the following form

$$\begin{cases} \frac{dx}{dt} = f(x, u, \varphi(x,u)) \\ \quad y = h(x, u, \varphi(x,u)) \end{cases} \tag{3.39}$$

If x denotes the state of the system, if $\varphi(x,u)$ is the unknown function, and $y(t)$ is the observed data, the identification problem is the problem of reconstructing the piece of the graph of $\varphi(.)$,visited during the experiment. That is, for an experiment of duration T, we want to determine the trajectories $(x(t), u(t), \varphi(x(t), u(t)))$, for all $t \in [0,T]$, using only the observed data $\{y(t), t \in [0,T]\}$. We say that a system is identifiable if this is possible, whatever the experiment.

An identifier is a device performing this task. We will be interested with "on-line identifiers" only, *i.e.* identifiers that estimate the graph of φ simultaneously to the experiment.

The two problems, of observation and identification, are of course strongly connected for two reasons:

1. we do not suppose that $x(0)$ is known. Hence, the identification problem include an observation problem: we want to estimate both $x(t)$ and $\varphi(.)$. It is the main difference with the right-inversion problem, also known as the input identification problem.
2. identification requires an identifiability study, and this study is closely related to observability study. Moreover, our main tools to perform identification are based on (high–gain) observers.

Let us explain the second point, in the uncontrolled case. We consider smooth (C^ω or C^∞, depending on the context) systems of the form Σ

$$\Sigma \begin{cases} \frac{dx}{dt} = f(x, \varphi(x)) \\ y = h(x, \varphi(x)) \end{cases} \tag{3.40}$$

where the state $x = x(t)$ lies in a n–dimensional analytic manifold[5] X, $x(0) = x_0$, the observation y is $\mathbb{R}^p$–valued, and f, h are respectively a smooth (parametrized) vector field and a smooth function. The function φ is an unknown function of the state. Each trajectory is supposed to be defined on some interval $[0, T_{x_0,\varphi}]$.

- If the number of outputs is three or more, then, identifiability is a generic property,
- If there is only one or two outputs, then, identifiability is a nongeneric property, so strong that it can be characterized by four very rigid normal forms.

Our goal is to estimate both state variable x and unknown function $\varphi : X \longrightarrow I$, I being a compact interval of $\mathbb{R}$ (the theory, developed in [10], clearly has extensions to higher dimension). More precisely, we want to reconstruct the piece of the graph of φ visited during experiment.

Let us recall some definitions and results from this last paper. For this introduction, we will only consider uncontrolled systems such as (3.40). Some results can be extended to controlled systems.

Let $\Omega = X \times L^\infty[I]$, where

$$L^\infty[I] = \{\hat{\varphi} : [0, T_{\hat{\varphi}}] \mapsto I,\ \hat{\varphi} \text{ measurable}\}$$

Then we can define the input/output mapping

$$\begin{aligned} P_\Sigma : \quad \Omega &\longrightarrow L^\infty\left[\mathbb{R}^{d_y}\right] \\ (x_0, \hat{\varphi}(\cdot)) &\longrightarrow y(\cdot) \end{aligned}$$

Definition 5. *Σ is said to be identifiable if P_Σ is injective.*

As for observability, we define an infinitesimal version of identifiability. Let us consider the first variation of the system (3.40) (where $\hat{\varphi}(t) = \varphi \circ x(t)$):

$$T\Sigma_{x_0,\hat{\varphi},\xi_0,\eta} \begin{cases} \frac{dx}{dt} = f(x, \hat{\varphi}) \\ \frac{d\xi}{dt} = T_x f(x, \hat{\varphi})\,\xi + d_\varphi f(x, \hat{\varphi})\,\eta \\ \hat{y} = d_x h(x, \hat{\varphi})\,\xi + d_\varphi h(x, \hat{\varphi})\,\eta \end{cases}$$

and the input/output mapping of $T\Sigma$

$$\begin{aligned} P_{T\Sigma,x_0,\hat{\varphi}} : T_{x_0}X \times L^\infty[\mathbb{R}] &\longrightarrow L^\infty\left[\mathbb{R}^{d_y}\right] \\ (\xi_0, \eta(\cdot)) &\longrightarrow \hat{y}(\cdot) \end{aligned}$$

Definition 6. *Σ is said to be infinitesimally identifiable if $P_{T\Sigma,,x_0,\hat{\varphi}}$ is injective for any $(x_0, \hat{\varphi}(\cdot)) \in \Omega$ i.e.* $\ker(P_{T\Sigma,x_0,\hat{\varphi}}) = \{0\}$ *for any $(x_0, \hat{\varphi}(\cdot))$.*

---

[5] Analytic manifold stands for analytic connected paracompact Hausdorf manifold.

Both identifiability and infinitesimal identifiability mean injectivity of some mapping. Clearly injectivity depends on the domain. Therefore, it seems that these notions are not well defined. In fact these notions do not depend on the domain. Indeed, if an analytic system Σ is not (infinitesimally) identifiable because there exists a L^∞ function which makes the system not (infinitesimally) identifiable, then there exists an analytic function which makes the system not (infinitesimally) identifiable.

We consider again a system Σ of the form (3.40). In [10], we have shown that identifiability is a generic property if and only if the number of observation p is greater or equal to 3. On the contrary, if p is equal to 1 or 2, identifiability is a very restrictive hypothesis (infinite codimension) and we have completely classified infinitesimally identifiable systems by giving certain geometric properties that are equivalent to the normal forms presented in Theorems 8 and 9 [10] below.

These theorems are the basis of our identifier construction: since every identifiable systems may be put, up to a change of coordinates, in one of these canonical form of identifiability, then it is sufficient to develop an identifier for these forms (exactly as observers for observable systems).

Theorem 8. *($p = 1$) If Σ is uniformly infinitesimally identifiable, then, there is a subanalytic closed subset Z of X, of codimension 1 at least, such that for any $x_0 \in X \backslash Z$, there is a coordinate neighborhood $(x_1, \dots, x_n, V_{x_0})$, $V_{x_0} \subset X \backslash Z$ in which Σ (restricted to V_{x_0}) can be written:*

$$\Sigma_1 \begin{cases} \dot{x}_1 &= x_2 \\ &\vdots \\ \dot{x}_{n-1} &= x_n \\ \dot{x}_n &= \psi(x, \varphi) \\ y &= x_1 \end{cases} \quad \text{and } \frac{\partial}{\partial \varphi}\psi(x, \varphi) \neq 0 \tag{3.41}$$

Theorem 9. *($p = 2$) If Σ is uniformly infinitesimally identifiable, then, there is an open-dense semi-analytic subset $\tilde{U}$ of $X \times I$, such that each point (x_0, φ_0) of $\tilde{U}$, has a neighborhood $V_{x_0} \times I_{\varphi_0}$, and coordinates x on V_{x_0} such that the system Σ restricted to $V_{x_0} \times I_{\varphi_0}$, denoted by $\Sigma_{|V_{x_0} \times I_{\varphi_0}}$, has one of the three following normal forms:*

- ***type 1 normal form***

$$\Sigma_{2,1} \begin{cases} y_1 = x_1 & y_2 = x_2 \\ \dot{x}_1 = x_3 & \dot{x}_2 = x_4 \\ \quad\vdots & \quad\vdots \\ \dot{x}_{2k-3} = x_{2k-1} & \dot{x}_{2k-2} = x_{2k} \\ \dot{x}_{2k-1} = f_{2k-1}(x_1, \dots, x_{2k+1}) & \\ \dot{x}_{2k} = x_{2k+1} & \\ \quad\vdots & \\ \dot{x}_{n-1} = x_n & \\ \dot{x}_n = f_n(x, \varphi) & \end{cases} \tag{3.42}$$

with $\frac{\partial f_n}{\partial \varphi} \neq 0$.

- ***type 2 normal form***

$$\Sigma_{2,2}\left\{\begin{array}{llll} y_1 &= x_1 & y_2 &= x_2 \\ \dot{x}_1 &= x_3 & \dot{x}_2 &= x_4 \\ & \vdots & & \vdots \\ \dot{x}_{2r-3} &= x_{2r-1} & \dot{x}_{2r-2} &= x_{2r} \\ \dot{x}_{2r-1} &= \psi(x,\varphi) & \dot{x}_{2r} &= F_{2r}(x_1,\ldots, \\ & & & \quad x_{2r+1},\psi(x,\varphi)) \\ & & \dot{x}_{2r+1} &= F_{2r+1}(x_1,\ldots, \\ & & & \quad x_{2r+2},\psi(x,\varphi)) \\ & & & \vdots \\ & & \dot{x}_{n-1} &= F_{n-1}(x,\psi(x,\varphi)) \\ & & \dot{x}_n &= F_n(x,\varphi) \end{array}\right. \tag{3.43}$$

with $\frac{\partial\psi}{\partial\varphi} \neq 0, \frac{\partial F_{2r}}{\partial x_{2r+1}} \neq 0, \ldots., \frac{\partial F_{n-1}}{\partial x_n} \neq 0$

- ***type 3 normal form***

$$\Sigma_{2,3}\left\{\begin{array}{llll} y_1 &= x_1 & y_2 &= x_2 \\ \dot{x}_1 &= x_3 & \dot{x}_2 &= x_4 \\ & \vdots & & \vdots \\ \dot{x}_{n-3} &= x_{n-1} & \dot{x}_{n-2} &= x_n \\ \dot{x}_{n-1} &= f_{n-1}(x,\varphi) & \dot{x}_n &= f_n(x,\varphi) \end{array}\right. \tag{3.44}$$

with $\frac{\partial}{\partial\varphi}(f_{n-1}, f_n) \neq 0.$

Theorem 10. *($p \geq 3$) If Σ is an infinitesimally identifiable generic system, then there is a connected open dense subset Z of X such that for any $x_0 \in X \backslash Z$, there exist a smooth C^∞-function F and a $\left(\breve{y}, \breve{y}', \ldots, \breve{y}^{(2n)}\right)$-dependent embedding $\Phi_{\breve{y},\ldots,\breve{y}^{(2n)}}(x)$ such that outside Z, trajectories of $\Sigma_{x_0,\varphi}$ are mapped via $\Phi_{\breve{y},\ldots,\breve{y}^{(2n)}}$ into trajectories of the following system*

$$\Sigma_{3+}\left\{\begin{array}{rl} \frac{dz_1}{dt} &= z_2 \\ \frac{dz_2}{dt} &= z_3 \\ & \vdots \\ \frac{dz_{2n}}{dt} &= z_{2n+1} \\ \frac{dz_{2n+1}}{dt} &= F\left(z_1,\ldots,z_{2n+1},\breve{y},\ldots,\breve{y}^{(2n+1)}\right) \\ \bar{y} &= z_1 \end{array}\right.$$

where z_i, $i = 1,\ldots,2n+1$ has dimension $p-1$, and with

$$\left\{\begin{array}{l} x = \Phi^{-1}_{\breve{y},\ldots,\breve{y}^{(2n)}}(z) \\ \varphi = \Psi(x,\breve{y}) \end{array}\right. \tag{3.45}$$

($\breve{y}$ is a selected output).

3.4.2 Identifiers

As explained before, we have to build an identifier for each canonical form of identifiability. The basic idea is the same for all these forms, and leads to the use of the nonlinear observers developed previously: we assume, along the trajectories visited, a local model for φ. For instance, a simple local model is: $\varphi^{(k)} = 0$.

This does not mean, at the end, that we will identify φ as a polynomial in t: the question is not that this polynomial models the function φ globally as a function of t, but only locally, on reasonable time intervals (reasonable w.r.t. the performances of the observer that we will use).

This idea is just an extension of the classical way to identify constant or slowly varying parameters m. In this case, one uses to add the parameter in the state variables and to add the equation $\frac{dm}{dt} = 0$. Therefore, the local model is a constant polynomial. In our case, such local model is too constrained (since φ is not supposed to vary slowly), so we add a polynomial local model.

Let us consider a system Σ in the identifiability normal form 3.41. Adding the local model for φ, we get the system:

$$
\begin{aligned}
y &= x_1, \qquad\qquad (3.46)\\
\dot{x}_1 &= x_2, ..., \dot{x}_{n-1} = x_n,\\
\dot{x}_n &= \Psi(x, \varphi_1), \dot{\varphi}_1 = \varphi_2, ..., \dot{\varphi}_{k-1} = \varphi_k, \dot{\varphi}_k = 0,\\
\frac{\partial \Psi}{\partial \varphi_1} &\neq 0 \text{ (never vanishes).} \qquad\qquad (3.47)
\end{aligned}
$$

This is a system on $\mathbb{R}^{n+k}$, which is not controlled (however, for the considerations that follow, Ψ could depend on a control u), and this system is under the normal form (3.17, 3.18).

Therefore, we may apply high gain Luenberger observer, or we may apply the trick in Section 3.3.2. Then, for instance, the observer of Sections 3.3.2, 3.3.3 and 3.3.4 may be applied to this system. It will provide estimations of $x(t), \varphi(t)$, that is, just an estimation of the piece of the graph of φ visited during the experiment.

The cases of normal forms (3.42), (3.43), (3.44), corresponding to Type 1 to 3 systems can be treated in a similar way to the single-output case, with some more or less easy adaptations of the methods of the previous sections. This exercise is left to the reader.

An application of this technique in a difficult case (the local polynomial model does not apply) is presented in Section 3.6. Some important remarks and practical considerations are discussed in this section.

3.5 Series-Connected DC Motor

In this first application we present (in simulation) the design of the adaptive-gain extended Kalman filter (AG–EKF, see Section 3.3.4) for a single input single output (SISO) system, namely a series-connected DC motor.

Basically, an electric motor converts electrical energy into mechanical energy. In a DC motor, the stator (also called field) is composed of an electromagnet, or a permanent magnet, that immerses the rotor in a magnetic field. The rotor (also called armature) is made of an electromagnet that once supplied with current creates a second magnetic field. The motion is then caused by the attraction/repelling behavior of magnets. As far as the magnetic field created by the stator remains fixed the rotor windings are connected to a commutator. The direction of the current flowing through the armature coils is then switched during the rotation and the polarity of the armature magnetic field is reversed. Successive commutations then maintain the rotating motion of the machine. A DC motor whose field circuit and armature circuit are connected in series, and therefore fed by the same power supply, is referred to as a series-connected DC motor [34].

3.5.1 Mathematical Model

The model of the series-connected DC motor is obtained from the equivalent circuit representation shown in Figure 3.1. We denote by I_f the current flowing through the field part of the circuit (between points A and C) and I_a the current through the armature circuit (between points C and B). When the shaft of the

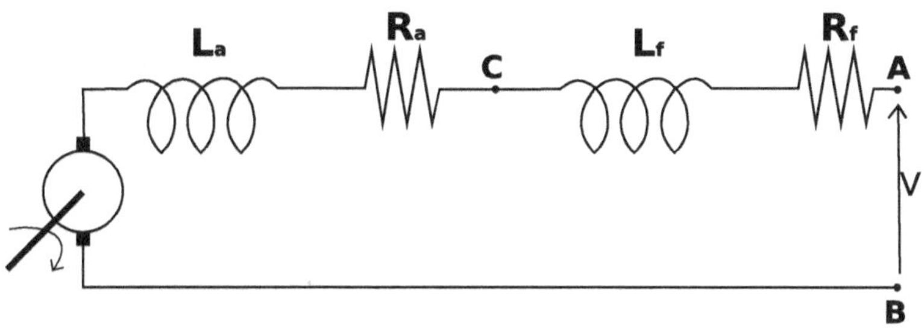

Fig. 3.1. Series-connected DC motor equivalent circuit representation

motor is turned by an external force, the motor acts as a generator and produces an electromotive force. In the case of the DC motor, this force will act against the current applied to the circuit and is then called *back or counter electromotive force* (BEMF or CEMF). The electrical balance leads to

$$L_f \dot{I}_f + R_f I_f = V_{AC}$$

for the field circuit, and to

$$L_a \dot{I}_a + R_a I_a = V_{CB} - E$$

where L_f and R_f are the inductance and the resistance of the field circuit, L_a and R_a are the inductance and the resistance of the armature circuit, and E denotes the Back EMF. Kirchoff's laws give us the relations

$$\begin{cases} I = I_a = I_f \\ V = V_{AC} + V_{CB} \end{cases}$$

which gives for the total electrical balance

$$L\dot{I} + RI = V - E$$

where $L = L_f + L_a$ and $R = R_f + R_a$. Now denoting by Φ the field flux, we have $\Phi = f(I_f) = f(I)$, and $E = K_m \Phi \omega_r$ where K_m is a constant and ω_r is the rotational speed of the shaft.

The second equation of the model is given by the mechanical balance of the shaft of the motor using the well known Newton's law. We consider that the only forces applied to the shaft are the electromechanical torque T_e, the viscous friction torque and the load torque T_a leading to

$$J\dot{\omega_r} = T_e - B\omega_r - T_a$$

where J denotes the rotor inertia, and B the viscous friction coefficient. The electromechanical torque is given by $T_e = K_e \Phi I$ with K_e denoting a constant parameter. We consider that the motor is operated **below saturation**: the field flux can be expressed by the linear expression $\Phi = L_{af} I$ where L_{af} denotes the mutual inductance between the field and the rotating armature coils. To conclude with the modeling of the DC motor we suppose the ideal hypothesis of 100% efficiency in the energy conversion expressed by $K = K_m = K_e$, and for notation simplicity we write L_{af} instead of KL_{af}. The voltage is the input of the system $u(t)$ and the current I is the measured output. We finally obtain the following SISO model for the series-connected DC motor

$$\begin{pmatrix} L\dot{I} \\ J\dot{\omega_r} \end{pmatrix} = \begin{pmatrix} u - RI - L_{af}\omega_r I \\ L_{af} I^2 - B\omega_r - T_a \end{pmatrix} \qquad (3.48)$$
$$y = I$$

This model will be used to simulate the DC motor by means of a Matlab/Simulink S-function.

3.5.2 Observability Canonical Form

Before implementing the observer in order to reconstruct the state vector of this system we test (quite easily) its observability property. We use the *differentiation* approach that is we verify the differential observability (Definition 3) which implies observability.

- $I(t)$ is known with time, then $\dot{I} = (1/L)(u - R.I - L_{af}\omega_r I)$ is known and as far as $u(t)$, R, and L_{af} are known then ω_r can be computed
- now that $\omega_r(t)$ is known, $\dot{\omega_r} = (1/J)(L_{af}I^2 - B\omega_r - T_a)$ can be computed and because of the knowledge we have of $I(t)$, L_{af}, B,and J, T_a can be estimated

We deduce from this that a third variable may be added to the state vector in order to reconstruct both the state of the system and the load torque applied to the shaft of the motor. We assume that the load torque is constant over time. Sudden changes of the load torque will then be considered as unmodeled perturbations. The observer we use is the adaptive-gain Kalman filter as described in

Section 3.3.4 because it has the classical EKF structure when no perturbations occur and the structure of a HG–EKF when the system faces a perturbation. Estimation of the load torque is made possible by the addition of the equation $\dot{T}_a = 0$ to (3.48) (see remarks in Section 3.4.2). We now need to find the coordinate transformation that puts this systems into the observability canonical form.

From the equation $y = I$, we choose $z_1 = I$ and then

$$\dot{z}_1 = \frac{1}{L}(u(t) - RI - L_{af} I \omega_r)$$

which by setting $z_2 = I\omega_r$ becomes

$$\dot{z}_1 = -\frac{L_{af}}{L} z_2 + \frac{1}{L}(u(t) - Rz_1) = \alpha_2(u) z_2 + b_1(z_1, u) \tag{3.49}$$

we now compute the time derivative of z_2

$$\dot{z}_2 = \dot{I}\omega_r + I\dot{\omega}_r = -\frac{1}{J} T_a I - \frac{B}{J} I\omega_r + \frac{L_{af}}{J} I^3 - \frac{L_{af}}{L}\omega_r^2 I + \frac{u(t)}{L}\omega_r - \frac{R}{L}\omega_r I$$

when $I > 0$ and consequently $z_1 > 0$ which sounds as a reasonable assumption as far as I is the current of the circuit which is equal to zero only when there is no power supplied to the engine (and therefore nothing to observe), we set $\omega_r = \frac{z_2}{z_1}$, and by setting $z_3 = T_a I$ this equation becomes

$$\dot{z}_2 = -\frac{1}{J} z_3 - \frac{B}{J} z_2 + \frac{L_{af}}{J} z_1^3 - \frac{L_{af}}{L}\frac{z_2^2}{z_1} + \frac{u(t)}{L}\frac{z_2}{z_1} - \frac{R}{L} z_2 = \alpha_3(u) z_3 + b_2(z_1, z_2, u) \tag{3.50}$$

and identical remark as above lead us to the expression $T_a = \frac{z_3}{z_1}$ and recalling that $\dot{T}_a = 0$ we obtain

$$\dot{z}_3 = -\frac{L_{af}}{L}\frac{z_2 z_3}{z_1} + \frac{u(t)}{L}\frac{z_3}{z_1} - \frac{R}{L} z_3 = b_3(z_1, z_2, z_3, u) \tag{3.51}$$

Thus the application from $\mathbb{R}^{*+} \times \mathbb{R} \times \mathbb{R} \to \mathbb{R}^{*+} \times \mathbb{R} \times \mathbb{R}$ defined by $(I, \omega_r, T_a) \to (I, I\omega_r, IT_a)$ with $(z_1, z_2, z_3) \to \left(z_1, \frac{z_2}{z_1}, \frac{z_3}{z_1}\right)$ as its inverse, is a change of coordinates that puts the system (3.48) into the observer canonical form defined by (3.49), (3.50) and (3.51). It is necessary to compute the coefficients of the matrix b^*.

3.5.3 Observer Implementation

We now recall the equations of the AG–EKF

$$\left\{\begin{aligned} \frac{dZ}{dt} &= A(u)Z + b(Z,u) + PC'R_\theta^{-1}(CZ - y(t)) \\ \frac{dS}{dt} &= P(A(u) + b^*(Z,u))' + (A(u) + b^*(Z,u)) \\ &\quad -PC'R_\theta^{-1}CP + Q_\theta \\ \frac{d\theta}{dt} &= \lambda(1 - s(\mathcal{I})).(1-\theta) + K.s(\mathcal{I}).(\theta_{max} - \theta) \end{aligned}\right. \tag{3.52}$$

where $R_\theta = \theta^{-1}R$ and $Q_\theta = \theta\Delta Q\Delta$ with $\Delta_\theta = diag\left(\theta, \theta^2, \ldots, \theta^n\right)$, $s(\mathcal{I}) = \left[1 + e^{-\beta(\mathcal{I}-m)}\right]^{-1}$ and

$$\mathcal{I} = \int_{t-T}^{t} \left\|y(s) - \bar{y}_{t-T}(s)\right\|^2 ds = \left\|y - \bar{y}_{t-T}\right\|^2_{L^2(t-T,t)} \tag{3.53}$$

In fact, these equations are a slight modification of (3.34): the function F has been modified in order to take into account noise effects, as we will explain below.

The simulation of the DC motor is straightforward, we then only comment the implementation of the observer. A Matlab/Simulink block diagram representing the DC machine and the observer is shown in Figure 3.2 (this figure is incomplete as far as one would surely want to plot errors between real and estimated states). As it may be seen from the simulink block diagram shown in Figure 3.3 the observer is decomposed into three parts: two level 1 S-functions and a transport delay block. As written on the diagram, the rightmost S-function is dedicated to the computation of the three main equations of the observer which are equations (3.52). This block has three type of inputs: the measured output of the observed system, the input delivered to the observed system and the innovation. The innovation is computed using a distinct S-function because unlike the main equations that may be processed continuously (or quasi-continuously), a

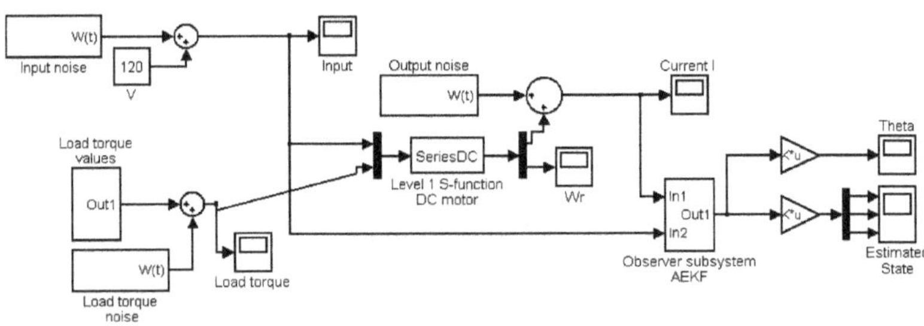

Fig. 3.2. Simulation and observation of the DC motor

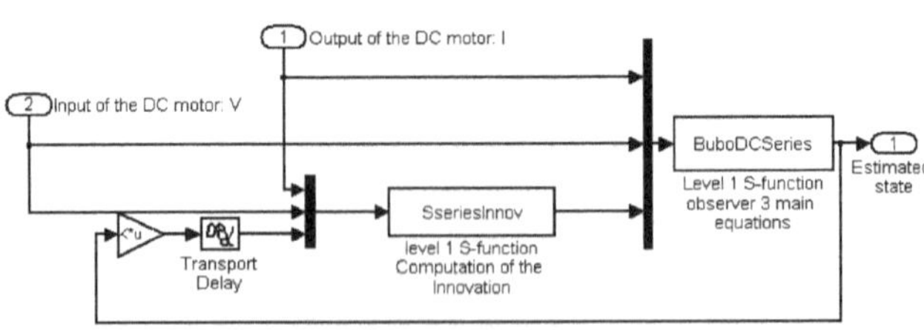

Fig. 3.3. Observer subsystem

discrete S-function is needed to compute the innovation. This choice was made because:

- the computation of the integral is made by means of a fixed step trapezoidal method
- we have to keep memory of the input and the output trajectories over a time interval $[0;T]$ where T is the delay of (3.53) which is easily done with a fixed step process.

The codes to implement those different functions may be downloaded from http://www.u-bourgogne.fr/monge/e.busvelle/springer/ or obtained from the authors if the link happens to be disabled.

3.5.4 Simulation Parameters and Observer Tuning

The parameters used to simulate the DC engine, motivated by measures made on a real system, are $L = 1.22\,H$, $Res = 5.4183\,\Omega$, $L_{af} = 0.0683\,N.m.Wb^{-1}.A^{-1}$, $J = 0.0044\,kg.m^2$, and $B = 0.0026\,N.m.s^{-1}.rad^{-1}$.

We now need to set the observer parameters d, Dt, R, Q, θ_{max}, λ, K, β, and m. Before explaining how those parameters may be tuned, we want to stress that the last four ones do not need to be reset for each new observer. Those parameters appear in the last equation in (3.52) and drive the evolution of the parameter θ. The values $\lambda = K = 500$, $\beta = 2000$, and $m = m_1 + m_2$ where $m_1 = 0.005$ (m_2 will be specific to each new process) may be kept each time a new observer is implemented. The procedure used to tune the parameters R, Q, θ_{max} is inspired by the one proposed in [9, part. 5.2.2].

1. As a first step, we determine the (symmetric positive definite) matrices R and Q by using an EKF. This observer can be obtained from the AG–EKF when the parameters of the adaptation function are set to 0 and $\theta(0) = 1$. Large perturbations are not considered and the observer is initialized to the proper (or previously estimated) values of the state vector.
2. As a second step, we set the R and Q matrices to the values previously found and use a HG-EKF in order to tune θ. As above the observer needed is obtained from the AG–EKF when the parameters of the adaptation function are set to 0. Then $\theta(0)$ is the value that is tuned. Here we will try to find a value for the high-gain parameter that allows fast and reasonable convergence (with respect to noise amplification) when large unmodeled perturbations are applied to the system. θ_{max} is then taken equal to the value estimated at this step.
3. As a last step we now set the parameters of the adaptation function. We remark that when $m = 0$ then $s(0) = 0.5$. Thus we need to shift the sigmoid function to the right if we want $s(0)$ to be close to zero. Choosing y_1 as small as we want and solving the equation $s(0) = y_1$ allows to obtain the parameter m. This solution is easily computed provided that the parameter β is known. As the sigmoid function is centered on $(0, 0.5)$ when $m = 0$, the computation of β is made by setting a length l for the transition part and solving the nonlinear equation (with $m = 0$): $s(l/2) - s(-l/2) = (1 - y_1) - y_1$. Of course, another

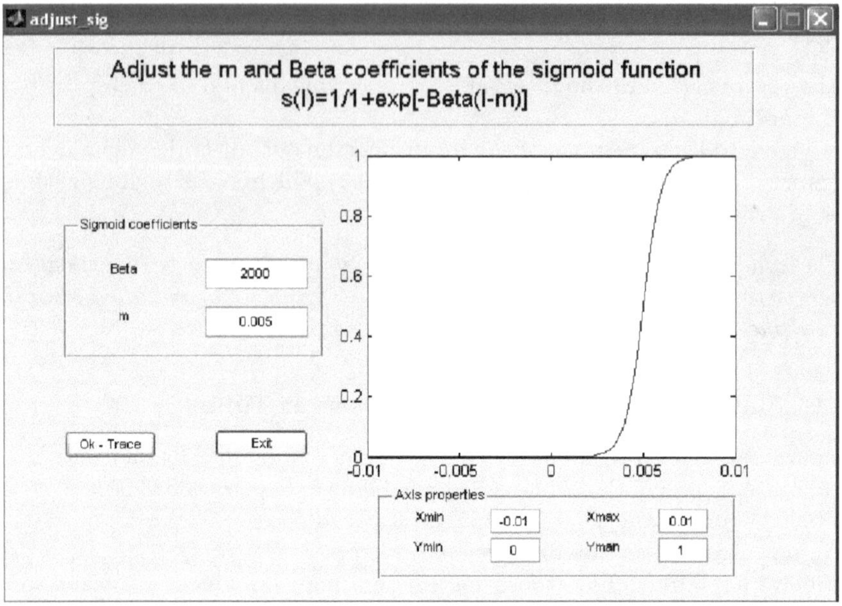

Fig. 3.4. Estimation of β and m_1 by trial and error

approach is to graphically define β and m from trial and error. Figure 3.4 shows a simple Matlab GUI implemented to ease this latter method (the result displayed is for the values of β and m_1 given above). The code of this GUI is also available from *http://www.u-bourgogne.fr/monge/e.busvelle/springer/*.

Now that the transition part is small, we want the gain to increase and decrease quickly. If we suppose that $\theta(t) = 1$ and that we want it to reach θ_{max} within a time τ then the equation $\dot{\theta} = \frac{\theta_{max}-1}{\tau} = K.(\theta_{max} - 1)$ allows the computation of K. As far as the equation used to compute K is only an approximation, a bigger value (e.g. twice the computed value) may be used. Finally, a reasonable choice for the last parameter remaining is $\lambda = K$.

The parameter T, the length of the window on which innovation is computed, is related to the rise time of the system when it is facing perturbations: it has to be sufficiently big so as to give an account of perturbations that occur on the system. The sample time Dt of the discrete S-function should ideally be chosen as small as possible, leading to a significant increase of the amount of time and of the memory needed to compute the innovation (we need to keep track of $\frac{T}{Dt} + 1$ system outputs and $\frac{T}{Dt}$ system inputs). $Dt = T/3$ or $Dt = T/4$ seems to be reasonable, fewer values will of course give more flexibility to the system.

Because of measurement noise the innovation will never be equal to zero and therefore the observer will stay in a high-gain mode. To avoid this problem, the parameter m is rewritten $m = m_1 + m_2$ where m_1 is the previously computed quantity and m_2 will represent the influence of the noise on the system. As a result, when $\mathcal{I} \leq m_2$ we will have $s(\mathcal{I}) \leq y_1$ and θ won't increase. We denote

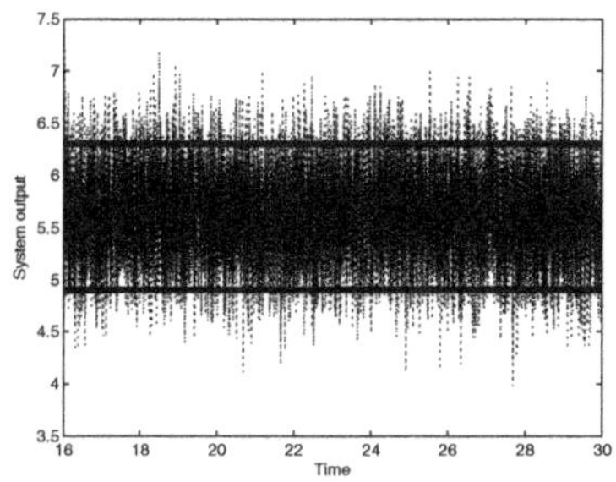

Fig. 3.5. Estimation of the standard deviation

by σ the standard deviation of the output noise, which can be estimated from output measurements, and then $m_2 = T.\sigma^2$ where T is the delay used in the definition of the innovation. Figure 3.5 shows the output of the simulated DC motor (with addition of noise) and that $\sigma = 0.7$ is a reasonable value for the standard deviation.

Finally all those steps allow us to set the parameters to $R = 1$, $Q = [1,0,0;0,5,0;0,0,5]$, $\theta_{max} = 3$, $\lambda = K = 500$, $\beta = 2000$, $T = 0.1$, $Dt = 0.01$, and $m = 0.005 + 0.049$.

3.5.5 Simulation Results

Figures 3.6 and 3.7 shows the performance of the designed observer, all the observers identify the values taken by the load torque but with different behaviors. The EKF rejects noise but converges slowly when the system faces unmodeled perturbations. We may add that in order to speed up a little bit the EKF the Q matrix was set to $[25,0,0;0,25,0;0,0,50]$ in this special case, it was kept to the value given in the previous chapter for all the other simulations.

The HG–EKF is on the contrary very sensitive to measurement noise but is very fast regarding convergence when a perturbation arises.

The AG–EKF presents both the advantages of the two previous filters, namely noise rejection and speed of convergence under perturbations. We observe that the adaptive-gain observer is a little bit slower than the fixed high-gain one. This is due to the delay induced by the computation of innovation, in fact the value chosen for Dt will have an impact on this delay as far as the behavior of θ (increasing towards θ_{max} or decreasing towards 1) will only change with the innovation. In all the parameters tuned for this last observer one will have a major impact, this is m_2. If indeed it is set to a too big value, then θ won't increase every time it is needed, which does not constitute a major drawback because the EKF rejects noise (this is true provided that m_2 is not such as big that it totally prevent θ from increasing). On the contrary, if m_2 is too small then θ will increase when it is not needed (only because of the noise) having the only effect to amplify noise. However as it can be seen from Figure 3.5, σ and therefore m_2 is not difficult to estimate from output measurements. To illustrate

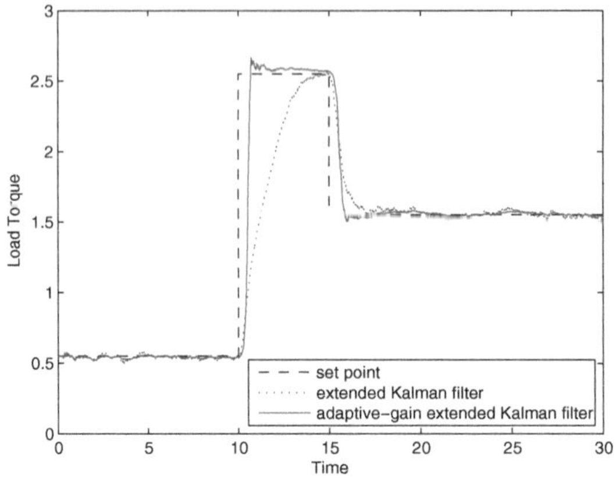

Fig. 3.6. EKF VS AEKF

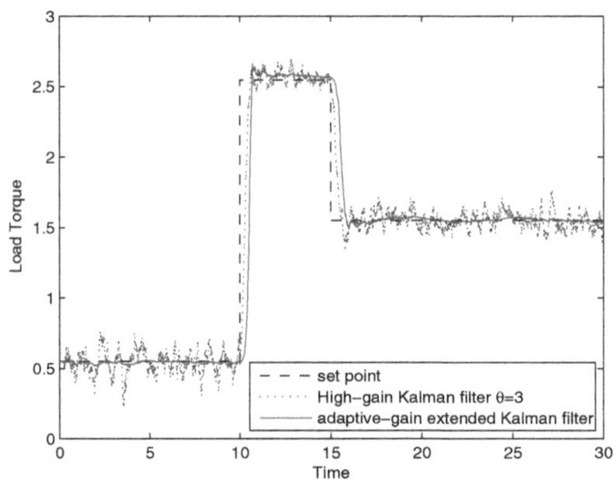

Fig. 3.7. HGEKF VS AEKF

this comment Figure 3.8 shows the evolution of θ for two different values of m_2 (the value 0.049 corresponds to the simulations which results are shown above).

3.6 Electronical Neuron Circuit

With this second application we illustrate how observers can assist system modeling and, in the case considered here, prototype assessment (as in Section 3.4). Identifiability study of this model has been presented in [5].

The modelization of neurons is extensively studied in neuroscience research. A large quantity of models of isolated neuron cells or of neuron cells networks are available in the literature each one of them presenting variable degrees in their accuracy. The model we use here, a modification of the model proposed by

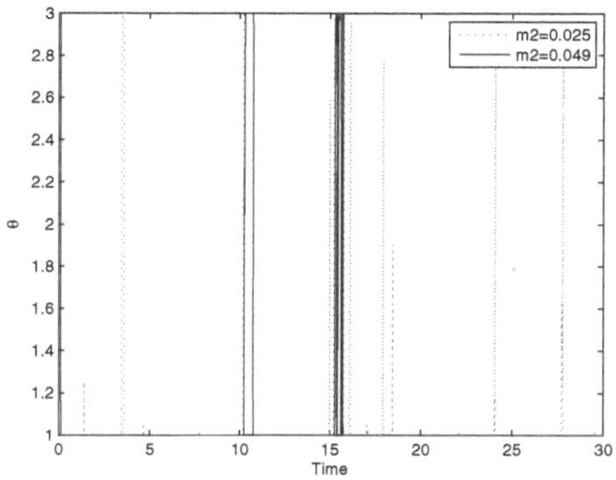

Fig. 3.8. Different values of m_2

Fitzhugh, Nagumo & al. in the early 1960's, is a simplification of the one of a single isolated biological neuron proposed by Hodgkin and Huxley [24]. Historical informations on the development of this model can be found in [28].

3.6.1 The Modified Fitzhugh-Nagumo Model (MFHN)

From the biological point of view this model is composed of two variables, V representing the membrane voltage and W that represents the recovery variable

$$\left\{ \begin{array}{l} \dot{V} = V - \frac{V^3}{3} - W \\ \dot{W} = \epsilon \left(g(V) - W - \eta \right) \end{array} \right. \tag{3.54}$$

where ϵ and η are constant parameters and g is the piecewise linear function

$$g(V) = \left\{ \begin{array}{l} \beta V \text{ if } V > 0 \\ \alpha V \text{ if } V \leq 0 \end{array} \right.$$

where α and β are constant parameters.

This model was implemented as an analogue circuit at LE2I laboratory (university of Burgundy), the exact description of this circuit is given in [6]. The analysis of this physical system is made by means of an observer based approach, real data being available.

3.6.2 Identifiability and Observability

From the analogue circuit point of view, V corresponds to a voltage and W to a current therefore both of them can be measured. Although in the case of a real biological system it will only be possible to measure V, the membrane voltage. Thus we will consider that only V is actually measured. The objective of this study is the identification of the function g (i.e. the part of its graph visited during the experiment) and the study of the identifiability property of the system constitutes a first step. In Section 3.4, we described an identifiability normal form for single output uncontrolled systems (normal forms for systems with more than one output are also given)

$$\begin{cases} \dot{x_1} &= x_2 \\ & \vdots \\ \dot{x_{n-1}} &= x_n \\ \dot{x_n} &= \psi(x,g) \\ y &= x_1 \end{cases} \tag{3.55}$$

We now want to find a change of coordinates that allow the MFHN equations to match this normal form. This coordinate transformation is easily found: set $x_1 = V$ and $x_2 = \dot{V}$.

$$\begin{cases} \dot{x_1} = \dot{V} \\ \quad = x_2 \\ \dot{x_2} = \dot{V} - \dot{V}V^2 - \dot{W} \\ \quad = (1 - x_1^2)x_2 - \epsilon\left(g(x_1) - x_1 + \frac{x_1^3}{3} + x_2 - \eta\right) \\ \quad = \psi(x,g) \end{cases} \tag{3.56}$$

Since $\epsilon \neq 0$, the system is clearly identifiable. We see that if the parameter η is unknown we have the possibility to redefine the unknown function g as $g(x_1) = g(x_1) - \eta$ with no change in the normal form.

In order to identify the function g, we extend the state vector by making g a state variable. As it is clear that g is not constant over time we model it as a local polynomial of time

$$g(V(t)) = g(t) = a_0 + a_1 t + \ldots + a_n t^n$$

which implies that $\frac{d^{n+1}g(t)}{dt^{n+1}} = 0$. The model is completed by the addition of n new state variables corresponding to the n first derivatives of g with respect to time (for a total of $n+1$ new variables). It appears that when the system defined by (3.56) is extended in that manner it is in the observability canonical form. However there exists a much more simpler way to obtain the canonical form that is not to do any change of variables. This latter form is the one we will consider so as to avoid change of variables while implementing the observer

$$\frac{d}{dt}\begin{pmatrix} V \\ W \\ h_0 \\ h_1 \\ \ldots \\ h_{n-1} \\ h_n \end{pmatrix} = \begin{pmatrix} 0 & -1 & 0 & 0 & \ldots & 0 & 0 \\ 0 & 0 & \epsilon & 0 & \ldots & 0 & 0 \\ 0 & 0 & 0 & 1 & \ldots & 0 & 0 \\ 0 & 0 & 0 & 0 & \ldots & 0 & 0 \\ \ldots & \ldots & \ldots & \ldots & \ldots & \ldots & \ldots \\ 0 & 0 & 0 & 0 & \ldots & 0 & 1 \\ 0 & 0 & 0 & 0 & 0 & 0 & 0 \end{pmatrix} \begin{pmatrix} V \\ W \\ h_0 \\ h_1 \\ \ldots \\ h_{n-1} \\ h_n \end{pmatrix} + \begin{pmatrix} V - \frac{V^3}{3} \\ -\epsilon(W - \eta) \\ 0 \\ 0 \\ \ldots \\ 0 \\ 0 \end{pmatrix}$$

where $h_i = \frac{d^i g(t)}{dt^i}$ for $i = 0, \ldots, n$ and with $\frac{d^0 g}{dt} = g$.

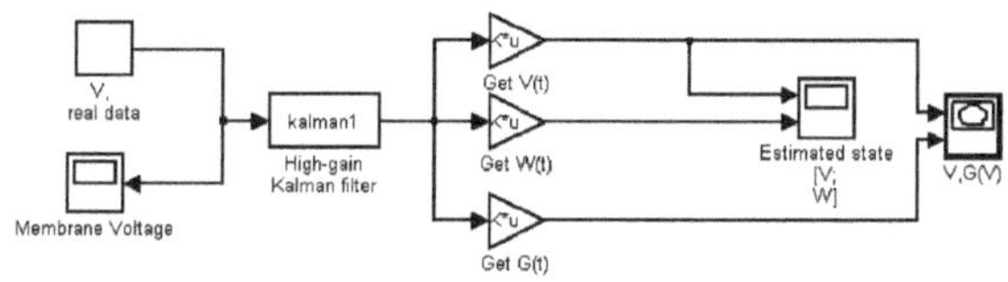

Fig. 3.9. Identification from real data

One could think that the choice of a local representation for the function g (here a polynomial of time) and the transformation of the model into the canonical observability form is enough to prove identifiability. It is in fact not the case. This subtle difference has been well illustrated in [9, part 6] where the authors exhibit the example

$$\begin{cases} \dot{x} = \quad \varphi(x) \\ y = x + \varphi(x) \end{cases} \qquad x \in \mathbb{R}$$

indeed, keeping the notations used above for the function g and setting $n = 1$, then the change of coordinates $(x, h_0, h_1) \to (z_1, z_2, z_3) = (x + h_0, h_0 + h_1, h_1)$ leads to an observability canonical form. However the authors showed that this system is not identifiable !

3.6.3 Implementation

The high-gain extended Kalman filter is adapted to the problem of identification of the unknown function g. The implementation of this observer is much more easy to carry on than the previous one: only one S-function is needed. Even if our objective is to use real data obtained from the analogue circuit mentioned above we use a continuous S-function. This is motivated by the fact that our data's sample time is smaller than the average time step used by the software to compute the continuous solutions (but a continuous–discrete observer (3.37)–(3.38) can be another possible choice). The corresponding Matlab/Simulink diagram is shown Figure 3.9.

Codes may be downloaded (together with a set of data) from http://www.u-bourgogne.fr/monge/e.busvelle/springer/.

3.6.4 Results

A first series of simulations of the MFHN model are done in order to tune the three parameters n, Q, and θ. The parameters for the MFHN model are set to $\alpha = 0.5$, $\beta = 1.96$, $\epsilon = 0.2966$, $\eta = 0.20531$ $V(0) = 1.0656$, and $W(0) = 2.6903$. Since we are using an observer that only has a high-gain behavior, Q is set to the identity matrix $Id_{(3+n)\times(3+n)}$. The high-gain parameter θ is then chosen to ensure an accurate identification of the function. Several simulations show that $\theta = 1$ (corresponding to an extended Kalman filter) does not lead to the

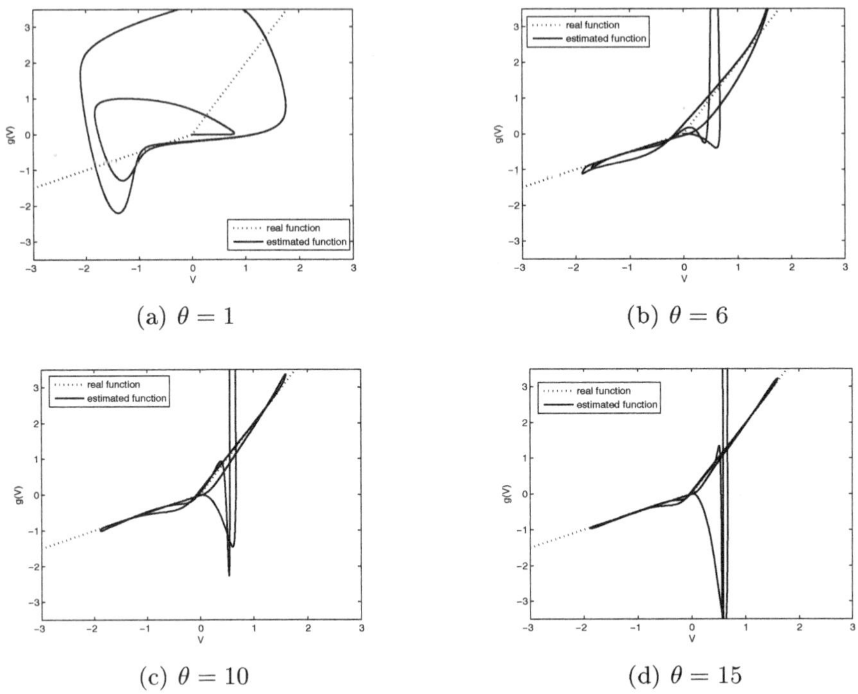

(a) $\theta = 1$

(b) $\theta = 6$

(c) $\theta = 10$

(d) $\theta = 15$

Fig. 3.10. Identification of g from simulations (without noise addition)

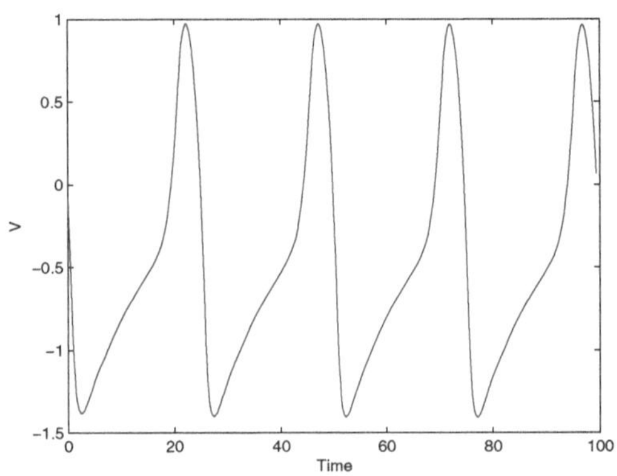

Fig. 3.11. Circuit voltage V

identification of the function. The identification is made possible when $\theta \in [5; 10]$, and is very accurate when $\theta > 10$. Figure 3.10 shows identification results for four different values of the high-gain parameter when the data fed to the observer

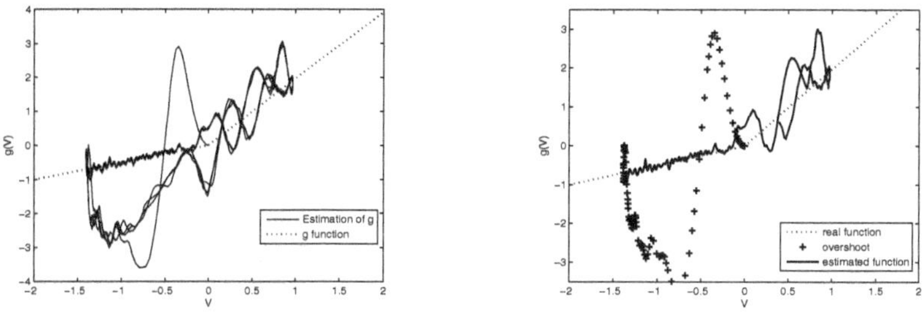

Fig. 3.12. Decomposed identification of the function g

are simulated. No noise has been added during those simulations and then even if $\theta = 15$ gives the best result, the trade-off between speed of convergence and sensibility to noise leads us to choose a smaller value.

The values for V got from the analogue circuit are shown Figure 3.11 and the result of the identification (with $\theta = 10$ and $n = 1$) is shown Figure 3.12(a). We see that the unknown function is identified as a loop and from the shape of the data used, we expect four of them.

We isolated the first values given by the observer in order to obtain the clearer graphic Figure 3.12(b) in which we highlighted the overshoot due to the inaccurate initialization of the observer. After this overshoot the observer converges to the values taken by the unknown function and while $V < 0$ the estimation is quite good. When V becomes positive the estimation is not that accurate anymore. Two reasons can be pointed out to explain this phenomenon: the real data do not correspond exactly to the output the theoretic model would give for the same set of parameters (which is analogous to modeling errors) and the fact that the function we want to identify is not differentiable in 0, a very specific property that is not reflected by our polynomial approximation.

We rewrite the model used to perform the identification so as to take this into consideration

$$\left\{\begin{array}{llll} \dot{x_1} = x_2 & \dot{x_2} = & \bar{\psi}(x, \hat{\alpha}, \hat{\beta}) \\ \dot{\hat{\alpha}} = \alpha_1 & \dot{\hat{\beta}} = & \beta_1 \\ \dot{\alpha_1} = \alpha_2 & \dot{\beta_1} = & \beta_2 \\ \dot{\alpha_2} = \alpha_3 & \dot{\beta_2} = & \beta_3 \\ \dot{\alpha_3} = 0 & \dot{\beta_3} = & 0 \end{array}\right. \tag{3.57}$$

The results of this new identification are shown Figure 3.13(a-b). This new estimation is very accurate after a few cycles. Small errors both for the positive and negative values of V are still visible, they can also be spotted when we trace the values taken by $\hat{\alpha}$ and $\hat{\beta}$ against time as in Figure 3.14. Those errors are due to the fact that real data differ from the ideal mathematical model.

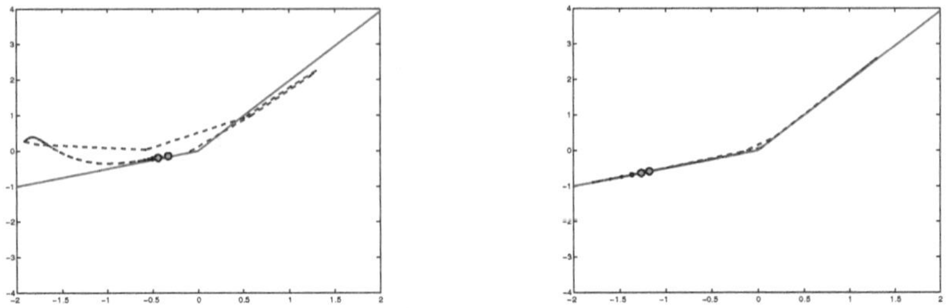

(a) after 4 cycles

(b) after 15 cycles

Fig. 3.13. Estimation of g

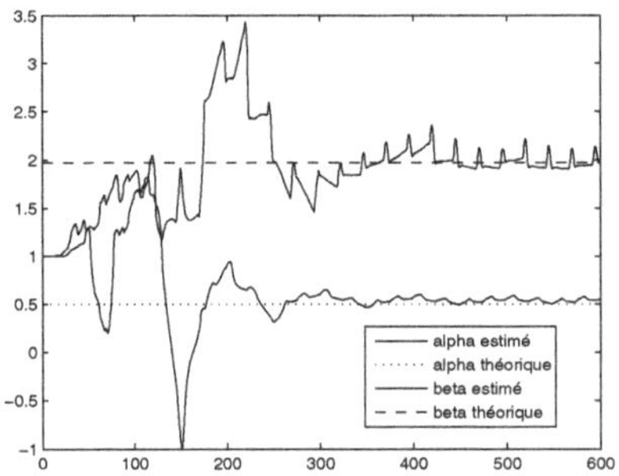

Fig. 3.14. $\hat{\alpha}$ and $\hat{\beta}$ against time

References

[1] Ammar S., Vivalda J.-C. (2005) On the genericity of the observability of controlled discrete-time systems, in ESAIM: Control, Optimisation and Calculus of Variations, vol. 11, 161–179

[2] Ammar S., Vivalda J.-C. (2004) On the preservation of observability under sampling, Systems & Control Letters, vol. 52, no 1, 7–15

[3] Balde M., Jouan P. (1988) Observability of control affine systems, ESAIM/COCV, Vol. 3, pp. 345-359, 1998.

[4] Baras J.S. , Bensoussan A. , James M.R. , (1988) Dynamic observers as asymptotic limits of recursive filters: special cases", SIAM J. Appl. Math., 48, 1147–1158

[5] Binczak S, Busvelle E, Gauthier JP, Jacquir S (2006) Identification of unknown functions in dynamic systems. 17th International Symposium on Mathematical Theory of Networks and Systems (MTNS 2006), Kyoto:2157–2162.

[6] Binczak S, Kazantsev V, Nekorkin V I, Bilbault JM (2003) Experimental study of bifurcations in modified Fitzhugh-Nagumo cell. Electronic letters 39,13:961–962.

[7] Boizot N, Busvelle E, Gauthier JP, Sachau J (2007) Adaptive-Gain Extended Kalman Filter: Application to a Series-Connected DC Motor. In Conference on Systems and Control (CSC'2007), Marrakech.

[8] Busvelle E., Gauthier J.-P. (2002) High-Gain and Non-High-Gain Observers for Nonlinear Systems, Contemporary Trends in Nonlinear Geometric Control Theory and its Applications, World Scientific, 233–256
[9] Busvelle E., Gauthier J.-P. (2005) Observation and Identification Tools for Nonlinear systems. Application to a Fluid Catalytic Cracker. International Journal of Control, Vol. 78, 3, 2005
[10] Busvelle E., Gauthier J.-P. (2003) On determining unknown functions in differential systems, with an application to biological reactors, ESAIM: COCV 9, 509–552
[11] Bucy R. , Joseph P. (1987) Filtering for stochastic processes with applications to guidance, Chelsea publishing company, second edition
[12] Busse F.D. , How J.P., Simpson J. (2003) Demonstration of Adaptive Extended Kalman Filter for Low-earth-orbit Formation Estimation using CDGPS Navigation. Journal of the Institute of Navigation, v 50, n 2, 79–93
[13] Deza F. (1991) Contribution to the synthesis of exponential observers, Phd thesis, INSA de Rouen, France, June 1991.
[14] Deza F., Busvelle E., Gauthier J.-P. (1992) High-gain estimation for nonlinear systems, Systems and Control Letters 18, 295–299
[15] Deza F., Busvelle E., Gauthier J.-P., Rakotopara D., (1992)A stability result on the continuous-continuous and continuous-discret extended Kalman filters, Compte-Rendus de l'Académie des Sciences de Paris, 314, Février 1992
[16] Gauthier J.-P., Hammouri H., Kupka I., (1991) Observers for nonlinear systems; IEEE CDC Conference, december 1991, Brighton, England, 1483–1489
[17] Gauthier J.-P., Hammouri H., Othman S. (1992) A simple observer for nonlinear systems. IEEE Trans. Aut. Control, 37, 875–880
[18] Gauthier J.P., Kupka I., (1994) Observability and observers for nonlinear systems. SIAM Journal on Control, vol. 32, N° 4, 975–994
[19] Gauthier J.P., Kupka I., (1996) Observability for systems with more outputs than inputs. Mathematische Zeitschrift, 223, pp. 47–78
[20] Gauthier, J.-P., and Kupka, I., (2001), Deterministic Observation Theory and Applications, Cambridge University Press
[21] Gikhman, I. I., Skorokhod, A.V. (1969), Introduction to the Theory of Random Processes, W.B. Saunders, Philadelphia
[22] Gevers M., Bastin G. (1986) A Stable Adaptive Observer for a class of Nonlinear Second-order Systems, Analysis and optimization of systems, ed. by Bensoussan, A., and Lions, J. L. Springer-Verlag, 143–155
[23] Hammouri, H., Farza, M., (2003), Nonlinear observers for locally uniformly observable systems, COCV, 9, 343-352.
[24] Hodgkin L A, Huxley A F (1952) A quantitative description of membrane current and its application to conduction and excitation in nerve. Journal of phisiology 117:500–544.
[25] Holland C.D. (1963) Multicomponent Distillation, Englewood Cliffs, New-Jersey, USA: Prentice Hall, 1963.
[26] Hu C., Chen W., Chen Y., Liu D. (2003) Adaptive Kalman filtering for vehicle navigation, Journal of Global Positioning Systems, Vol. 2, No 1, 42–47
[27] Huillet T., Salut G. (1989) Interprétation des équations du filtrage non linéaire, GdR Automatique du CNRS, Paris, 8 Novembre 1989
[28] Jacquir S (2006) Systèmes dynamiques non-linéaires, de la biologie à l'électronique. Thèse de doctorat, Université de Bougogne, France.
[29] Jaswinski A., (1970) Stochastic processes and filtering theory, Academic Press, New York

[30] Johansson A., Medvedev A. (2003), An observer for systems with nonlinear output map, Automatica 39, pp 909–918
[31] Jouan P., (1995) Singularités des systèmes non linéaires, observabilité et observateurs, PHD thesis, Université de Rouen,
[32] Jouan P., J.P. Gauthier J.P., (1996) Finite singularities of nonlinear systems. Output stabilization, observability and observers. Journal of Dynamical and Control Systems, vol. 2, N° 2, 255–288.
[33] Kent K. C. Yu, N. R. Watson, J. Arrillaga, (2005) An adaptive Kalman filter for dynamic harmonic state estimation and harmonic injection tracking, IEEE Trans. on power delivery, Vol. 20, No 2
[34] Krause P C, Wasunczuk O, Sudhoff S D (2002) Analysis of Electric Machinery and Drive Systems, 2nd edition. Wiley-interscience, IEEE series on power engineering.
[35] La Moyne L., Porter L. L., Passino K. M. , (1995) Genetic adaptive observers, Engng Applic. Artif. Intell. Vol. 8, No 3, 261–269
[36] Mehta S,Chiasson J (1998) Nonlinear Control of a Series DC Motor: Theory and Experiment. IEEE transactions on industrial electronics 45, 1.
[37] E. Pardoux (1991) Filtrage non linéaire et équations aux dérivées partielles stochastiques associées, Ecole d'Eté de Probabilités de Saint-Flour XIX-1989, Lecture Notes in Math., 1464, Springer, 67–163
[38] Picard J. (1991), "Efficiency of the extended Kalman filter for nonlinear systems with small noise", SIAM J. Appl. Math., 51, No3, 843–885.
[39] Rigal G. (1993) Filtrage non-linéaire, résolution particulaire et applications au traitement du signal, Thèse de l'Université de Toulouse, LAAS.
[40] Rosenbrock H. (1962), A Lyapunov function with applications to some nonlinear physical systems, Automatica, 1, 31–53
[41] Ropuchon P. (1990), Simulation dynamique et commande non linéaire des colonnes à distiller, Thèse de l'école des mines de Paris,
[42] Shen L., Wang H. (2003), Adaptive observer design for general nonlinear systems with linear output structure, 4th International Conference on Control and Automation, June 2003. ICCA '03. Proceedings, 48–52
[43] Shim, H., Son, Y. I., and Seo, J. H., (2001), Semi-global observer for multi-output nonlinear systems, Systems Control Lett. 42 233–244.
[44] Stubberud S., Lobbia R., Owen M., (1998) An adaptive extended Kalman filter using artificial neural networks, The international journal on smart system design, Vol. 1, pp 207-221
[45] Viel, F (1994) Stabilité des systèmes non linéaires controlés par retour d'état estimé. Application aux réacteurs de polymérisation et aux colonnes à distiller, Thèse de l'université de Rouen
[46] Viel F., Busvelle E., Gauthier J.P. (1997) A stable control structure for binary distillation columns, International Journal on Control, Vol 67, N°4, 475–505

4

Immersion-Based Observer Design

Alexandru Ţiclea[1] and Gildas Besançon[2]

[1] Universitatea Politehnica Bucureşti, Romania
ticlea@indinf.pub.ro

[2] Control Systems Department (former Laboratoire d'Automatique de Grenoble)
GIPSA-lab, ENSIEG, BP 46 38402 Saint-Martin d'Hères, France
gildas.besancon@inpg.fr

4.1 Introduction

In this chapter we present *immersion transformations* of nonlinear systems for observer synthesis. A transformation through immersion is a generalization of an equivalence transformation to the extent that the dimension of the state space is not necessarily preserved. The immersion of a system for estimation purposes involves in fact the immersion (in the differential geometry sense) of the state space into a space of larger dimension, leading to a *dynamical extension* of the system.

The idea of dynamical extension is actually quite natural when solving estimation problems. For a simple illustration, consider the linear system

$$\begin{aligned} \dot{x} &= -x + u \\ y &= x + v \end{aligned}$$

where v is an unknown measurement bias, assumed constant. The usual approach for the estimation of the state in this case is to extend the state vector to include the bias, which leads to an immersion into a second order system:

$$\begin{aligned} \dot{x}_1 &= -x_1 + u \\ \dot{x}_2 &= 0 \\ y &= x_1 + x_2. \end{aligned}$$

As for a common example in a nonlinear setting, dynamical extension is frequently performed with the intention of estimating the parameters of a system through a state observation technique.

The immersion transformations discussed in this chapter go beyond the extension of the state vector with variables that can be assimilated with constants inasmuch as the objective is to transform nonlinear systems—possibly obtained through preliminary dynamical extension—into systems that suit the design of observers, through (further) dynamical extension. To that aim, we present precise immersion conditions with respect to several structures that are particularly

G. Besançon (Ed.): Nonlinear Observers and Applications, LNCIS 363, pp. 115–138, 2007.
springerlink.com

interesting from the observer design viewpoint, with an emphasis on the conditions that can be easily translated into immersion algorithms.

Section 4.2 recalls some standard definitions for the basic concepts used throughout this chapter without aiming at a rigorous treatment but rather at setting out the notation. Section 4.3 presents results concerning the immersion into state affine structures, while section 4.4 presents results for immersion into linear structures. Finally, section 4.5 presents an observer design based on immersion into a nonlinear system satisfying some specific structural constraints.

4.2 Notation and Definitions

4.2.1 Nonlinear Systems

The general description of nonlinear systems used in this chapter assumes that the state space is a C^∞ manifold of dimension n, denoted by M, the input space, denoted by E is a subset of $\mathbb{R}^m$, and the output space, denoted by S is a subset of $\mathbb{R}^p$. As far as the inputs as functions of time $u : [0, t_u) \to E$ are concerned, we assume that they are elements of the set of measurable and bounded functions defined on $\mathbb{R}^+$ with values in E.

We consider that the dynamics of the system are characterized by a family of vector fields parameterized by the input u, $f_u = \{f_c \mid c \in E\}$, such that the application $M \times E \to TM$ (tangent structure of M), $(x, c) \mapsto f(x, c)$, is C^∞. We also consider that the output of the system is given by a C^∞ output map $h : M \to S$.

For convenience, we assume that M admits a global system of coordinates, which leads to a global representation of the system,

$$\begin{aligned} \dot{x} &= f_u(x) = f(x, u) \\ y &= h(x), \end{aligned} \tag{4.1}$$

and very often we consider that $M = \mathbb{R}^n$.

We denote by $x_{x^\circ,u}$ the internal trajectory of the system under the action of the input u starting from x°, and by $y_{x^\circ,u}$ the corresponding output trajectory. We assume that $y_{x^\circ,u} : [0, t_{x^\circ,u}) \to S$ is a measurable function.

4.2.2 Observability

The only observability-related concept that will be needed in this chapter is the observability rank condition relative to the observation space of the system. In that respect, let us recall the following:

Definition 1 (Observation space). *Given a system (4.1), we denote by $\mathcal{O}(h)$ and we call the* observation space *of the system the smallest vector space over $\mathbb{R}$ of functions defined on M with values in $\mathbb{R}$ that contains the set $\{h_1, \ldots, h_p\}$ and is invariant under the action (through Lie derivation) of the vector fields in f_u.*

We denote by $\mathrm{d}\mathcal{O}(h)$ the space of the differentials of the elements of $\mathcal{O}(h)$ and by $\mathrm{d}\mathcal{O}(h)(x)$ the (finite dimensional) vector space over $\mathbb{R}$ obtained through evaluation at x of the elements of $\mathrm{d}\mathcal{O}(h)$. The system satisfies the *observability rank condition at a point* x° if

$$\dim \mathrm{d}\mathcal{O}(h)(x^\circ) = n.$$

We note that the observability rank condition characterizes the local weak observability of the system [18].

4.2.3 Immersion

This section gives the precise mathematical characterization of the notion of *subsystem* towards the situations in which the input-output behavior of a system is reproduced in the input-behavior of another system. As suggested by the term, a system has lower order than the system of which it is subsystem. We usually refer to subsystems as *immersed* systems. Conversely, if a system admits one or several subsystems, then it can be *submersed* into such system.

Let us first recall the definition of the immersion as concept in differential geometry.

Definition 2 (Immersion and submersion of manifolds [6]). *An application $\tau : M \to M'$ is an* immersion *(*submersion*) if its rank is $n = \dim M$ ($n' = \dim M'$) everywhere. If τ is an injective immersion, then it establishes a one-to-one correspondence of M and the subset $M'' = \tau(M)$ of M'.*

We also recall the concept of *embedding*.

Definition 3 (Embedding [6]). *An embedding is a one-to-one immersion $\tau : M \to M'$, which is a homeomorphism of M into M', that is, τ is a homeomorphism of M onto its image, $M'' = \tau(M)$, with its topology as a subspace of M'. Every one-to-one immersion is locally an embedding.*

Since the rank of τ is less than $\min(n, n')$ at every point, if τ is an immersion then $n \leq n'$, while if τ is a submersion, $n \geq n'$. As far as the observer design is concerned, we are mainly concerned with immersions.

Definition 4 (Immersion of dynamical systems). *Consider two C^∞ systems $S = (M, f_u, h)$ and $S' = (M', f'_u, h')$ such that every input that is admissible for one of them is also admissible for the other. The system S is* immersible *into system S' if there exists a C^∞ map $\tau : M \to M'$ such that*

(i) *For every pair $(x^\circ, x^\bullet) \in M \times M$, $h(x^\circ) \neq h(x^\bullet)$ implies $h'(\tau(x^\circ)) \neq h'(\tau(x^\bullet))$,*
(ii) *For every pair (x, u), the domain of definition of $y'_{\tau(x),u}$ includes the domain of definition of $y_{x,u}$ and on the intersection of their domains, $y_{x,u}$ and $y'_{\tau(x),u}$ coincide.*

In this situation, τ is an immersion of dynamical systems *and S can be represented as* subsystem *of S'.*

It turns out that an immersion of dynamical systems for observer design purposes has to be at least an immersion of manifolds. This guarantees that the internal trajectories of the two systems initialized respectively at x° and $\tau(x^\circ)$ are (at least locally) in one-to-one correspondence, making possible the inverse transformation of estimated trajectories in order to recover the original variables, and ensuring that the systems share (locally) the same observability properties. Moreover, when the immersion is an embedding, S is subsystem of a uniquely determined C^∞ system of order n' defined on M' and their internal trajectories are in one-to-one correspondence.

Sometimes, the system we are dealing with does not possess the required observability properties. In some cases it is possible to perform a submersion through the canonical map obtained by factoring the state space by the relation of indistinguishability [18] and obtain a system which possesses a certain observability property. We shall see in Section 4.5.3 an example of such a submersion.

4.3 Immersion in a State-Affine Structure

One important difficulty when designing observers for nonlinear systems arises from the fact that in general the observability properties of such systems depend on the applied input. The main advantage of state affine systems is that they lend themselves to the characterization of the quality of the applied input through the observability grammian specific to linear time varying systems. The observation of state-affine system can be then easily performed through Kalman [21], or Kalman-like [16] observers (as recalled in chapter).

4.3.1 Immersion Without Output Injection

The first result on the immersion of continuous-time nonlinear systems into a state-affine structure

$$\begin{aligned}\dot{z} &= A(u)z + \varphi(u)\\ y &= Cz\end{aligned}$$

has been presented by Fliess [11] for input-affine analytic systems as an application to the theory of nonlinear causal functionals he had previously introduced in [10].

The immersion transformation leads in this case to a bilinear representation and the necessary and sufficient condition for immersion is that the observation space of the system to be immersed has a finite dimension. The proof uses on the one hand the fact that the output of an analytic system is a causal functional of the input u, generated by a power series with coefficients in the observation space of the system, and on the other hand the fact that the observation space of any bilinear system has a finite dimension.

It turns out that the finiteness condition with respect to the observation space also applies to general C^∞ systems.

Theorem 1 (Immersion into a state-affine system [12])

(i) *If the observation space of a system (4.1) has a finite dimension, then the system can be immersed into a state-affine system.*
(ii) *If in addition the system is originally control-affine, then the immersion leads to a bilinear system.*
(iii) *If the class of admissible inputs contains the piecewise-constant inputs, then the immersion condition is also necessary.*

The main characteristic of the immersion performed under the conditions of this theorem is that the order of the system obtained after immersion coincides with the dimension of the observation space of the original system. Some proof elements for the "if" part of the theorem turn out to be useful for the construction of the immersion.

First, the finiteness of $\mathcal{O}(h)$ implies that every element of this space can be expressed as $\mathbb{R}$-linear combination of basis elements. Let $\mathcal{L}(f_u)$ denote the Lie algebra generated by the vector fields of the family f_u. The observation space is invariant under the action (through the Lie derivative) of the elements of $\mathcal{L}(f_u)$, so this action can be given an $\mathbb{R}$-linear representation through a map $\theta : \mathcal{L}(f_u) \to \mathrm{End}(\mathcal{O}(h))$.

On the other hand, every linear endomorphism of $\mathcal{O}(h)$ determines a unique linear endomorphism of the dual space $\mathcal{O}(h)^\star$. This means that there is a "natural" action of $\mathcal{L}(f_u)$ on $\mathcal{O}(h)^\star$ with $\mathbb{R}$-linear representation given by a map $\rho : \mathcal{L}(f_u) \to \mathrm{End}(\mathcal{O}(h)^\star)$ such that, if $f \in \mathcal{L}(f_u)$, then $\rho(f)$ is the dual map of $\theta(f)$, defined as follows: if $l \in \mathcal{O}(h)$ and $\lambda \in \mathcal{O}(h)^\star$, then

$$(\rho(f)\lambda)[l] = \lambda(\theta(f)l).$$

In this equality, l can also be seen as an element of the double dual $\mathcal{O}(h)^{\star\star} \equiv \mathcal{O}(h)$, so the endomorphism $\rho(f)$ defines a linear vector field on $\mathcal{O}(h)^\star$. It is then possible to define a dynamical system on $\mathcal{O}(h)^\star$ with dynamics given by the family of vector fields $\rho(f_u)$ and output map h' defined as follows: if $\lambda \in \mathcal{O}(h)$, then $h'(\lambda) = \left[\lambda(h_1) \cdots \lambda(h_p)\right]^T$.

The original system is subsystem of the above defined system, with immersion map $\tau : M \to \mathcal{O}(h)^\star$ defined as follows: if $x \in M$ and $l \in \mathcal{O}(h)$, then $\tau(x)[l] = l(x)$. According to this definition, the components of $\tau(x)$ in the dual basis are the basis elements of $\mathcal{O}(h)$ evaluated in x, so knowledge of a basis of $\mathcal{O}(h)$ is sufficient to construct the immersion. Next, given a vector field f of the family f_u, in order to compute $\rho(f)$ (which is also the image of f through the tangent map $\tau_\star$ [6]), it is enough to compute the directional derivative of the components of τ in the direction of f.

Example 1 (Immersion of an input-affine system [11, 12]). Consider the following SISO system defined on $\mathbb{R} \setminus \{0\}$:

$$\begin{aligned} \dot{x} &= f_0(x) + f_1(x)u = ax - bx^\alpha + xu \\ y &= h(x) = \frac{1}{x^{\alpha-1}}, \end{aligned}$$

with $a, b \in I\!R$, $\alpha \in I\!N$ and $\alpha \geq 2$. A basis of $\mathcal{O}(h)$ is $\{1, \frac{1}{x^{\alpha-1}}\}$, which defines the immersion map $\tau : I\!R \setminus \{0\} \to I\!R^2$ as

$$x \mapsto \begin{bmatrix} 1 \\ \frac{1}{x^{\alpha-1}} \end{bmatrix}.$$

In order to compute the projections of the vector fields f_0 and f_1 through the immersion τ, we proceed as indicated above:

$$\begin{bmatrix} L_{f_0}1 \\ L_{f_0}\frac{1}{x^{\alpha-1}} \end{bmatrix} = \begin{bmatrix} 0 \\ b(\alpha-1) + \frac{a(1-\alpha)}{x^{\alpha-1}} \end{bmatrix} = \begin{bmatrix} 0 & 0 \\ b(\alpha-1) & a(1-\alpha) \end{bmatrix} \begin{bmatrix} 1 \\ \frac{1}{x^{\alpha-1}} \end{bmatrix},$$
$$\begin{bmatrix} L_{f_1}1 \\ L_{f_1}\frac{1}{x^{\alpha-1}} \end{bmatrix} = \begin{bmatrix} 0 \\ \frac{1-\alpha}{x^{\alpha-1}} \end{bmatrix} = \begin{bmatrix} 0 & 0 \\ 0 & 1-\alpha \end{bmatrix} \begin{bmatrix} 1 \\ \frac{1}{x^{\alpha-1}} \end{bmatrix}.$$

The original system is therefore subsystem of the bilinear system

$$\begin{aligned} \dot{z} &= \begin{bmatrix} 0 & 0 \\ b(\alpha-1) & a(1-\alpha) \end{bmatrix} z + u \begin{bmatrix} 0 & 0 \\ 0 & 1-\alpha \end{bmatrix} z \\ y &= \begin{bmatrix} 0 & 1 \end{bmatrix} z. \end{aligned}$$

4.3.2 Immersion with Output Injection

The finiteness condition with respect to the observation space is fairly strong and seldom satisfied in practice, even for apparently "simple" systems, like the one in the following example.

Example 2. It can be easily verified, through successive differentiation of the output application along the direction of the vector field that describes the dynamics, that the observation space of system

$$\begin{aligned} \dot{x}_1 &= x_2 \\ \dot{x}_2 &= x_1 x_2 \\ y &= x_1 \end{aligned} \tag{4.2}$$

has infinite dimension.

In some situations, it is however possible to immerse a system with infinite dimensional observation space into a state-affine system, by resorting to output injection.

We shall first present an approach by Hammouri and Celle [15], which uses an idea that does not differ much from the idea in the previous section as the immersion condition is still a finiteness condition related to the observation space. In concrete terms, the system to immerse is an autonomous, single-output, analytical system defined on $I\!R^n$:

$$\begin{aligned} \dot{x} &= f(x) \\ y &= h(x), \end{aligned} \tag{4.3}$$

and the immersion condition is that the elements of the observation space are linear combinations in a finite basis with coefficients in the space of the functions of h, more precisely in $\mathcal{R}(h)$—the ring of the functions of the type $l \circ h$, where $l \in C^{\omega}(S)$ (the set of analytic functions defined on the output space of the system with values on the real axis). The system is then immersed in a system of the form

$$\dot{z} = \begin{bmatrix} 0 & 1 & 0 & \cdots & 0 \\ \vdots & \ddots & \ddots & \ddots & \vdots \\ \vdots & & \ddots & 1 & 0 \\ 0 & \cdots & \cdots & 0 & 1 \\ a_1(y) & & \cdots & & a_N(y) \end{bmatrix} z + \begin{bmatrix} 0 \\ \vdots \\ 0 \\ 0 \\ \varphi_N(y) \end{bmatrix} \qquad (4.4)$$
$$y = z_1$$

with $N \geq n$. However, there is no a priori information towards the value, nor the existence of N. In fact, the available result can be stated as follows:

Theorem 2 (Immersion into a state-affine system [15]). *If a system (4.3) satisfies the conditions:*

(i) *There is an integer N such that $h, L_f h, \ldots, L_f^{N-1} h$ are $\mathcal{R}(h)$-linearly independent;*
(ii) *$L_f^N h$ is element of the $\mathcal{R}(h)$-module generated by the $\mathbb{R}$-vector space $\mathbb{R} \oplus \mathrm{span}(h, L_f h, \ldots, L_f^{N-1} h)$,*

then it can be immersed into a system (4.4). Conversely, if N is the smallest integer for which a given system (4.3) can be immersed in a system (4.4), then (i) *and* (ii) *hold for that system.*

Example 3. For the system (4.2), we have

$$h = x_1, \qquad L_f h = x_2, \qquad L_f^2 h = h(x) x_2,$$

so in this particular case we obtain an *equivalence* with the state-affine system

$$\dot{z} = \begin{bmatrix} 0 & 1 \\ 0 & y \end{bmatrix} z$$
$$y = z_1.$$

Remark 1. Since the construction of the immersion resembles to a certain extent the construction performed when the observation space has finite dimension, the result can be easily extended to non-autonomous and multiple-output systems.

Again, just like in the case of the immersion without output injection, there are systems to which the above result does not apply, but that can still be immersed into a state-affine structure.

Example 4. It can be easily verified that the observation space of the system

$$\begin{aligned}\dot{x}_1 &= x_1 + x_1 x_2 \\ \dot{x}_2 &= x_1 \\ y &= x_1\end{aligned}$$

is not finite-dimensional, nor can it be given the structure of a module over $\mathcal{R}(h)$. Nevertheless, if the dependence on y is made explicit in the nonlinearity $x_1 x_2$, then the system can be written as

$$\begin{aligned}\dot{x} &= \begin{bmatrix} 1 & y \\ 1 & 0 \end{bmatrix} x \\ y &= x_1.\end{aligned}$$

However, the conditions for immersion into a state-affine structure through such "general" output injection are very difficult to characterize. This fact is equally true for diffeomorphism transformations into systems of the form

$$\begin{aligned}\dot{z} &= A(u, y)z + \varphi(u, y) \\ y &= Cz,\end{aligned} \tag{4.5}$$

a problem which is treated for instance in [17], where generic transformation conditions are given, which are very difficult to check in practice and do not offer any guidelines for constructing the transformation. Specific cases have been considered in [4]. The difficulties originate in the fact that in general there are multiple ways to parameterize the vector fields that describe the dynamics of the system by the measured output y. A more detailed discussion on this subject is available in [27].

Here, we shall present an idea to immerse a nonlinear system into a state affine structure by means of output injection which is more along the lines of the idea in Section 4.3.1. More precisely, the idea consists in fixing an explicit dependence on y of the dynamics in the original description of the system and then try to immerse the system using the finiteness condition towards the observation space by considering the extended input $\left[\begin{smallmatrix} u \\ y \end{smallmatrix}\right]$. We summarize this result in the following proposition, which provides a sufficient condition for immersion.

Proposition 1. *Given a system (4.1), if the family of vector fields $f(x, u)$ can be parameterized by the output y such that the observation space is finite dimensional when considering the extended input $\left[\begin{smallmatrix} u \\ y \end{smallmatrix}\right]$, then the system can be immersed in a state-affine structure (4.5).*

Remark 2. In general, there is no link between the observation space described in the above proposition and the real observation space of the system. In the above proposition, this object is created only for immersion construction and should not be used for other purposes, such as testing the observability rank condition.

Example 5. Consider the system

$$\begin{aligned}\dot{x}_1 &= x_1 + x_1 x_2^2 \\ \dot{x}_2 &= x_1 \\ y &= x_1\end{aligned}$$

and write the dynamics as

$$\dot{x} = f_0(x) + y f_1(x) = \begin{bmatrix} x_1 \\ 0 \end{bmatrix} + y \begin{bmatrix} x_2^2 \\ 1 \end{bmatrix}.$$

Then,

$$L_{f_0} h = x_1 = h, \qquad L_{f_1} h = x_2^2, \qquad L_{f_1}^2 h = 2x_2, \qquad L_{f_1}^3 h = 1,$$

so the system is immersible into a state affine structure through the immersion map

$$x \mapsto \begin{bmatrix} x_1 \\ x_2^2 \\ 2x_2 \\ 1 \end{bmatrix}.$$

The new representation is obtained by proceeding as in Example 1:

$$\dot{z} = \begin{bmatrix} 1\,0\,0\,0 \\ 0\,0\,0\,0 \\ 0\,0\,0\,0 \\ 0\,0\,0\,0 \end{bmatrix} z + y \begin{bmatrix} 0\,1\,0\,0 \\ 0\,0\,1\,0 \\ 0\,0\,0\,2 \\ 0\,0\,0\,0 \end{bmatrix} z.$$

Example of application

For a more practical example of application of Proposition 1, consider the model of an induction motor with state variables

$i_{s\alpha}, i_{s\beta}$ – the components of the stator current phasor
$\phi_{s\alpha}, \phi_{s\beta}$ – the components of the stator flux phasor
ω_r – the mechanical speed

and inputs the components of the stator voltage phasor $u_{s\alpha}, u_{s\beta}$:

$$\begin{aligned}
\tfrac{\mathrm{d}}{\mathrm{d}t} i_s &= [-(\tfrac{R_r}{\sigma L_r} + \tfrac{R_s}{\sigma L_s})I + p\omega_r J] i_s + [\tfrac{R_r}{\sigma L_s L_r} I - p\tfrac{1}{\sigma L_s}\omega_r J]\phi_s + \tfrac{1}{\sigma L_s} u_s \\
\tfrac{\mathrm{d}}{\mathrm{d}t} \phi_s &= -R_s i_s + u_s \\
\tfrac{\mathrm{d}}{\mathrm{d}t} \omega_r &= -\tfrac{f_v}{J_m}\omega_r + p\tfrac{1}{J_m}(i_{s\beta}\phi_{s\alpha} - i_{s\alpha}\phi_{s\beta}) - \tfrac{1}{J_m}\tau_l \\
y &= \begin{bmatrix} i_s \\ \omega_r \end{bmatrix}.
\end{aligned}$$

In the electrical part, L stands for inductance, R stands for resistance, $\sigma = 1 - \frac{M^2}{L_s L_r}$ with M the maximum mutual inductance between one stator and one rotor winding, the indexes s and r refer respectively to the stator and the rotor, and

$$I = \begin{bmatrix} 1 & 0 \\ 0 & 1 \end{bmatrix}, \qquad J = \begin{bmatrix} 0 & -1 \\ 1 & 0 \end{bmatrix}.$$

As far as the mechanical part is concerned, τ_l denotes the load torque, J_m the total inertia momentum (rotor plus load) and f_v the viscous friction coefficient.

Finally, in both electrical and mechanical parts, p represents the number of pairs of poles.

For such a system, the control problems generated by uncertainties in the electrical parameters as well as by the absence of flux, torque and sometimes even speed transducers are well-known and have motivated a lot of work. As a result, it can be shown that the above model extended with the unknown electrical parameters and load torque can be immersed into a form (4.5) for any considered set of such unknown parameters, even when the speed is not measured [28, 29]. Notice however that in these works the immersion is performed through a heuristic approach in which a suitable parametrization of the dynamics by the measured output is chosen at each step of the construction with the objective of obtaining an affine structure. Here, we shall show that it is possible to achieve the same result—at least towards the simultaneous estimation of the electrical parameters, load torque and state variables when the speed is measured—by using the approach of Proposition 1.

First, a preliminary dynamical extension is performed such that the electrical parameters $\frac{R_r}{\sigma L_r}$, $\frac{R_s}{\sigma L_s}$, $\frac{R_r}{\sigma L_s L_r}$, $\frac{1}{\sigma L_s}$, R_s and the load torque τ_l are included in the state vector, denoted from now on by x. Resistances usually vary with temperature, but the dynamics are in general unknown and this is true for the load torque as well. For this reason, it is assumed that these dynamics are slow compared to the dynamics of the employed observer and the corresponding variables are assimilated with constants.

Let us now define the following vector fields:

$$
\begin{aligned}
f_0(x) &= \left[\frac{R_r}{\sigma L_s L_r}\phi_{s\alpha}\ \frac{R_r}{\sigma L_s L_r}\phi_{s\beta}\ 0\ 0\ -(\frac{f_v}{J_m}\omega_r + \frac{1}{J_m}\tau_l)\ 0 \cdots 0\right]^T \\
f_1(x) &= \left[-(\frac{R_r}{\sigma L_r} + \frac{R_s}{\sigma L_s})\ 0\ -R_s\ 0\ -p\frac{1}{J_m}\phi_{s\beta}\ 0 \cdots 0\right]^T
\end{aligned}
$$

$$
\begin{aligned}
f_2(x) &= \left[0\ -(\frac{R_r}{\sigma L_r} + \frac{R_s}{\sigma L_s})\ 0\ -R_s\ p\frac{1}{J_m}\phi_{s\alpha}\ 0 \cdots 0\right]^T \\
f_3(x) &= \left[-p(i_{s\beta} - \frac{1}{\sigma L_s}\phi_{s\beta})\ p(i_{s\alpha} - \frac{1}{\sigma L_s}\phi_{s\alpha})\ 0\ 0\ 0\ 0 \cdots 0\right]^T \\
f_4(x) &= \left[\frac{1}{\sigma L_s}\ 0\ 1\ 0\ 0\ 0 \cdots 0\right]^T \\
f_5(x) &= \left[0\ \frac{1}{\sigma L_s}\ 0\ 1\ 0\ 0 \cdots 0\right]^T
\end{aligned}
$$

and the output map

$$
h(x) = \left[i_{s\alpha}\ i_{s\beta}\ \omega_r\right]^T.
$$

Then the original system can be represented as

$$
\begin{aligned}
\dot{x} &= f_0(x) + y(1)f_1(x) + y(2)f_2(x) + y(3)f_3(x) + u(1)f_4(x) + u(2)f_5(x) \\
y &= h(x)
\end{aligned}
$$

and one can compute:

$$
\begin{aligned}
L_{f_0}h_1 &= \tfrac{R_r}{\sigma L_s L_r}\phi_{s\alpha} \\
L_{f_0}h_2 &= \tfrac{R_r}{\sigma L_s L_r}\phi_{s\beta} \\
L_{f_1}L_{f_0}h_1 = L_{f_2}L_{f_0}h_2 &= -\tfrac{R_s R_r}{\sigma L_s L_r} \\
L_{f_4}L_{f_0}h_1 = L_{f_5}L_{f_0}h_2 &= \tfrac{R_r}{\sigma L_s L_r} \\
L_{f_1}h_1 = L_{f_2}h_2 &= -\tfrac{R_r}{\sigma L_r} - \tfrac{R_s}{\sigma L_s} \\
L_{f_3}h_1 &= -p(i_{s\beta} - \tfrac{1}{\sigma L_s}\phi_{s\beta}) \\
L_{f_3}h_2 &= p(i_{s\alpha} - \tfrac{1}{\sigma L_s}\phi_{s\alpha}) \\
L_{f_1}L_{f_3}h_1 = -L_{f_2}L_{f_3}h_2 &= -pL_{f_1}h_1 - \tfrac{R_s}{\sigma L_s} \\
L_{f_4}h_1 = L_{f_5}h_2 &= \tfrac{1}{\sigma L_s} \\
L_{f_0}h_3 &= -(\tfrac{f_v}{J_m}\omega_r + \tfrac{1}{J_m}\tau_l) \\
L_{f_1}h_3 &= -p\tfrac{1}{J_m}\phi_{s\beta} \\
L_{f_2}h_3 &= p\tfrac{1}{J_m}\phi_{s\alpha} \\
L_{f_2}L_{f_1}h_3 = -L_{f_1}L_{f_2}h_3 &= p\tfrac{1}{J_m}R_s \\
L_{f_5}L_{f_1}h_3 = -L_{f_4}L_{f_2}h_3 &= -p\tfrac{1}{J_m},
\end{aligned}
$$

from where a basis of the "observation space" is obtained: $i_{s\alpha}$, $i_{s\beta}$, ω_r, $\frac{R_r}{\sigma L_s L_r}\phi_{s\alpha}$, $\frac{R_r}{\sigma L_s L_r}\phi_{s\beta}$, $\frac{R_s R_r}{\sigma L_s L_r}$, $\frac{R_r}{\sigma L_s L_r}$, $\frac{R_r}{\sigma L_r}$, $\frac{R_s}{\sigma L_s}$, $\frac{1}{\sigma L_s}\phi_{s\alpha}$, $\frac{1}{\sigma L_s}\phi_{s\beta}$, $\frac{1}{\sigma L_s}$, τ_l, $\phi_{s\alpha}$, $\phi_{s\beta}$, R_s and 1, so the system can be immersed into a state affine structure with output injection and an exponential forgetting factor observer (Kalman-like) can be employed for estimation.

We shall illustrate the effectiveness of this method through some estimation results obtained for a real data set collected from a 7.5 kW induction motor available at the control systems department of GIPSA-lab. The set contains terminal voltage, terminal current and mechanical speed measurements corresponding to the response of the system to a change in the speed set-point from 0 to 75 rad/s.

The evolution of the speed, both real and estimated is presented in fig. 4.1. Since flux measurements were unavailable, the performances towards the estimation of these variable are illustrated through the norm of the flux phasor, plotted against the reference imposed to the flux controller in fig. 4.2. Finally, the evolution of the parameter estimates is presented in fig. 4.3 where it can be seen that the obtained steady-state values are very realistic. They actually are in a good accordance with the a priori available information on those parameters.

4.4 Immersion into a Linear Structure

Linear structures are obviously very appealing for observer synthesis, which in this case is trivial; in particular, linear systems can be observed through Luenberger-like observers [26] (the errors dynamics are linear, with freely assignable spectrum under the observability condition).

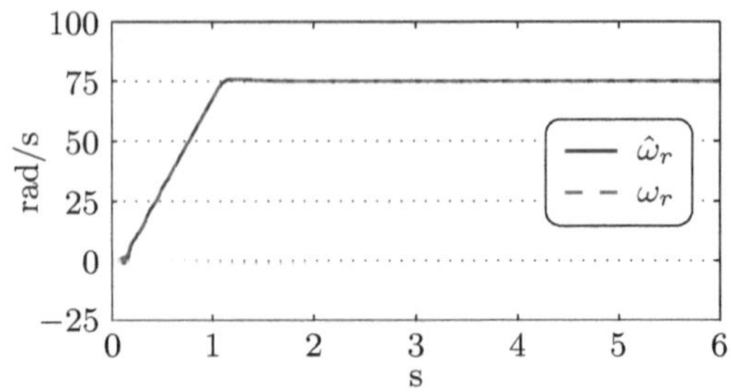

Fig. 4.1. Real and estimated mechanical speed

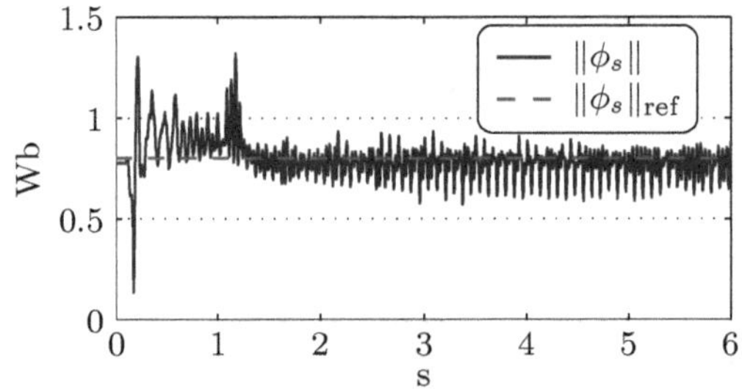

Fig. 4.2. Real and estimated flux norm

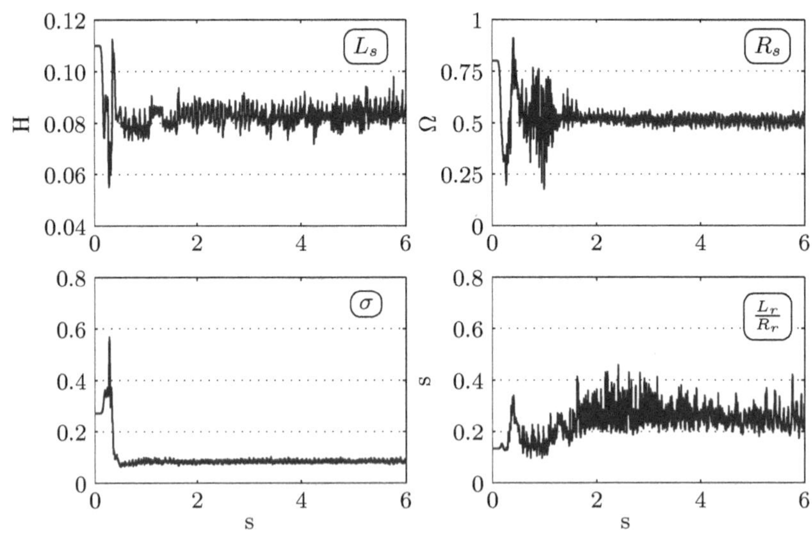

Fig. 4.3. Estimated values of electrical parameters

4.4.1 Extensions of the Immersion into a State-Affine Structure

The first approach to the immersion into a linear structure represents a natural extension of the result in Section 4.3.1 on the immersion into a state-affine structure, in the particular case of input-affine systems [9, 25]. More precisely, if we relate to Example 1, the way the vector fields are defined in the new representation shows that for an input-affine system

$$\begin{aligned} \dot{x} &= f_0(x) + \sum_{i=1}^{m} f_i(x)u_i \\ y &= h(x), \end{aligned} \tag{4.6}$$

if, besides the finiteness of the observation space, the Lie derivatives of the basis elements of this space along the vector fields $f_1, \dots, f_m$ are constant, then the system obtained through immersion is linear.

Just as in the case of the immersion into a state-affine structure, it is possible to weaken the immersion condition by resorting to output injection. This idea is considered for instance in [8], where the objective is to immerse a state-affine system (4.6) into a linear (up to output injection) structure:

$$\begin{aligned} \dot{z} &= Az + \varphi(u, y) \\ y &= Cz. \end{aligned} \tag{4.7}$$

Theorem 3 (Immersion into a linear system [8]). *The necessary conditions for a system (4.6) to be locally immersed around a point x° in a system (4.7) are*

(i) $\dim \mathcal{O}(h) < \infty$,
(ii) $\forall \lambda \in \mathcal{O}(h)$, $\mathrm{d}L_{f_i} \wedge \mathrm{d}h_1 \wedge \cdots \wedge \mathrm{d}h_p = 0$, $i = 1, \dots, m$.

Conversely, if $\mathrm{d}h_1 \wedge \cdots \wedge \mathrm{d}h_p\big|_{x^\circ} \neq 0$, these conditions are also sufficient.

Remark 3. Condition (ii) of the theorem translates the requirement that the Lie derivatives of the basis elements of $\mathcal{O}(h)$ along the vector fields $f_1, \dots, f_m$ can be written as functions of the type $l \circ h$, with $l \in C^\omega(S)$ (see also Lemma 1).

4.4.2 Observer Linearization Approach

This second approach to the immersion into a linear structure is more recent and has stemmed from the works on the transformations that allow the synthesis of observers with linear error dynamics, initiated by Krener and Isidori [23] (a detailed exposition of these results is also available in [19]). The considered problem in [23] is the transformation of an autonomous, single-output system of the form (4.4) into a linear (up to output injection) structure

$$\begin{aligned} \dot{z} &= Az + \varphi(y) \\ y &= Cz \end{aligned} \tag{4.8}$$

which is also observable, or, without loss of generality, in the observability canonical form:

$$A = \begin{bmatrix} 0 & 1 & 0 & \cdots & 0 \\ & & \ddots & \ddots & \vdots \\ \vdots & & \ddots & \ddots & 0 \\ & & & & 1 \\ 0 & & \cdots & & 0 \end{bmatrix}, \qquad C = \begin{bmatrix} 1 & 0 & \cdots & 0 \end{bmatrix}. \tag{4.9}$$

The integrability of a certain distribution is a necessary condition for the existence of the transformation, which is then obtained as solution of a first-order partial differential equation. An extension to multiple-output and non autonomous systems under the assumption that the input functions are piecewise constant is available in [24], where the transformation of the state space can also be combined with a diffeomorphism of the output space.

An alternative approach to the construction of the transformation is based on the expression of the nth derivative of the output of a system of the form (4.8–4.9). The *characteristic equation* is then obtained [22], which, in the general case where a diffeomorphism ψ is considered in the output space, can be written as

$$L_f^n\tilde{h} - L_f^{n-1}(\varphi_1 \circ \tilde{h}) - L_f^{n-2}(\varphi_2 \circ \tilde{h}) - \ldots - L_f(\varphi_{n-1} \circ \tilde{h}) - \varphi_n \circ \tilde{h} = 0,$$

where $\tilde{h} = \psi \circ h$. Under the assumption that ψ is known, this a partial differential equation of order $n-1$ in n unknowns (the components of φ).

There are however situations where the characteristic equation has no solution. The idea to use immersion transformations in such situations appears in the independent works of Jouan [20] and Back and Seo [1], which explore the possibility to solve the characteristic equation when its order is $N-1$, with $N > n$. Both references are, however, mainly concerned with the computational aspects of the problem, as it is still very difficult to check the existence of a finite N and even more difficult to find the value of N such that the immersion can be performed. Basically, the only solution is to try applying the immersion algorithm for successive values of N, starting with $N = n+1$.

Besides the solution of the characteristic equation, supplementary immersion conditions are specified in the non autonomous case, the latter restricted, in addition, to input-affine systems.

Theorem 4 (Immersion into a linear system [20]). *A input-affine system (4.6) can be immersed into a system (4.7) if and only if the following conditions hold:*

(i) *The autonomous part of the system is immersible in a system (4.8);*
(ii) *If τ is the immersion of the autonomous part, eventually combined with a diffeomorphism ψ of the output space, then there exist some functions $\gamma_{i,j}$, $i = 1, \ldots, m$, $j = 1, \ldots, N$, such that*

$$L_{f_i}\tau_j = \gamma_{i,j} \circ (\psi \circ h).$$

Moreover, around regular points of the codistribution dh*, condition* (ii) *is equivalent to*

$$\mathrm{d}L_{f_i}\tau_j \wedge \mathrm{d}h = 0.$$

Note that the conditions of Theorem 4 are weaker than those of Theorem 3 to the extent that in the latter the non autonomous part must be immersible into a linear structure *without output injection.*

4.5 Immersion into a Constrained Nonlinear Structure

In this section we discuss the immersion of a nonlinear system into another nonlinear system that satisfies particular structural constraints. Our interest towards this structure is justified on the one hand by the fact that it is suited, under appropriate excitation conditions, to observer design and on the other hand by the fact that it can be obtained through immersion—as far as input-affine systems are concerned—under quite mild conditions. The observer features are first given (extending a basic case presented in chapter), and then is discussed the immersion procedure.

4.5.1 A Triangular Structure for Observer Design

We shall first discuss the synthesis of an observer for nonlinear, single-output systems of the form

$$\begin{aligned} \dot{z} &= A(u,y)z + \varphi(u,z) \\ y &= C(u)z + \eta(u) \end{aligned} \tag{4.10}$$

where the involved matrices have particular structures

$$\begin{aligned} A(u,y) &= \begin{bmatrix} 0 & A_{1,2}(u,y) & 0 & \cdots & 0 \\ & & \ddots & \ddots & \vdots \\ \vdots & & \ddots & \ddots & 0 \\ & & & & A_{q-1,q}(u,y) \\ 0 & & \cdots & & 0 \end{bmatrix} \\ \varphi(u,z) &= \begin{bmatrix} \varphi_1(u,z_1) \\ \varphi_2(u,z_1,z_2) \\ \cdots \\ \varphi_{q-1}(u,z_1,\ldots,z_{q-1}) \\ \varphi_q(u,z) \end{bmatrix} \\ C(u) &= \begin{bmatrix} C_1(u) & 0 & \cdots & 0 \end{bmatrix}, \end{aligned} \tag{4.11}$$

with $z = \mathrm{col}(z_1,\ldots,z_q) \in I\!R^N$, $z_i \in I\!R^{N_i}$ for $i = 1,\ldots,q$, $A_{i-1,i} \in I\!R^{N_{i-1}\times N_i}$ for $i = 2,\ldots,q$ and $C_1(u) \in I\!R^{1\times N_1}$.

Note that particular cases of this structure have already been considered for observer design. When the $A_{i-1,i}$'s are scalars different from zero we get the structure for classical high gain design from [14] when the scalars are also independent of u and y, or from [13] when the dependence is allowed. Some high-gain-based observer for the case when $A_{i-1,i}(u,y) = A_{i-1,i}(u) \in I\!R$ was proposed in [7], the same problem being reconsidered in [5].

Finally, notice that when $\varphi(u,z) = \varphi(u,y)$ we get the so-called state affine structure for which a Kalman-like observer can be designed under appropriate excitation conditions [16, 3]. All these suggest that an observer for a system (4.10–4.11) should combine ingredients required for both high gain design and Kalman-like design.

In particular, as it is often the case with nonlinear systems, the observability of a system (4.10–4.11) typically depends on the inputs. When further aiming at a high gain observer design, one needs a guarantee of observability at arbitrarily short times. This can be characterized as follows:

Definition 5 (Locally regular inputs [7, 5]). *An input function u is said to be locally regular for a system (4.10–4.11) if for any initialization z° of the system there exist $\alpha > 0$, $\lambda_0 > 0$ such that*

$$\int_{t-\frac{1}{\lambda}}^{t} \Phi_{u,y}(\tau,t)^T C^T(u) C(u) \Phi_{u,y}(\tau,t) \mathrm{d}\tau \geq \alpha\lambda\Lambda^{-2}(\lambda) \tag{4.12}$$

for all $\lambda \geq \lambda_0$ and $t \geq \frac{1}{\lambda}$, where

$$\Lambda(\lambda) = \begin{bmatrix} \lambda I_{N_1} & & & 0 \\ & \lambda^2 I_{N_2} & & \\ & & \ddots & \\ 0 & & & \lambda^q I_{N_q} \end{bmatrix}$$

and $\Phi_{u,y}(\tau,t)$ satisfies

$$\frac{\mathrm{d}\Phi_{u,y}(\tau,t)}{\mathrm{d}\tau} = A(u(\tau),y(\tau))\Phi_{u,y}(\tau,t), \qquad \Phi(t,t) = I_N.$$

With such an excitation, and under the usual technical (Lipschitz) assumption for high gain observer design, one can obtain asymptotic estimation of the state:

Theorem 5. *If a system (4.10–4.11) is such that the following hold:*

(i) *The input u is bounded, locally regular and making $A(u,y)$ bounded,*
(ii) *The nonlinearity φ is Lipschitz globally in z and uniformly in u,*

then for every $\sigma > 0$ there exist $\lambda, \gamma > 0$ such that the system:

$$\begin{aligned} \dot{\hat{z}} &= A(u,y)\hat{z} + \varphi(u,\hat{z}) - \Lambda(\lambda)S^{-1}C(u)^T[C(u)\hat{z} + \eta(u) - y] \\ \dot{S} &= \lambda(-\gamma S - A(u,y)^T S - SA(u,y) + C^T C) \end{aligned}$$

with any initial condition $\hat{z}(0) \in \mathbb{R}^N$ and $S(0) \geq 0$, ensures for all $t \geq \frac{1}{\lambda}$

$$\|z(t) - \hat{z}(t)\| \leq \mu e^{-\sigma t}$$

with $\mu > 0$.

Proof. The idea of this observer originates in the work of Besançon [5], where the introduction of the two tuning parameters λ and γ helps to accomplish two objectives:

(i) It can be shown that, regardless of the choice of $\lambda \geq \lambda_0 > 0$, if the input is locally regular, then there exist $\gamma > 0$ and $\alpha_1, \alpha_2 > 0$ such that $\alpha_1 I_N \leq S(t) \leq \alpha_2 I_N$ for all $t \geq \frac{1}{\lambda}$;

(ii) Using the fact that $S(t)$ has finite, positive bounds, it can be checked through typical high-gain arguments (such as in [14] for instance) that for sufficiently large λ, if $\varepsilon := \hat{z} - z$ the candidate Lyapunov function

$$V(t) := \varepsilon(t)^T \Lambda_\lambda^{-1} S(t) \Lambda_\lambda^{-1} \varepsilon(t)$$

satisfies $\dot{V} \leq -\beta(\lambda) V \leq 0$ along the trajectories of $\varepsilon(t)$, where $\beta(\lambda)$ is a strictly positive increasing function.

Details for establishing these two points can be found in [27] or in [2]. The conclusion clearly follows from standard Lyapunov arguments. □

4.5.2 Immersion of Rank-Observable Systems

Once an observer is designed for systems of the form (4.10–4.11), we would obviously like to know to what extent an arbitrary nonlinear system can be put in that form. We will show that every input-affine, single-output system of order n that satisfies the observability rank condition at a point x° can be immersed around this point into a system (4.10–4.11) of order N, with $N \geq n$ and $A(u, y) = A(u)$. This result can be easily extended:

- to include the dependence on y,
- to systems that do not satisfy the observability rank condition,
- to systems that are not input-affine.

Formally, we consider control-affine systems of the general form:

$$\dot{x} = f_0(x) + \sum_{i=1}^{m} f_i(x) u_i$$
$$y = h_0(x) + \sum_{i=1}^{m} h_i(x) u_i,$$

but for convenience we shall use the condensed representation:

$$\begin{aligned} \dot{x} &= \sum_{i=0}^{m} f_i(x) u_i = f(x) u \\ y &= \sum_{i=0}^{m} h_i(x) u_i = h(x) u \end{aligned} \tag{4.13}$$

where $u_0 := 1$, and:

$$f(x) = \left[f_0(x)\ f_1(x)\ \cdots\ f_m(x)\right],$$
$$h(x) = \left[h_0(x)\ h_1(x)\ \cdots\ h_m(x)\right],$$
$$u = \left[u_0\ u_1\ \cdots\ u_m\right]^T.$$

Given the structure (4.13), we will be interested in immersions into systems (4.10–4.11) with a control-affine structure, but also with the temporary restriction that $A(u,y) = A(u)$, namely systems of the following form:

$$\begin{aligned}\dot{z} &= \sum_{i=0}^{m} u_i A_i z + \sum_{i=0}^{m} b_i(z) u_i = A(u)z + B(z)u \\ y &= \sum_{i=0}^{m} u_i C_i z + \sum_{i=1}^{m} d_i u_i = C(u)z + Du\end{aligned} \tag{4.14}$$

where $A(u)$, $C(u)$ and the vectors $b_i(z)$ are like in (4.11) (the structure of each b_i is identical to the structure of φ with respect to z).

Theorem 6. *A system (4.13) that satisfies the observability rank condition at some x° can be immersed around such point into a system (4.14) with structure (4.11).*

The proof of the theorem relies on a straightforward result from the theory of exterior differential forms, which will be stated here without proof.

Lemma 1. *Let $\mathrm{d}\phi_1,\ldots,\mathrm{d}\phi_k$ be independent, exact 1-forms on an open set $U \subset \mathbb{R}^n$. If $\mathrm{d}\psi$ is an exact 1-form such that*

$$d\phi_1 \wedge \cdots \wedge d\phi_k \wedge d\psi = 0$$

on U, then $\psi = \psi(\phi_1, \ldots, \phi_k)$ on U.

Proof (of Theorem 6). We will show that

(i) under the condition of the theorem the immersion procedure given hereafter applied to the system (4.13) yields a system (4.14) with structure (4.11),
(ii) the corresponding transformation is an immersion in the sense of Definition 4.

Immersion procedure

1. *At the first step, build the vector $z_1(x)$ of all h_i, $0 \le i \le m$, that depend on x,*
2. *At step $k+1$, assume that the vectors $z_1, \ldots, z_k$ were constructed in the previous steps and choose among the differentials of their components the maximum number of independent differentials that generate a regular codistribution around x°. Let $\{\mathrm{d}\phi_1, \ldots, \mathrm{d}\phi_{\nu_k}\}$ denote the set of these differentials.*
   - *If $\nu_k = n$, end the procedure,*
   - *If not, assume that $z_k = \mathrm{col}(z_k^1, \ldots, z_k^{N_k})$ and construct the vector z_{k+1} of all functions $L_{f_i} z_k^k$, $i = 0, \ldots, m$, $j = 1, \ldots, N_k$ that do not satisfy*

   $$\mathrm{d}\phi_1 \wedge \cdots \wedge \mathrm{d}\phi_{\nu_k} \wedge \mathrm{d}L_{f_i} z_k^j = 0$$

   around x°.

(i) Notice that by construction, the components of the vectors $z_1, z_2, \ldots$ belong to the observation space of the system, $\mathcal{O}(h)$, which means that their differentials are elements of $\mathrm{d}\mathcal{O}(h)$. Therefore, the construction will continue until a basis of $\mathrm{d}\mathcal{O}(h)$ is obtained.

In order to see that the construction cannot stop unless $\nu_k = n$, we note that $\nu_k = n$ if and only if the vector z_{k+1} is empty. To prove this affirmation, suppose that $\nu_k < n$ and the resulting z_{k+1} is empty. This means that each covector $L_{f_i} z_k^j$, $0 \leq i \leq m$, $1 \leq j \leq N_k$ can be written as linear combination of $\mathrm{d}\phi_1, \ldots, \mathrm{d}\phi_{\nu_k}$ in a neighborhood of x°. Combined with the manner in which the vectors $z_1, \ldots, z_k$ have been defined, this means that the codistribution $\mathrm{span}(\mathrm{d}\phi_1, \ldots, \mathrm{d}\phi_{\nu_k})$, which contains the codistribution $\mathrm{span}(\mathrm{d}h_0, \ldots, \mathrm{d}h_m)$, is invariant under the Lie derivative along the vector fields $f_0, \ldots, f_m$. This codistribution has dimension less than n around x°, which contradicts the assumption $\dim \mathrm{d}\mathcal{O}(h)(x^\circ) = n$.

The assumption $\dim \mathrm{d}\mathcal{O}(h)(x^\circ) = n$ also guarantees that the construction ends in a finite number of steps, since a situation in which $\nu_k \to n$ when $k \to \infty$ would obviously lead to a contradiction.

Suppose now that $\nu_q = n$. We claim that the dynamical system having the components of $z = \mathrm{col}(z_1, \ldots, z_q)$ as state variables can be put into the form (4.14) with structure given by (4.11). Since indeed for an arbitrary element z_k^j, $1 \leq j \leq N_k$,

$$\dot{z}_k^j(x) = \sum_{i=0}^{m} L_{f_i} z_k^j(x) u_i,$$

from the condition used for the construction of z_{k+1} and using Lemma 1, if $k < q$, one can write:

$$\dot{z}_k = A_{k,k+1}(u) z_{k+1} + B_k(z_1, \ldots, z_k) u.$$

When $k = q$, the map $\phi = \mathrm{col}(\phi_1, \ldots, \phi_n)$ is a diffeomorphism of a neighborhood V° of x°. Therefore, all functions of x, which include the iterated Lie derivatives of the functions $h_0(x), \ldots, h_m(x)$ along vector fields in the set $\{f_0, \ldots, f_m\}$, can be expressed on this neighborhood as functions of n elements of z. Thus,

$$\dot{z}_q = B_q(z) u.$$

(ii) For the proof of this part, it is enough to note that the map $z(x)$ can stand for the immersion map $\tau(x)$. In this case, if we express the system (4.14) in the form

$$\dot{z} = \sum_{i=0}^{m} \tilde{f}_i(z) u_i$$

$$y = \sum_{i=0}^{m} \tilde{h}_i(z) u_i$$

then, by construction,

$$\frac{\partial \tau}{\partial x} f_i(x) = \tilde{f}_i(\tau(x)) \tag{4.15}$$

$$h_i(x) = \tilde{h}_i(\tau(x)), \tag{4.16}$$

with $i = 0 \dots m$, for all $x \in V^\circ$. Equation (4.15) translates the fact that the flows $\Phi_t^{f_i}(x)$ and $\Phi_t^{\tilde{f}_i}(z)$ of the vector fields f_i and $\tilde{f}_i$ satisfy

$$\tau(\Phi_t^{f_i}(x)) = \Phi_t^{\tilde{f}_i}(\tau(x))$$

for all $x \in V^\circ$ and all $t \geq 0$ such that $\Phi_t^{f_i}(x) \in V^\circ$, which implies that

$$\tau(x_{x^\bullet,u}(t)) = z_{\tau(x^\bullet),u}(t)$$

for all $x \in V^\circ$ and all $t > 0$ such that $x_{x^\bullet,u}([0,t)) \subset V^\circ$. Combined with (4.16), this shows that the two systems have the same input-output map when initialized respectively at $x^\bullet$ and $\tau(x^\bullet)$ for all $x^\bullet \in V^\circ$. □

Remark 4. Note that this transformation is indeed of interest for observer design since an estimation of z can be obtained through the observer in Theorem 5 and then a (unique) estimation of x can be computed by inverting the diffeomorphism ϕ.

An example of application of the immersion procedure and subsequent use of the observer in Theorem 5 can be found in [2].

4.5.3 Extensions

Non-rank-observable systems

The immersion procedure described in the proof of Theorem 6 can also be applied to certain input-affine systems that do not satisfy the observability rank condition, but whose dynamics can be decomposed in two parts such as the output is related to only one of these parts, which is also locally weakly observable. In such a situation, under the assumption that x° is a regular point of $\mathrm{d}\mathcal{O}(h)$, if $\dim \mathrm{d}\mathcal{O}(h) = d < n$, the immersion procedure ends when d independent covectors exist among the differentials of the elements of z.

Recall first that if there exists a codistribution Ω with the properties that

a) it is spanned around a regular point x° by d' *exact* covector fields;
b) it contains the codistribution $\mathrm{span}(\mathrm{d}h_0, \dots, \mathrm{d}h_m)$;
c) it is invariant under the vector fields $f_0, \dots, f_m$;

then a coordinates transformation $\tilde{z} = \tilde{\phi}(x)$ such that

$$\mathrm{span}(\mathrm{d}\tilde{\phi}_{n-d'+1}, \dots, \mathrm{d}\tilde{\phi}_n\} = \Omega,$$

puts the system (4.6) in the following form around x°:

$$\dot{\xi}_1 = f_{10}(\xi_1, \xi_2) + \sum_{i=1}^{m} f_{1i}(\xi_1, \xi_2)u_i$$

$$\dot{\xi}_2 = f_{20}(\xi_2) + \sum_{i=1}^{m} f_{2i}(\xi_2)u_i$$

$$y = h_0(\xi_2) + \sum_{i=1}^{m} h_i(\xi_2)u_i$$

where $\xi_1 = \mathrm{col}(\tilde{z}_1, \ldots, \tilde{z}_{n-d'})$ and $\xi_2 = \mathrm{col}(\tilde{z}_{n-d'+1}, \ldots, \tilde{z}_n)$ [19]. In this representation, the output is only related to the last d' components of the state vector, which correspond to the components of the coordinates transformation whose differentials span $\mathrm{d}\mathcal{O}(h)$. It is obvious that $x \mapsto \xi_2(x)$ is a submersion of the considered neighborhood of x°, since the rank of this application is d' around x°.

Notice now that the codistribution $\mathrm{d}\mathcal{O}(h)$ with a basis of exact covector fields generated around a regular point x° by the immersion procedure meets the required properties for such a decomposition of the system. Moreover, $\mathrm{d}\mathcal{O}(h)$ is the *minimal* codistribution with these properties, i.e. there is no other distribution with the same properties and dimension $d' < d$ such that in the resulting decomposition the output be affected by d' elements of the state vector. Therefore, the estimation of ξ_2 by means of a state observer is the maximum that can be obtained around x° from the input-output map of the considered system.

The immersion can be performed directly on the original system (4.13), which is the same as first decomposing the system and then immersing the observable part (which now satisfies the observability rank condition around $\xi_2(x^\circ)$). However, in contrast with the immersion performed under the conditions of Theorem 6, there no longer exists a one to one correspondence between the internal trajectories of the two systems.

The $A(u, y)$ case

One important effect of the immersion procedure is a (sometimes significant) increase of the order of the system. This effect could be reduced by output injection combined with a slightly different construction strategy. Nevertheless, the presence of y makes the immersion procedure to no longer be systematic. We obtain a heuristic procedure, with no guarantees as to its effectiveness.

More precisely, instead of considering as state variables the iterated Lie derivatives of the functions $h_0, \ldots, h_m$ along vector fields in the set $\{f_0, \ldots, f_m\}$, one could proceed as follows: starting with the same set of variables as in the original immersion procedure, write at each step of the construction the Lie derivatives of the functions considered as state variables at the previous step as sums, by conveniently isolating the terms that can be expressed around x° through already defined state variables and separating the ones that are to become new state variables. The construction ends when there exist n independent covectors fields among the differentials of the functions considered as state variables. This necessarily happens at some point, as it can be easily seen that the codistribution spanned by these covectors around x° coincides with $\mathrm{d}\mathcal{O}(h)$.

Sometimes it may happen that for a term that has to be considered as a new state variable, say ζ, the explicit dependence on y leads to a representation $\zeta = \bar{\zeta}(y)\tilde{\zeta}(x)$. In this case, the new state variable could be chosen to be $\tilde{\zeta}(x)$, making the matrix A dependent on y through $\bar{\zeta}(y)$. In this way, a simpler expression may be obtained for the derivatives of the new state variables, which may also lead in the end to a system of lower order.

Non-input-affine systems

A similar approach to the one in the preceding discussion may yield results in some non control-affine cases. More precisely, one can handle situations in which the output map can be suitably written as a sum of terms in the form $\bar{\eta}(u)\tilde{\eta}(x)$. The immersion procedure is then initialized with the terms generically denoted by $\tilde{\eta}(x)$. Then, constraints have to be put on the expressions of their *time* derivatives in order to carry on the construction and it goes the same for all subsequently created state variables. More precisely, just like for the output map, a suitable arrangement of these expressions as sums has to exist such that the terms that cannot be expressed around x° as functions of already defined state variables (and eventually u) are in the form $\bar{\eta}(u)\tilde{\eta}(x)$, with the exception that the dependence $\bar{\eta}(u, y)$ is also allowed.

Example 6. Consider the system

$$\begin{aligned}
\dot{x}_1 &= u\alpha(x_2) + \beta(x_1)x_2 \\
\dot{x}_2 &= \gamma(x_1, x_2, u) \\
y &= x_1.
\end{aligned}$$

If for instance $u\frac{\partial\alpha(x_2)}{\partial x_2} + \beta(x_1) \geq a > 0$ for all x_1, x_2, one could design a high gain observer as in [13]. But when this condition is not fulfilled, one can yet immerse the system into the form studied in this section.

First, set $z_1^1 = x_1$. By making the dependence on y explicit in the time derivative of this first state variable,

$$\dot{z}_1^1 = u\alpha(x_2) + \beta(y)x_2,$$

new state variables are obtained: $z_2^1 = \alpha(x_2)$ and $z_2^2 = x_2$. Their differential equations are

$$\begin{aligned}
\dot{z}_2^1 &= \frac{\partial\alpha(x_2)}{\partial x_2}\gamma(x_1, x_2, u) = \delta(z_2^2)\gamma(z_1^1, z_2^2, u) \\
\dot{z}_2^2 &= \gamma(x_1, x_2, u) = \gamma(z_1^1, z_2^2, u)
\end{aligned}$$

which shows that the construction has ended.

4.6 Conclusion

In this chapter, some results directly based on the idea of increasing the dimension of the state for a given representation, so as to obtain a new representation with a form well suited to observer design, have been reviewed. It can here be emphasized how such an approach has been shown to provide an estimation tool for the widely studied problem of simultaneous state and parameter estimation in induction motors. It has also been shown in this chapter how such a method can be thought of for a wide class of nonlinear systems.

References

[1] J. Back and J. H. Seo. Immersion of nonlinear systems into linear systems up to output injection: characteristic equation approach. *International Journal of Control*, 77(8):723–734, 2004.
[2] G. Besançon and A. Ţiclea. An immersion-based observer design for rank-observable nonlinear systems. *IEEE Transactions on Automatic Control*, 52(1): 83–88, 2007.
[3] G. Besançon, G. Bornard, and H. Hammouri. Observer synthesis for a class of nonlinear control systems. *European Journal of Control*, 3(1):176–193, 1996.
[4] G. Besançon. On charactering classes of observer forms for nonlinear systems. In *Proceedings of European Control Conf*, 1997.
[5] G. Besançon. Further results on high gain observer for nonlinear systems. In *Proceedings of the 38th IEEE Conference on Decision and Control*, pages 2904–2909, 1999.
[6] William M. Boothby. *An Introduction to Differentiable Manifolds and Riemannian Geometry*. Academic Press, San Diego, 2nd edition, 2003.
[7] G. Bornard, F. Celle-Couenne, and G. Gilles. Observabilité et observateurs. In A. J. Fossard and D. Normand-Cyrot, editors, *Systèmes Non Linéaires*, volume 1. Modélisation - Estimation, pages 177–221. Masson, 1993.
[8] D. Bossane, D. Rakotopara, and J. P. Gauthier. Local and global immersion into linear systems up to output injection. In *Proceedings of the 28th IEEE Conference on Decision and Control*, pages 2000–2004, 1989.
[9] D. Claude, M. Fliess, and A. Isidori. Immersion, directe et par bouclage, d'un système non linéaire dans un linéaire. *Comptes Rendus des Séances de l'Académie de Sciences de Paris. Série 1, Mathématique*, 296(4):237–240, 1983.
[10] M. Fliess. Fonctionnelles causales non linéaires et indéterminées non commutatives. *Bulletin de la Société Mathématique de France*, 109:3–40, 1981.
[11] M. Fliess. Finite-dimensional observation-spaces for nonlinear systems. In D. Hinrichsen and A. Isidori, editors, *Feedback Control of Linear and Nonlinear Systems*, volume 39 of *Lecture Notes in Control and Information Sciences*, pages 73–77. Springer-Verlag, 1982.
[12] M. Fliess and I. Kupka. A finiteness criterion for nonlinear input-output differential systems. *SIAM Journal of Control and Optimization*, 21(5):721–728, 1983.
[13] J. P. Gauthier and I. A. K. Kupka. Observability and observers for nonlinear systems. *SIAM Journal on Control and Optimization*, 32(4):975–994, 1994.
[14] J. P. Gauthier, H. Hammouri, and S. Othman. A simple observer for nonlinear systems — applications to bioreactors. *IEEE Transactions on Automatic Control*, 37(6):875–880, 1992.

[15] H. Hammouri and F. Celle. Some results about nonlinear systems equivalence for the observer synthesis. In G. Conte, A. M. Perdon, and B. Wyman, editors, *New Trends in Systems Theory*, volume 7 of *Progress in Systems and Control Theory*, pages 332–339. Birkhäuser, 1991.
[16] H. Hammouri and J. de Leon Morales. Observer synthesis for state-affine systems. In *Proceedings of the 29th IEEE Conference on Decision and Control*, pages 784–785, 1990.
[17] H. Hammouri and J. de Leon Morales. On systems equivalence and observer synthesis. In G. Conte, A. M. Perdon, and B. Wyman, editors, *New Trends in Systems Theory*, volume 7 of *Progress in Systems and Control Theory*, pages 340–347. Birkhäuser, 1991.
[18] Robert Hermann and Arthur J. Krener. Nonlinear controllability and observability. *IEEE Transactions on Automatic Control*, 22(5):728–740, 1977.
[19] A. Isidori. *Nonlinear Control Systems*. Communications and Control Engineering. Springer-Verlag, London, 3rd edition, 1995.
[20] Philippe Jouan. Immersion of nonlinear systems into linear systems modulo output injection. *SIAM Journal on Control and Optimization*, 41(6):1756–1778, 2003.
[21] R. E. Kalman and R. S. Bucy. New results in linear filtering and prediction theory. *Journal of Basic Engineering*, 83:95–108, 1961.
[22] H. Keller. Nonlinear observer by transformation into a generalized observer canonical form. *International Journal of Control*, 46(6):1915–1930, 1987.
[23] A. J. Krener and A. Isidori. Linearization by output injection and nonlinear observers. *Systems and Control Letters*, 3(1):47–52, 1983.
[24] A. J. Krener and W. Respondek. Nonlinear observers with linearizable error dynamics. *SIAM Journal on Control and Optimization*, 23(2):197–216, 1985.
[25] J. Levine and R. Marino. Nonlinear system immersion, observers and finite-dimensional filters. *Systems and Control Letters*, 7(2):133–142, 1986.
[26] D. G. Luenberger. Observing the state of a linear system. *IEEE Transactions on Military Electronics*, 8:74–80, 1964.
[27] A. Ţiclea. *Techniques d'immersion pour l'estimation non linéaire. Application aux systèmes de puissance.* PhD thesis, Institut National Polytechnique de Grenoble, 2006.
[28] A. Ţiclea and G. Besançon. Observer scheme for state and parameter estimation in asynchronous motors with application to speed control. *European Journal of Control*, 12(4):400–412, 2006.
[29] A. Ţiclea and G. Besançon. Observer design for state and parameter estimation in induction motors with simulation and experimental validation. In *Proceedings of the 32nd Annual Conference of the IEEE Industrial Electronics Society*, 2006.

5

Nonlinear Moving Horizon Observers: Theory and Real-Time Implementation

Mazen Alamir

Centre National de la Recherche Scientifique (CNRS)
GIPSA-Lab. Control Systems Department. SysCo-Team
BP 46, Rue de la Houille Blanche,
Domaine Universitaire, 38400 Saint Martin d'Hères, France
mazen.alamir@inpg.fr

5.1 Definitions and Notation

In this chapter, the concept of Moving-Horizon Observer (**MHO**) is recalled and some related topics are discussed and illustrated through dedicated examples. Throughout this chapter, interest is focused on nonlinear systems that may be described by the following equations:

$$x(t) = X(t, t_0, x_0), \tag{5.1}$$

$$y(t) = h(t, x(t)), \tag{5.2}$$

where $X : \mathbb{R}_+ \times \mathbb{R}_+ \times \mathbb{R}^n \to \mathbb{R}^n$ is a map that gives the state $x(t)$ of the system at instant t based on the knowledge of the state $x(t_0) = x_0$ at some instant t_0. The map $X(\cdot, \cdot, \cdot)$ may be obtained by using an appropriate system model (Ordinary Differential Equations (ODE's), Differential Algebraic Equations (DAE's) or even a quite sophisticated hybrid simulator). Some of the results presented hereafter may need a particular system model. This is indicated when needed. $y(t) \in \mathbb{R}^{n_y}$ denotes the measured output at instant t. Using similar notations as for the state, the output trajectory is denoted hereafter by:

$$Y(\cdot, t_0, x_0) \doteq h(X(\cdot, t_0, x_0)) \tag{5.3}$$

Note also that in (5.1)-(5.2), dependency w.r.t measured variables such as control input, time varying parameters with known time evolution is implicitly handled through the argument t of the map X. When unmeasured disturbances $w \in \mathbb{R}^{n_w}$ and measurement noise $v \in \mathbb{R}^{n_y}$ are to be considered, equations (5.1)-(5.2) are replaced by the following ones:

$$x(t) = X(t, t_0, x_0, w_{t_0}^t), \tag{5.4}$$

$$y(t) = h(t, x(t)) + v(t), \tag{5.5}$$

where $w_{t_0}^t$ denotes the disturbance profile $\{w(\tau)\}_{\tau \in [t_0, t]}$. Note that these disturbances may also represent model discrepancies. The same notation $v_{t_0}^t$ are used to denote measurement noise profiles.

G. Besançon (Ed.): Nonlinear Observers and Applications, LNCIS 363, pp. 139–179, 2007.
springerlink.com

In the present chapter, it is assumed that some knowledge is available on the admissible sets of states, disturbances and measurement noise. Namely, there are known compact sets maps $\mathbb{X}(\cdot)$, $\mathbb{W}(\cdot)$ and $\mathbb{V}(\cdot)$ such that the following inclusions hold at each instant t:

$$x(t) \in \mathbb{X}(t) \subset \mathbb{R}^n \quad ; \quad w(t) \in \mathbb{W}(t) \subset \mathbb{R}^{n_w} \quad ; \quad v(t) \in \mathbb{V}(t) \subset \mathbb{R}^{n_y}. \tag{5.6}$$

These constraints enable the following definition to be stated:

Definition 1 (Measurements-compatible configurations)
Consider some time interval $[t-T, t]$ and a corresponding measurement profile y_{t-T}^t. A pair $(\xi, \mathbf{w}) \in \mathbb{X}(t-T) \times [\mathbb{R}^{n_w}]^{[t-T,T]}$ is said to be (y_{t-T}^t)-compatible if the following conditions hold for all $\sigma \in [t-T, t]$:

1. $w(\sigma) \in \mathbb{W}(\sigma)$,
2. $X(\sigma, t-T, \xi, \mathbf{w}) \in \mathbb{X}(\sigma)$,
3. $y_{t-T}^t(\sigma) - Y(\sigma, t-T, \xi, \mathbf{w}) \in \mathbb{V}(\sigma)$.

When these conditions hold, the following short notation is used:

$$(\xi, \mathbf{w}) \in \mathbb{C}(t, y_{t-T}^t) \tag{5.7}$$

to denote the set of (y_{t-T}^t)-compatible pairs. ♡

Roughly speaking, a (y_{t-T}^t)-compatible pair $(\xi, \mathbf{w})$ is a pair of initial state (at instant $t-T$) and a disturbance profile $\mathbf{w}$ defined on $[t-T, t]$ such that the resulting trajectory obtained by (5.4) meets the constraints (5.6) over $[t-T, t]$.

5.1.1 Technical Definitions

In this section, some technical definitions that are needed in the remainder of this chapter are successively given:

✓ A function $\alpha : \mathbb{R}_+ \to \mathbb{R}_+$ is a **K-function** if it is positive definite, continuous, strictly monotonic increasing and proper ($\lim_{x\to\infty} \alpha(x) = \infty$).

✓ Given some closed subset $\mathbb{S}$ of an Euclidian space $\mathbb{E}$, the projection map $P_{\mathbb{S}}$ is defined as follows:

$$P_{\mathbb{S}} : \mathbb{E} \to \mathbb{S} : P_{\mathbb{S}}(e) = \min_{\sigma \in \mathbb{S}} d(e - \sigma) \tag{5.8}$$

where d is some distance that is to be understood from the context.

✓ For all matrix A, $\underline{\sigma}(A)$ denotes the smallest singular value of A.

✓ Given a piece-wise continuous function $g(\cdot)$ defined over some time interval I and some integer i, the following notations are used:

$$\|g(\cdot)\|_{L_i} = \int_I \|g(\tau)\|^i d\tau \quad ; \quad \|g(\cdot)\|_\infty = \sup_{\tau \in I} \|g(\tau)\|$$

✓ Given a multi-variable function $f(x_1, x_2, \dots)$, the partial derivative of G w.r.t x_i is shortly denoted by $f_{x_i}(x_1, x_2, \dots)$.

5.2 The Constrained Observation Problem

Based on the above definitions, the observation problem can be stated as follows:

Definition 2 (The finite horizon observation problem)
The finite horizon observation problem amounts to choose some observation horizon length $T > 0$ and to use at each instant t, the available information, namely:

1. *the system equations (5.4)-(5.5)*
2. *the past measurements y_{t-T}^t,*
3. *the constraints (5.6) and*
4. *some additional exogenous knowledge.*

in order to produce an estimation $\hat{x}(t)$ of the current state $x(t)$. ♡

The need for some *additional Knowledge* comes from the fact that the first three available information (system equations, measurements and constraints) are of no help to choose between all the candidate states that belong to the following subset:

$$\Omega_t = \Big\{X(t, t-T, \xi, \mathbf{w}) \quad | \quad (\xi, \mathbf{w}) \in \mathbb{C}(t, y_{t-T}^t)\Big\}. \tag{5.9}$$

Indeed, all states in Ω_t belong to trajectories that respect the constraints and the noise level and are therefore equally valuable candidates to *explain* the output measurements. Therefore, there is indeterminism unless one of the following conditions holds:

✓ $\Omega_t = \{x(t)\}$ or
✓ some additional criteria is considered.

The first case $\Omega_t = \{x(t)\}$ occurs in particular when no disturbances nor measurement noises are present ($\mathbb{W} = \{0\}$ and $\mathbb{V} = \{0\}$) provided that the system is observable in the following trivial sense:

Definition 3 (Uniform Observability of nominal systems)
The system (5.1)-(5.2) is uniformly observable if there is some $T > 0$ and a **K-function** *α such that the following inequality holds:*

$$\int_{t-T}^{t} \|Y(\sigma, t-T, x^{(1)}) - Y(\sigma, t-T, x^{(2)})\|^2 d\sigma \geq \alpha(\|x^{(1)} - x^{(2)}\|) \tag{5.10}$$

for all $t \geq 0$ and all $(x^{(1)}, x^{(2)}) \in \mathbb{X}(t-T) \times \mathbb{X}(t-T)$. ♡

Indeed, under disturbance and noise free assumption and for uniformly observable nominal systems, if $X(t, t-T, \xi)$ belongs to Ω_t for some ξ [see (5.9)], then one has according to condition 3 of definition 1: $y_{t-T}^t(\sigma) = Y(\sigma, t - T, x(t-T)) = Y(\sigma, t-T, \xi)$ for all $\sigma \in [t-T, t]$ and this implies according to (5.10) that $\alpha(\|x(t-T) - \xi\|) = 0$ which simply means by definition of α that $\xi = x(t-T)$. Consequently, under the above assumptions, the only element in Ω_t is $X(t, t-T, x(t-T)) = x(t)$.

It is important to underline that definition 3 involves state constraints since inequality (5.10) has to be satisfied only on the set $\mathbb{X}(t-T)\times\mathbb{X}(t-T)$ of admissible pairs. The following example shows a system that is uniformly observable on some restricted region of admissible states but not in the whole state space.

Example 1. Consider the nominal nonlinear system given by:

$$\dot{x}_1 = -x_1 + x_2 \quad ; \quad \dot{x}_2 = 0 \quad ; \quad y = x_1 x_2$$

This system is observable on the subset $\mathbb{X} = \{x \in \mathbb{R}^2 \mid x_2 > 0\}$ but not on $\mathbb{R}^2$. This is because any pair of states $(x^{(1)}, x^{(2)})$ such that $x^{(1)} = -x^{(2)}$ leads to identically the same output profile. This would contradict (5.10) if *global* uniform observability is checked. ♢

In the general uncertain and noisy situations, Ω_t may not be a singleton and one needs to add some additional requirement in order to *make the best choice* between all the pairs $(\xi, \mathbf{w}) \in \mathbb{C}(t, y_{t-T}^t)$. Once such a criterion is defined, the *best choice* denoted by $(\hat{\xi}(t), \hat{\mathbf{w}}(t))$ is used to compute the best state estimate $\hat{x}(t)$ according to:

$$\hat{x}(t) = X(t, t-T, \hat{\xi}(t), \hat{\mathbf{w}}(t)) \tag{5.11}$$

Typically, one way to define a choice criterion is to look for $(\xi, \mathbf{w}) \in \mathbb{C}, (t, y_{t-T}^t)$ that minimizes some functional:

$$J(t, \xi, \mathbf{w}) := \Gamma(t, \xi - \xi^*(t)) + \int_{t-T}^{t} L\Big(\mathbf{w}(\sigma), \varepsilon_y(\sigma)\Big) \quad \text{where} \tag{5.12}$$

$$\varepsilon_y(\sigma) \; = \; y_{t-T}^t(\sigma) - Y(\sigma, t-T, \xi, \mathbf{w}). \tag{5.13}$$

More precisely, the *best choice* $(\hat{\xi}(t), \hat{\mathbf{w}}(t))$ is obtained by solving the following optimization problem:

$$P(t) \quad : \quad \min_{(\xi, \mathbf{w}) \in \Omega_t} J(t, \xi, \mathbf{w}) \tag{5.14}$$

Note that the definition of the performance index J introduces an *additional knowledge* through the relative weights on the disturbance term w and the output prediction error term ε_y. The resulting trade-off recalls the one introduced in the Kalman filter by using penalties equal to the inverses of the corresponding covariance matrices, namely:

$$L(w, \varepsilon_y) = w^T Q^{-1} w + \varepsilon_y^T R^{-1} \varepsilon_y$$

Moreover, the weighting term $\Gamma(\cdot,\cdot)$ in (5.12) enables to penalize the distance between ξ and some particular value $\xi^*(t)$ that may condense the past knowledge on the most likely value of the state at instant $t-T$. The value of $\xi^*(t)$ is generally induced by the *past estimation*.

Remark 1. The formulation given above can be viewed as the generalization of the Kalman filter equations that hold only for linear unconstrained systems with particular statistical properties of the uncertainty (w) and the measurement noise v (white Gaussian signals). In this case, the penalty term writes $\Gamma(t,\eta) = \eta P(t)\eta$ and $\xi^*(t)$ is induced by the past estimation giving rise to the Kalman filter updating rules in the discrete or in the continuous case (see [3] for more details).

5.2.1 About Temporal Parametrization of Uncertainties

In this section, attention is focused on the need for temporal parametrization of uncertainties when using (5.11)-(5.14) to design a nonlinear observer. Indeed, the decision variable $(\xi, \mathbf{w})$ involved in the optimization problem $P(t)$ is infinite dimensional as $\mathbf{w}$ is the uncertainty profile over the time interval $[t-T,t]$. Consequently, any concrete implementation of the above scheme needs a finite dimensional approximation of candidate profiles $\mathbf{w}$.

In many academic texts (see for instance [3]), a discrete time version of the system model is used and a piece-wise constant structure is implicitly used with the unknowns

$$p_w := \{\mathbf{w}(k\tau)\}_{k=k_0}^{k_0+N-1} \in \mathbb{W}(k_0) \times \cdots \times \mathbb{W}(k_0+N-1) \subset \mathbb{R}^{n_w \cdot N},$$

where $\mathbb{W}(k)$ is a short writing of $\mathbb{W}(k\tau)$ and where the observation horizon length is $T = N\tau$. This leads to a decision variable (ξ, p_w) of dimension $n + N \cdot n_w$.

This choice although apparently natural shows the following major drawbacks:

1. First, the piece-wise constant structure is very often too rich when compared to realistic uncertainties that are often due to badly identified rather constant parameters, slowly drifting variables or even periodic disturbances. This excess of spectral content enlarges the *size* of the set Ω_t of candidate paires [see (5.9)] and hence lead to noisy estimation even in presence of small physical measurement noise.
2. In addition to the drawback mentioned above, the piecewise constant structure leads to a high dimensional decision variable $(n+N\cdot n_w)$ with a generally badly conditioned optimization problem. This is because the high spectral content of the resulting $\mathbf{w}$ leads to too many possible *interpretations* of the past measurements.
3. In case of continuously varying uncertain signals, the piecewise constant parametrization implies a small sampling time leading again to even higher dimensional problem for the same observation horizon (buffer length).

One way to overcome these drawbacks is to choose a parametrization of $\mathbf{w}$ that reflects in a more realistic way what would be the time evolution of this uncertainty vector. This can be denoted generically by:

$$w(t) = \mathcal{W}(t, p_w) \quad ; \quad p_w \in \mathbb{P}.$$

Note that here, the dimension of the unknown disturbance parameter p_w is no more directly related to the dimension of the disturbance vector w nor to the length of the observation horizon. The cost function to be minimized can then be rewritten as a function of the new decision variable (ξ, p_w):

$$J(t, \xi, p_w) = J(t, \xi, \mathcal{W}(\cdot, p_w)). \tag{5.15}$$

Example 2. A typical example of reduced dimensional parametrization of an uncertainty vector that evolves smoothly in time is to use time polynomial approximations:

$$\mathcal{W}_i\big(t, \underbrace{(p_w^{(1)}, \dots, p_w^{(n_w)})}_{p_w}\big) = P_{\mathbb{W}(t)}\Big[(1, t, \dots, t^{n_w^{(i)}}) \cdot p_w^{(i)}\Big] \quad ; \quad p_w^{(i)} \in \mathbb{R}^{n_w^{(i)}}. \tag{5.16}$$

where $P_{\mathbb{W}(t)}(\cdot)$ is the projection map on the admissible set $\mathbb{W}(t)$.

The order of the polynomial development for the i-th component of p_w, namely $n_w^{(i)}$ is to be chosen according to what could be a realistic evolution of this component during the observation horizon $[t - T, t]$. The resulting optimization problem shows a decision variable (ξ, p_w) of dimension

$$n_p := n + \sum_{i=1}^{n_w} n_w^{(i)}.$$

It goes without saying that other time parameterizations can be used in order to be closer to any available information about the uncertainty evolution. ♢

When such parametrization is used, the following straightforward notation is adopted to denote the corresponding state trajectory:

$$X(t, t_0, x_0, p_w) = X\big(t, t_0, x_0, \mathcal{W}(\cdot, p_w)\big).$$

The *best* estimate of the state is then given by

$$\hat{x}(t) = X(t, t - T, \hat{\xi}(t), \hat{p}_w(t))$$

where the pair $(\hat{\xi}(t), \hat{p}_w(t))$ minimizes the cost function $J(t, \xi, p_w)$ defined by (5.15).

Note that by using the extended state:

$$\bar{x} = \big(x^T \; p_w^T\big)^T \in \mathbb{R}^n \times \mathbb{R}^{n_p}, \tag{5.17}$$

together with the trivial dynamic $\dot{p}_w = 0$ on the *additional state* vector, the uncertain observation problem is put in a deterministic uncertainty free context with a higher dimensional extended system. Note however that for the new extended uncertainty-free system, the admissible set Ω_t is generally not reduced to $\{x(t)\}$ and the result is still dependent on the additional knowledge that are introduced through the weighting parameters of the cost function $J(t, \xi, p_w)$.

Heuristic approaches can also be used to avoid time structured model of the uncertainties evolution that are not discussed here. See [13] for more details.

5.2.2 Optimization Based vs Analytic Observers

Recall that the Kalman filter equations are originally derived based on optimal design considerations (maximum likelihood under white Gaussian signals assumption). One nice feature of the observer equations is that in the absence of disturbances and measurement noise, the estimation error:

$$e := x - \hat{x}$$

shows a comprehensively asymptotically stable dynamic behavior with a stable closed loop matrix. The generalization of the optimization based formulation that underlines the Kalman filter to general nonlinear systems leads to generally non convex and hard to solve optimization problems.

This fact together with the relatively limited computational facilities in the 80's motivated researches on nonlinear observers that are based on the study of the resulting estimation error's dynamic and that can be expressed in analytic form without the use of on-line computations. However, the possibility to derive observer equations such that the induced dynamics on the estimation error is *provably asymptotically stable* is quite limited. Indeed, given a general nonlinear system expressed in ODE's form

$$\dot{x} = f(x) \quad ; \quad y = h(x), \tag{5.18}$$

and a candidate consistent observer equation:

$$\dot{\hat{x}} = f(\hat{x}) + K(\hat{x}, y)$$

the explicit observer design problem amounts to find a function $K(\cdot,\cdot)$ of the observer's internal state and the measured output such that the induced estimation error equation that is involved in the extended resulting ODE's:

$$\begin{aligned} \dot{x} &= f(x) \\ \dot{e} &= f(x) - f(x-e) - K(h(x), x-e) \end{aligned}$$

can be proved to be asymptotically stable. This is clearly a hard task as long as a high level of genericity is required.

To overcome this difficulty, researchers imagined conditions on the maps f and h involved in the system and measurement equations (5.18) in order for a correction map $K(\cdot,\cdot)$ to be found. High gain observers [20, 2] and sliding mode observers [4, 10, 23] resulted from this approach.

Almost twenty years of this *state estimation error* (**SEE**)-based observer design enforced the idea according to which, the very basic notion of observability expressed in definition 3 is largely insufficient to derive a concrete state estimation scheme. Additional (generally structural) properties are still needed in order for a state observer to exist. Moreover, these additional conditions are constructive in the sense that they are needed not only to guarantee the convergence of the estimation error but they are needed for the observer design itself.

It goes without saying that, faced with these difficulties even in the nominal case, studies on nonlinear observers were essentially directed towards nominal state estimation problems (without uncertainty nor measurement noises). The robustness issues are generally viewed as a by-side product or tackled through an even more restrictive structural properties that are expressed for some extended systems in the spirit of what is presented in section 5.2.1.

Contrary to analytic observers that use the explicit study of the state estimation error in order to design the observer correction term, optimization based observers use the very definition of observability in order to derive the state estimation algorithm. The idea is to use the fact that as long as the nominal system is considered, estimating the state of an observable system is equivalent to minimizing $J(t, \xi)$: the integral of the squared output prediction error over some observation horizon (see definition 3).

Consequently, if an algorithm can guarantee that this quantity converges asymptotically to 0, then there is no need for additional proof of convergence. The convergence of the state estimation error is a direct consequence of the convergence of the cost function $J(t, \xi(t))$ since this proves that $\xi(t)$ converges to $x(t-T)$ and that $\hat{x}(t) = X(t, t-T, \xi(t))$ converges to $x(t)$.

Unfortunately, there is no such algorithms with guaranteed convergence properties for general non convex optimization problems. The keywords *Global convergence* that is widely used in scientific papers refer to global convergence to *some local minimum.* Convergence results still need dedicated sufficient conditions. However, these sufficient conditions are not constructive unlike the ones used in analytic observers design. These conditions are not needed in the construction of optimization based state estimation algorithms. More clearly, even if one cannot guarantee the convergence of the resulting state estimation scheme, one can always *investigate* the performance of an optimization based observer on his own system. It is likely that the resulting scheme works quite correctly even so there is no convergence proof.

Another difficulty arises when using optimization based nonlinear observer. This concerns the real-time implementability issue. Indeed, the number of iterations that would be needed for a solver to *find* the optimal solution of $P(t)$ may exceed the available computation time that would be compatible with the necessary updating rate. This difficulty is made worst by the fact that each evaluation of the cost function needs the evolution of the system to be *simulated* during the observation horizon which may be heavy to perform.

In a word, optimization based nonlinear observers offer several advantages such as constraints handling and independence w.r.t the mathematical model of the system. However, there are still several bottlenecks in their implementation and reliability. Despite these difficulties, these observers are very often the only available choice. Consequently, investigating implementation issues that enable to (at least partially) overcome the above mentioned difficulties is certainly a *profitable investment*. This is the aim of this chapter.

Typically, two main issues are to be considered when implementing moving-horizon observers:

- The first one is related to the presence of local valleys that may attract the optimization process leading to bad estimation of the state. As long as generic observer design is concerned, this problem is unavoidable in constrained non convex optimization. However, one can use a very particular feature of the state estimation induced optimization problem to derive singularities avoidance heuristic scheme. This is depicted in section 5.3 with an experimental validation on a terpolymerization processes.
- The second implementation issue is related to the computation time the iterative process would need to achieve the optimization task. This time may be prohibitive when compared to the necessary updating rate. In this chapter, two different approaches to address this problem are discussed:
  - In the first, a differential formulation of moving horizon observer is proposed. In this formulation, the observer equations take a rather standard form (the observer equations is obtained by copying the system equation and adding a correction term). The only difference is that the correction term uses an integral norm of the output prediction error rather than a point-wise output prediction error. This formulation and the related techniques enabling to reduce the computational burden are discussed in section 5.4.
  - In the second approach to address the real-time implementation issue, the optimization process is *distributed over the system life-time*. A concrete derivative-free iterative scheme is proposed that may address discontinuous (hybrid) behavior of the dynamic system. This scheme is presented in section 5.5

As it is discussed in section 5.2.2, in the forthcoming developments, the robustness issue is addressed indirectly by extending the state vector or by a posteriori validating tests.

5.3 Singularities Avoidance Heuristic Scheme

In this section, we consider a nominal system given by (5.1)-(5.2). The observer design is developed on the nominal system under the uniform observaility assumption (see definition 3). The robustness of the state estimation algorithm is then checked under modeling errors and measurement noise as well as experimentally on a real terpolymerization reactor[1].

[1] The experimental part of this section is a result of a joint work with Nida Sheibat-Othman and Sami Othman for the Laboratoire d'Automatique et du Génie des Procédés (LAGEP, Lyon, France). See [17].

5.3.1 Expression of the Moving Horizon Observer

Using the notations of section 5.2 in the nominal context, consider a sampled receding-horizon observer with observation horizon $T = N\tau_s$ that updates the estimated state at instants $t_k = k\tau_s$ according to:

$$\hat{x}(t_k) = X(t_k, t_{k-N}, \hat{\xi}(t_k)) \tag{5.19}$$

$$\hat{\xi}(t_k) = arg \min_{\xi \in \mathbb{X}(t_{k-N})} \Big[J(t_k, \xi)\Big] := \sum_{i=k-N}^{k} \|y(t_i) - Y(t_i, t_{k-N}, \xi)\|^2_{Q_i(k)} \tag{5.20}$$

where for all $i \in \{1, \dots, N\}$, $Q_i(k) \in \mathbb{R}^{n_y \times n_y}$ is a positive definite weighting matrix. Note here that there is no more integrals used to define the cost function as the measurements are assumed to be acquired with the sampling period τ_s.

Reference to uniform observability is therefore implicitly based on a slight adaptation of definition 3 to the case of sampled measurement acquisition. This would lead to what could be referred to as uniform observability under τ_s-sampling. The corresponding definition is identical to definition 3 with the l.h.s of (5.10) being replaced by the r.h.s of (5.20)

Recall that under the uniform observability assumption, the optimization problem (5.20) admits a unique global minimum $\hat{\xi}(t_k) = x(t_{k-N})$. Moreover, this global minimum correspond to à 0 optimal cost value.

The solution of the constrained generally non convex optimization problem (5.20) is ideally obtained as the asymptotic output of some iterative subroutine $\mathcal{S}$ (see figure 5.1), namely:

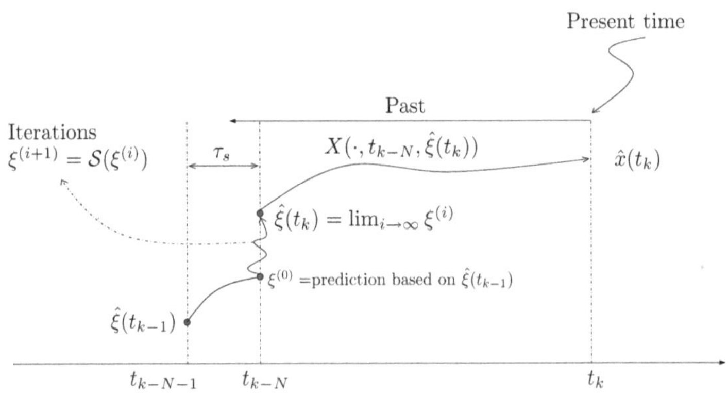

Fig. 5.1. Ideal computation scheme to solve the optimization problem (5.20). The iterative process $\mathcal{S}$ is initialized at $\xi^{(0)}$ that is obtained using the past estimated value $\hat{\xi}(t_{k-1})$. Then, the iterations ideally lead asymptotically to the solution $\hat{\xi}(t_k)$. Integrating the system equations enables the computation of the current estimation $\hat{x}(t_k)$.

$$\hat{\xi}(t_k) \leftarrow \lim_{i\to\infty} \xi^{(i)} \tag{5.21}$$

$$\xi^{(i+1)} = \mathcal{S}(\xi^{(i)}, t_k, y_{t_{k-N}}^{t_k}) \quad ; \quad \xi^{(0)} = X(t_{k-N}, t_{k-N-1}, \hat{\xi}(t_{k-1})) \tag{5.22}$$

More precisely, an initial guess $\xi^{(0)}$ for the iterative process is computed based on the past estimation $\hat{\xi}(t_{k-1})$ by integrating the system equations one sampling period ahead. The iterations defined by (5.22) can then be performed to yield $\hat{\xi}(t_k)$ after some iterations and the estimation $\hat{x}(t_k)$ is obtained according to (5.19).

Note that the initialization of the iterative process by $\xi^{(0)}$ that is based on the past estimation represents in some way an exogenous knowledge that is *injected* in addition to the past measurements. Note also that this can be used explicitly in the definition of the cost function to play the role of ξ^* invoked in section 5.2. More precisely, one can replace the cost function used in (5.20) by:

$$J^*(t_k, \xi) := \|\xi - \xi^{(0)}\|_{Q_0} + \sum_{i=k-N}^{k} \|y(t_i) - Y(t_i, t_{k-N}, \xi)\|_{Q_i(k)}^2 \tag{5.23}$$

where $\xi^{(0)}$ is given by (5.22).

It goes without saying that in practice, the number of iterations of the process $\mathcal{S}$ that can be performed within a sampling period τ_s is necessarily limited and the assignment in (5.21) must be replaced by:

$$\hat{\xi}(t_k) = \xi^{(N_{max})} = \mathcal{S}^{N_{max}}(\xi^{(0)}, t_k, y_{t_{k-N}}^{t_k}) \tag{5.24}$$

where $\mathcal{S}^j(\cdot)$ denotes the results of j successive applications of the map $\mathcal{S}$ starting from the initial guess $\xi^{(0)}$. More precisely, the number of iterations depends on the required precision $\varepsilon > 0$ used in the solver and N_{max} is just an upper bound on this number. Consequently, the number of effective iterations $N_{eff}(t_k, \varepsilon) \leq N_{max}$ varies in time with the parameters $x(t_k)$, $y_{t_k - N}^{t_k}$ that contribute to the definition of the optimization problem and its related complexity for a given required precision ε. It results that using an a priori given bound N_{max}, the required precision is no more guaranteed even regardless the problem of local minima.

Now, regardless the iterative process $\mathcal{S}$ used to perform the optimization task, local minima may exist that potentially prevent the iterate $\xi^{(i)}$ from converging to the global minimum $x(t_{k-N})$. This is afforded by using multiple initial guesses unless some other characterization of the global minimum is available.

Fortunately, when dealing with the nominal state estimation problem for uniformly observable systems, the global minimum $x(t_{k-N})$ one looks for when trying to solve the optimization problem (5.20) can be strongly characterized by the following property:

> $x(t_{k-N})$ is the unique global minimum of ALL the optimization problems (5.20) that may be obtained by changing the positive definite weighting matrices $Q_i(k)$.

This makes the state estimation problem a very particular optimization problem since the global minimum one is looking for is THE global minimum of an infinite number of KNOWN functons. A subset of this set of functions sharing $x(t_{k-N})$ as global minimum can be generated by choosing the following family of weighting matrices:

$$Q_i(k) = \gamma^{k-i} \cdot q_i \cdot \mathbb{I}_{n_y} \quad \text{s.t} \quad q_i > 0 \quad \text{and} \quad \sum_i q_i = 1 \tag{5.25}$$

where $\gamma \in [0,1]$ is some forgetting factor while $\mathbb{I}_{n_y}$ is the identity matrix in $\mathbb{R}^{n_y \times n_y}$. Note that since the vector of weights:

$$\bar{q} = \begin{pmatrix} q_1 \; q_2 \; \dots q_{n_y} \end{pmatrix}^T \tag{5.26}$$

is involved in the definition of the cost function (5.20), the iterative process (5.24) can be worth rewritten in the following form:

$$\hat{\xi}(t_k) = \mathcal{S}_{\bar{q}}^{N_{max}}(\xi^{(0)}, t_k, y_{t_{k-N}}^{t_k}) \tag{5.27}$$

The idea is then to notice that a local minimum for (5.20) in which some weighting vector $\bar{q}^{(1)}$ is used may probably not remain a local minimum for another randomly chosen value of the weighting vector $\bar{q}^{(2)}$ since it seems reasonable to admit that only the true global minimum $x(t_{k-N})$ is a singular point for all possible values of the weighting vector $\bar{q}$. Following this intuition, the *one trials* updating rules (5.27) is replaced by the following *multiple trials* updating rule:

$\bar{q} \leftarrow \dfrac{1}{n_y}(1\;1\;\dots\;1)$; $\hat{\xi}(t_k) =\leftarrow X(t_{k-N}, t_{k-N-1}, \hat{\xi}(t_{k-1}))$

for $(i = 1 : N_{\text{trials}})$

  $\hat{\xi}(t_k) \leftarrow \mathcal{S}_{\bar{q}}^{N_{max}}(\hat{\xi}(t_k), t_k, y_{t_{k-N}}^{t_k})$

  Generate randomly new $\bar{q}$ satisfying (5.26)

end

$\hat{x}(t_k) \leftarrow X(t_k, t_{k-N}, \hat{\xi}(t_k))$

Note that when $N_{\text{trials}} = 1$, the *multiple trials* updating rule defined above gives the classical *one trial* updating rule (5.27) in which the initial guess ξ^0 is given by (5.22).

Note that in the algorithm described above, the quantities t_k, $y_{t_{k-N}}^{t_k}$ and $\hat{\xi}(t_{k-1})$ are inputs while the resulting estimated values $\hat{x}(t_k)$ and $\hat{x}(t_k)$ are outputs while $N_{\text{trials}} \in \mathbb{N}$ is a parameter. This can be shortly written as follows:

$$\hat{\xi}(t_k) = \mathcal{A}_{N_{\text{trials}}}\Big(\hat{\xi}(t_{k-1}), t_k, y_{t_{k-N}}^{t_k}\Big) \tag{5.28}$$

$$\hat{x}(t_k) = X(t_k, t_{k-N}, \hat{\xi}(t_k)) \tag{5.29}$$

which clearly defines a dynamic observer with internal state $\hat{\xi}$ that delivers the estimated state $\hat{x}(t_k)$ as output.

It is worth noting that according to the definition of $\mathcal{A}_{N_{\text{trials}}}(\cdot)$, one need to perform $N_{\text{trials}} \times N_{max}$ iteration of the process $\mathcal{S}$. Denoting by τ_{iter} the time needed to perform a single iteration, the following constraint has to be satisfied:

$$N_{trial} \times N_{max} \times \tau_{iter} \leq \tau_s \tag{5.30}$$

in order for the above moving horizon observer to be real time implementable with τ_s as updating period.

It is worth noting that in the real-time implementability constraint (5.30), a trade-off is clearly to be found that is probably problem dependent. Examples may be found in which it is worth increasing N_{trials} and reducing N_{max} and vice-versa. On the other hand, the updating period τ_s may be quite larger than the acquisition rate in order to leave time for convergence.

5.3.2 Application to a Terpolymerization Batch Process

In this section[2], the moving horizon state observer defined in the preceding section is applied to the state estimation of a terpolymerization batch process.

Multimonomer systems are usually used to produce polymeric materials with suitable final properties. Terpolymerization systems usually allow producing high performance materials. In order to control the final polymer properties, such as the polymer composition, it is of high importance to model and monitor such processes. In particular, monitoring the number of each one of the three monomers is a key issue in controlling the final product quality.

In this section, we will be interested in estimating the polymer composition in emulsion terpolymerization. A complete description of the state estimation results presented in this section can be obtained in [17]. Here, only a sketch of the result are given to illustrate the estimation process described above.

While several estimators have been proposed for polymerization processes (see for instance [22, 21, 7] and the references therein), as long as emulsion terpolymerization is concerned, only two applications could be found in the literature. In [12], an open loop observer is designed to estimate the polymer composition using calorimetric measurements combined to the process model. In [19], a closed loop high gain observer is proposed to estimate the polymer composition and it has been shown by simulation and experimentally that the system can be observable if the total amounts of monomers are measured. However, because of the model complexity (see below), the design of such a high gain observer and the tuning of its gain in order to cope with the system constraints remains a quite involved task and the high gain observer has been obtained at the price of tremendous simplification of the dynamic model that lead to rather poor estimation performance.

In the remainder of this section, the process model is first described, then simulations as well as experimental validations are discussed.

---

[2] The process description given in this section is basically borrowed from [17] and is due to Nida Sheibat-Othman to whom I am deeply indebted for our fruitful and exciting collaborations.

Process Model

Assuming that monomers are not soluble in the aqueous phase and that the reaction takes place mainly in the polymer particles, the material balances of monomers are given by:

$$\dot{N}_i = Q_i - R_{Pi} \quad i - 1, 2, 3 \tag{5.31}$$

The reaction rate in the polymer particles R_{Pi} is proportional to the concentration of monomer in the polymer particles ($[M_i^P]$) and the number of moles of radicals in the polymer particles (μ):

$$R_{Pi} = \mu[M_i^P](k_{p1i}P_1^P + k_{p2i}P_2^P + k_{p3i}P_3^P) \tag{5.32}$$

The time averaged probabilities (P_i^P) that an active chain be of ultimate unit of type i are defined by:

$$P_1^P = \frac{\alpha}{\alpha+\beta+\gamma} \quad ; \quad P_2^P = \frac{\beta}{\alpha+\beta+\gamma} \quad ; \quad P_3^P = 1 - P_1^P - P_2^P \tag{5.33}$$

where

$$\begin{aligned}
\alpha &= [M_1^P](k_{p21}k_{p31}[M_1^P] + k_{p21}k_{p32}[M_2^P] + k_{p31}k_{p23}[M_3^P]) \\
\beta &= [M_2^P](k_{p12}k_{p31}[M_1^P] + k_{p12}k_{p32}[M_2^P] + k_{p13}k_{p32}[M_3^P]) \\
\gamma &= [M_3^P](k_{p13}k_{p21}[M_1^P] + k_{p21}k_{p23}[M_2^P] + k_{p13}k_{p23}[M_3^P])
\end{aligned}$$

In emulsion polymerization, it is well known that the reaction can be divided into three intervals. In interval I, the polymer particles are produced. Modelling of this interval allows the calculation of the particle size distribution and the average number of radicals per particle which allows to calculate the total number of moles of radicals in the polymer particles (μ) in (5.32). This part of the model will not be considered since it adds a lot of complexity to the process model besides the fact that it remains very sensitive to impurities. μ will therefore be considered as a parameter in the process model to be estimated without modelling. It is important to outline that μ can undergo important changes during the reaction since it is affected by the gel effect phenomena.

In interval II, the particle number is supposed to be constant. Polymer particles are saturated with monomer and the excess of monomer is stored in the monomer droplets. During interval III, monomer droplets disappear and all the residual monomer is supposed to be in the polymer particles. Therefore, the concentration of monomer in the polymer particles can be calculated by the following system:

$$[M_i^P] = \begin{cases} \dfrac{(1-\phi_p^p)N_i}{\sum_j \dfrac{N_j MW_j}{\rho_j}}, & \text{(Phase II)} \\[2ex] \dfrac{N_i}{\sum_j MW_j(\dfrac{N_j^T - N_j}{\rho_{j,h}} + \dfrac{N_j}{\rho_j})} & \text{(Phase III)} \end{cases} \tag{5.34}$$

Table 5.1. Parameter values of the terpolymerization of BuA/MMA/VAc (used in the experimental validation)

Parameter	Value	Unit
ϕ_p^p	0.4	
MW_1	128.2	(g/mol)
MW_2	100.12	(g/mol)
MW_3	86.09	(g/mol)
ρ_1	0.89	(g/cm^3)
ρ_2	0.94	(g/cm^3)
ρ_3	0.93	(g/cm^3)
$\rho_{1,h}$	1.08	(g/cm^3)
$\rho_{2,h}$	1.15	(g/cm^3)
$\rho_{3,h}$	1.17	(g/cm^3)
k_{p11}	4.5×10^5	$(cm^3/mol/s)$
k_{p22}	1.28×10^6	$(cm^3/mol/s)$
k_{p33}	4.26×10^6	$(cm^3/mol/s)$
r_{12}	0.355	
r_{21}	1.98	
r_{13}	6.635	
r_{31}	0.037	
r_{23}	22.21	
r_{32}	0.07	

The condition for the existence of monomer droplets and therefore for determining if the reaction is in interval II, is governed by the following equation:

$$N_1\delta_1 + N_2\delta_2 + N_3\delta_3 - \frac{(1-\phi_p^p)}{\phi_p^p}\sigma > 0 \tag{5.35}$$

where

$$\delta_i = MWi(\frac{1}{\rho_i} + \frac{(1-\phi_p^p)}{\rho_{i,h}\phi_p^p}) \ , \quad i = 1,2,3 \tag{5.36}$$

and

$$\sigma = \sum_{j=1}^{3} \frac{MW_j N_j^T}{\rho_j, h} \tag{5.37}$$

The overall monomer conversion that can be measured easily online by calorimetry is defined by:

$$y = \frac{\sum_{i=1}^{3} MW_i(N_i^T - N_i)}{\sum_{j=1}^{3} MW_j N_j^T} \tag{5.38}$$

Parameters used for the experimental validation of the model are given in table 5.1 where $k_{pij} = k_{pii}/r_{ij}$. The recipe used for the experimental validation of the observer is given by table 5.2 [19].

Table 5.2. Recipe of the terpolymerization of BuA/MMA/VAc

Component	Charge (g)
Butyl acrylate	300
Methyl methacrylate	300
Vinyl acetate	60
Sodium dioctyl sulfosuccinate	3
Potassium persulfate	2
Water	2380

Simulation-based Validation of the Moving-Horizon Observer

In order to apply the moving-horizon estimation scheme proposed in the preceding section to reconstruct the value of $N := (N_1, N_2, N_3)$ and μ, a constant evolution of μ is assumed (over the prediction horizon) and the general state equation is built up with the state vector being defined by :

$$x := \begin{pmatrix} N_1 & N_2 & N_3 & \mu \end{pmatrix} \in \mathbb{R}^4_+ \quad ; \quad \dot{\mu} = 0$$

Recall however that despite this constant behavior during the prediction horizon, the resulted *closed-loop* estimation of μ may show dynamic behavior thanks to the moving horizon technique (see figures 5.6 and 5.7).

Note that this is a concrete example of how dynamically unmodelled uncertain parameters can be tackled by the state extension technique that is described in section 5.2.1 [see equation (5.17)].

Considering global relative uncertainties d_1, d_2 and d_3, the following model is obtained to be used by the observer:

$$\dot{N} = \begin{pmatrix} 1+d_1 & 0 & 0 \\ 0 & 1+d_2 & 0 \\ 0 & 0 & 1+d_3 \end{pmatrix} \cdot f(x,u) \tag{5.39}$$

$$\dot{\mu} = 0 \tag{5.40}$$

$$y = (1+\nu) \cdot h(x) \tag{5.41}$$

Namely, relative uncertainties are introduced directly on the r.h.s of the system ODE's through the variables d_i's. This can gather all sources of model discrepancy. On the other hand measurement noises are introduced through the variable ν used in the measurement equation (5.41). More precisely, the following definitions of d and μ are used in the simulations:

$$d_i(k) = d_{max} \cdot r_i(k) \tag{5.42}$$

$$\nu(k) = \nu_{max} \cdot r_\nu(k) \tag{5.43}$$

where the $r_i(k)$'s and $\nu(k)$ are chosen randomly in $[-1, 1]$.

The results are shown on figures 5.2 and 5.3 (respectively without and in the presence of measurement noises) where up to 10% relative errors are introduced on the r.h.s of the system's model.

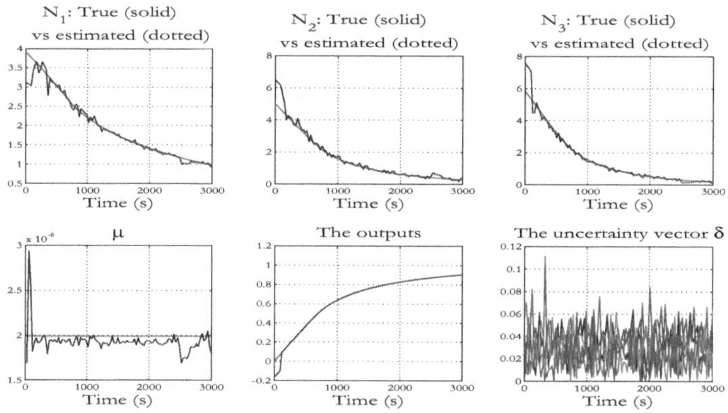

Fig. 5.2. Observer behavior under model uncertainty given by (5.39)-(5.43) with $d_{max} = 10\%$ and no measurement noise ($\nu_{max} = 0$). The observation horizon is $N = 10$ and the number of trials for the singularity crossing scheme is $N_{\text{trials}} = 4$. Initial state of the observer is $\hat{x}(0) = diag(0.8, 1.3, 1.3) \cdot x(0)$ and $\mu_{obs}(0) = 0.8\mu_{model}$.

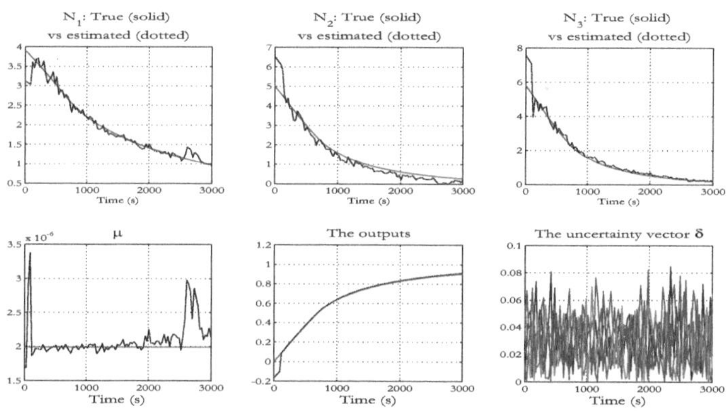

Fig. 5.3. Observer behavior under model uncertainty given by (5.39)-(5.43) with $d_{max} = 10\%$ and in the presence of measurement noise ($\nu_{max} = 0.01$). The observation horizon is $N = 15$ and the number of trials for the singularity crossing scheme is $N_{\text{trials}} = 4$. Initial state of the observer is $\hat{x}(0) = diag(0.8, 1.3, 1.3) \cdot x(0)$ and $\mu_{obs}(0) = 0.8\mu_{model}$. Note that concerning the output, only the true output and the estimated one are shown, measurement noise is not presented. This scenario uses a tolerance $\varepsilon = 10^{-8}$ for the optimization subroutine.

In order to show the benefit from the *singularity crossing* mechanism introduced in section 5.3, simulations with $N_{\text{trials}} = 1$ and $N_{\text{trials}} = 4$ are compared. The results are shown on Figure 5.4. The scenario being used is the same as the one depicted on figure 5.3.

Finally, to end this simulation based validation section, let us check the real time implementability of the moving-horizon observer. The computation times that lead to the results of figure 5.3 are given on Figure 5.5. Note that an explicit upper bound is imposed on the number of function evaluations. More precisely, the internal loop of the optimizer stops as soon as the computation time exceeds the sampling period (30 seconds). Note that all the results shown above use a tolerance threshold $\varepsilon = 10^{-8}$ for the optimization subroutine. It is shown in the following section illustrating the experimental validation results that this precision is unnecessarily high and quite similar results can be obtained using a lower precision (for instance $\varepsilon = 10^{-3}$) while reducing dramatically the computation time (see figures 5.6 and 5.7 hereafter). This is especially true under the multiple trials technique proposed above.

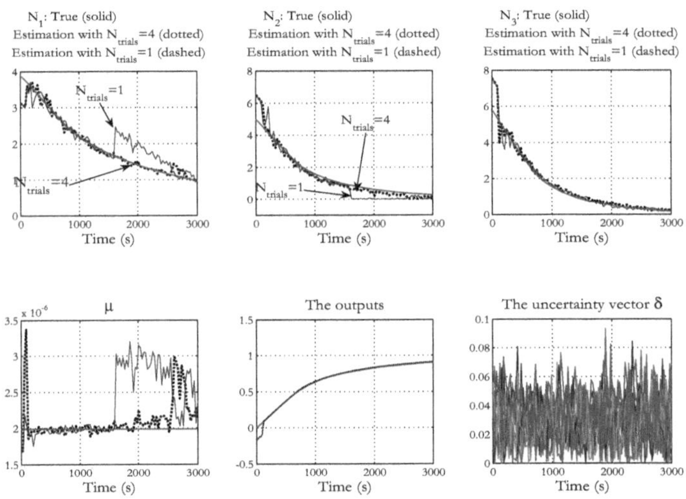

Fig. 5.4. Comparison between the observer behavior when $N_{trials} = 1$ and $N_{trials} = 4$ under the scenario depicted on figure 5.3. Note how the singularity cross mechanism enables to avoid drops in the estimation quality when the observer encounters a singular situation. This scenario uses a tolerance $\varepsilon = 10^{-8}$ for the optimization subroutine.

Experimental Validation of the Moving-Horizon Observer

In this section, the ability of the proposed state observer to reconstruct the individual values of N_1, N_2 and N_3 as well as the unmeasured and dynamically unmodeled variable μ is shown. Note that in order to experimentally measure the values of the N_i's, Samples are withdrawn during the reaction and an inhibitor is added to stop the reaction. The latex is then diluted in a solvent and injected in a gas chromatograph to measure the residual amount of monomer. By doing so, the true values of the N_i can be obtained. This has been done only during the 80 first minutes of the Batch where only 9 samples have been analyzed. The dots (*) on figures 5.6 and 5.7 indicate the corresponding measurements.

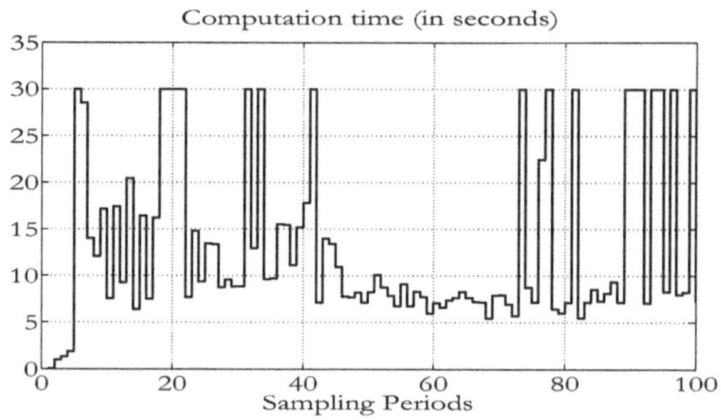

Fig. 5.5. Computation times needed to achieve the state estimation depicted on figure 5.3. Note that an explicit upper bound has been imposed in the internal loop of the optimizer in order to deliver the best estimation that can be obtained within the available computation time defined by the sampling period (30 seconds). This scenario uses a tolerance $\varepsilon = 10^{-8}$ for the optimization subroutine.

These figures clearly show the efficiency of the proposed pair (model,observer) in retrieving with an astonishing precision the values of the N_i's despite the unmodelled dynamic of μ. The rather short computation times (less than 5 seconds compared to the computation times obtained under high precision tolerance) underlines how real-time implementability depends on such parameters that are difficult to set a priori. Finally, it is worth underlying that the times needed to perform $N_{trials} = 10$ (figure 5.6) is much less than 10 times the mean computation time for $N_{trials} = 1$. This strengthens that the proposed singularity cross technique is different from the multiple initial guess technique in the sense that each trials starts from the best result achieved from the previous trial, only the weighting parameter vector $\bar{q}$ is randomly modified.

Figure 5.6 clearly shows an interesting (though expected) feature according to which the *closed-loop* dynamic of the additional state μ is much more *rich* than the dynamic used in the *prediction* algorithm. Indeed, while the supposed dynamic is $\dot{\mu} = 0$, the estimated evolution of μ shows realistic dynamic that is typical for this variable as it can be attested by polymerization experts. This asserts the efficiency of the *extended state* technique invoked in section 5.2.1 in handling the uncertainties using nominal uncertainty free framework even for uncertainties showing important dynamics.

5.4 Differential Form of Moving Horizon Observers

Throughout this section, the system model is assumed to be given in the following ODE form:

$$\dot{x}(t) = f(t, x(t)) \tag{5.44}$$

$$y(t) = h(t, x(t)) \tag{5.45}$$

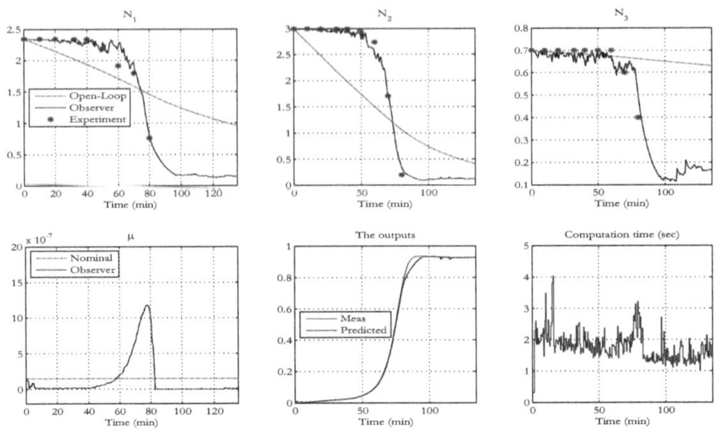

Fig. 5.6. Experimental validation with $N_{trials} = 10$ and tolerance threshold $\varepsilon = 10^{-3}$ for the optimization subroutine. Note how the dynamic behavior of μ is recovered despite the constant behavior assumption used in the receding horizon observer model. The dashed lines show what would be obtained if an open-loop simulator is used to obtain an on-line estimation of the N_i's. Note the excellent matching between the experimentally measured values of the N_i's and those recovered by the observer. The same scenario is depicted on figure 5.7 where $N_{trials} = 1$ is used. Note also the quite rich estimated dynamic for μ despite the over simplified (constant) dynamic used in the definition of the extended state. This asserts the efficiency of using the extended state formalism in handling uncertainties using nominal uncertainty free framework.

This is because the differential form of the moving-horizon observer needs the time evolution of the system to be continuously differentiable. Consequently, under this assumption, there is no clear advantage from using the general form adopted in the preceding sections. The state and the output trajectories related notations, namely $X(t, t_0, x_0)$ and $Y(t, t_0, x_0)$ are however maintained unchanged.

Throughout this section, it is assumed that the cost function $J(t, \xi(t))$ used in the receding-horizon estimation scheme is given by:

$$J(t, \xi) = \int_{t-T}^{t} \|Y(\tau, t-T, \xi) - y(\tau)\|^2 d\tau \tag{5.46}$$

In addition to the continuous differentiability of the r.h.s of (5.44), the following technical assumption is needed for the convergence result of the present section:

Assumption 1 (Uniform global regularity)
There is a **K-function** $\Upsilon : \mathbb{R}_+ \to \mathbb{R}_+$ *such that the following inequality holds:*

$$\|J_\xi(t, \xi)\|^2 \geq \Upsilon(J(t, \xi)) \tag{5.47}$$

for all (t, ξ) ♡

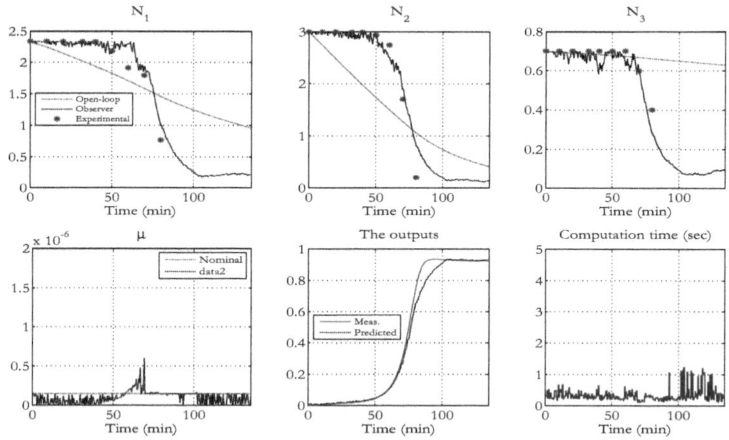

Fig. 5.7. Results under the same experimental validation scenario as figure 5.6 with $N_{trials} = 1$ and tolerance threshold $\varepsilon = 10^{-3}$. Note the slight drop in the estimation quality (particularly on N_2) compared to figure 5.6 where the singularity cross technique is used.

This assumption simply means that regardless the state $x(t-T)$ that holds at instant $t-T$, the corresponding cost function $J(t,\cdot)$ has a unique singular point which is precisely the unique global minimum $\xi = x(t-T)$. This is clearly a strong assumption that may not be necessary for the success of the estimation task in practical situations but that is mandatory to obtain a provably convergent state estimation scheme *in the large* (regardless the initial state estimation error). Locally, this property is clearly satisfied for systems (5.44)-(5.45) having an observable linearization (see [15] for more details). Further discussion on how to tackle the case where this assumption is not rigorously satisfied can be found in [13]. In this general survey on moving-horizon nonlinear observers, we restrict the presentation to the original basic framework. Another (although quite closely related) viewpoint leading to differential form of moving-horizon observer is based on continuation approach and can br found in [24].

The moving-horizon observer described in section 5.3 follows the standard scheme of the early formulation of [11] except that a singularities avoidance heuristic has been introduced. In particular, this formulation leaves aside the detailed description of the optimization process $\mathcal{S}$ used in (5.22) to perform the optimization task.

Regardless the particular choice of the optimizer $\mathcal{S}$, the classical scheme leads to a dynamic process on the observer's *internal state* $\xi(t)$ that is precisely given by (5.28) which is reproduced here for clarity:

$$\xi(t_k) = \mathcal{A}_{N_{\text{trials}}}\Big(\xi(t_{k-1}), t_k, y_{t_{k-N}}^{t_k}\Big)$$
$$\hat{x}(t_k) = X(t_k, t_{k-N}, \xi(t_k))$$

But there is clearly a more direct way to induce a dynamic on the internal state $\xi = \hat{x}(t-T)$ that is oriented towards the decrease of the cost function $J(t,\xi(t))$. Indeed, taking the time derivative of J, one may write[3]:

$$\dot{J}(t,\xi(t)) = J_t(t,\xi(t)) + \left[J_\xi(t,\xi(t))\right]\dot{\xi} \tag{5.48}$$

Note that the cost function $J(t,\zeta(t))$ implicitly depends on the value of the state $x(t-T)$ at the past instant $t-T$. It is worth emphasizing however that this dependence involves only the past measurements y_{t-T}^t over the time interval $[t-T,t]$.

The evolution of ξ has to satisfy two conditions:

1. It must lead to a consistent observer in the absence of modeling errors and measurement noise. This means that if $\xi(t_0) = x(t_0 - T)$ at some instant t_0, then $\xi(t) = x(t-T)$ for all $t \geq t_0$. This implies the following *structure* for $\dot{\xi}$:
$$\dot{\xi}(t) = f(t-T,\xi(t)) + \underbrace{c(t,\xi(t))}_{\text{correction term}} \quad . \tag{5.49}$$
Note that the first term in the r.h.s of (5.49) is the nominal time derivative of $\xi(t)$ (i.e. when $\xi(t) = x(t-T)$) while $c(\cdot,\cdot)$ is a correction function that is such that:
$$\Big\{J(t,\xi(t)) = 0\Big\} \quad \Rightarrow \Big\{c(t,\xi(t)) = 0\Big\}.$$
This enables to recover the nominal behavior as soon as $J(t,\xi(t)) = 0$, or equivalently as soon as $\xi(t) = x(t-T)$ under the observability condition in the sense of definition 3.
2. The correction term must be oriented towards the decrease of the cost function J.

Note that by injecting (5.49) in (5.48), the dynamic of J becomes:

$$\dot{J} = J_t(t,\xi(t)) + \Big[J_\xi(t,\xi(t))\Big] \cdot \Big[f(t-T,\xi(t)) + c(t,\xi(t))\Big] \tag{5.50}$$

To go further, the following two lemmas are needed:

Lemma 1. *The correction-free time derivative of J satisfies:*

$$\frac{dJ}{dt}|_{c(\cdot,\cdot)\equiv 0} \leq |\epsilon_y(t,\xi(t)) - \epsilon_y(t-T,\xi(t))| + \Big[\phi(t,\xi(t))\Big] \cdot \sqrt{J}$$

where

$$\epsilon_y(\tau,\xi(t)) = Y(\tau,t-T,\xi(t)) - y(\tau) \quad \forall \tau \in [t-T,t]$$

♡

[3] Assuming that the necessary regularity conditions are satisfied.

Proof. *Taking the time derivative of (5.46) when no correction is used gives:*

$$\frac{dJ}{dt}|_{c(\cdot,\cdot)\equiv 0} = \epsilon_y(t,\xi(t)) - \epsilon_y(t-T,\xi(t)) + \tag{5.51}$$

$$\int_{t-T}^{t}\Big[Y(\tau,t-T,\xi(t)) - y(\tau)\Big]^T\Big[\tilde{\phi}(\tau,\xi(t))\Big]d\tau \tag{5.52}$$

where $\tilde{\phi}(t,\xi(t))$ *is given by:*

$$\tilde{\phi}(\tau,\xi(t)) := \frac{dY}{dt}\Big|_{c\equiv 0}(\tau,t-T,\xi(t)) - \dot{y}(\tau) \tag{5.53}$$

Using appropriate upper-bounding inequalities, equation (5.52) gives:

$$\frac{dJ}{dt}|_{c(\cdot,\cdot)\equiv 0} \le |\epsilon_y(t,\xi(t)) - \epsilon_y(t-T,\xi(t))| + \tag{5.54}$$

$$\underbrace{\sup_{\tau\in[t-T,t]}\Big|\tilde{\phi}(\tau,\xi(t))\Big|}_{=:\phi(t,\xi(t))}\cdot\underbrace{\int_{t-T}^{t}\|Y(\tau,t-T,\xi(t)) - y(\tau)\|\,d\tau}_{\le\sqrt{J}} \tag{5.55}$$

which clearly gives the result. □

Note that lemma 1 states that a function ϕ exists. The following lemma gives the conditions under which an upper bound i=of this function can be obtained to be used in the definition of the observer dynamic.

Lemma 2. *If it is possible to estimate an upper bound* $\rho(t)$ *satisfying:*

$$\forall\tau\in[t-T,t] \quad ; \quad \|\dot{y}(\tau)\| \le \rho(t) \tag{5.56}$$

then there is a known computable function $\bar{\phi}_\rho(t,\xi(t))$ *satisfying:*

$$0 \le \phi(t,\xi(t)) \le \bar{\phi}_\rho(t,\xi(t)) \tag{5.57}$$

Proof. This is a direct consequence of (5.53) from which it can be inferred that:

$$\phi(t,\xi(t)) \le \sup_{\tau\in[t-T,t]}\Big[\|\frac{dY}{dt}\Big|_{c\equiv 0}(\tau,t-T,\xi(t))\| + \rho(t)\Big]$$

But for given τ, the time derivative of $Y(\tau,t-T,\xi(t))$ is given by:

$$\dot{Y}(\tau,t-T,\xi(t)) = Y_{t_2}(\tau,t-T,\xi(t)) + Y_\xi(\tau,t-T,\xi(t))f(t-T,\xi(t))$$

where $Y_{t_2}(\cdot)$ is the partial derivative of Y w.r.t its second argument. The fact that the partial derivative terms Y_{t_2} and Y_ξ can be computed by classical sensitivity related ODE's ends the proof of the lemma. □

Based on lemmas 1 and 2, equation (5.50) leads to:

$$\dot{J} \leq \left|\Delta_{t-T}^{t}(\epsilon_y(\cdot,\xi(t)))\right| + \left[\phi_\rho(t,\xi(t))\right] \cdot \sqrt{J} + J_\xi(t,\xi(t)) \cdot c(t,\xi(t))$$

where the following short notation has been used:

$$\Delta_{t-T}^{t}(\epsilon_y(t,\xi(t))) = \epsilon_y(t,\xi(t)) - \epsilon_y(t-T,\xi(t))$$

This suggests the following expression for the correction term $c(t,\xi)$:

$$c(t,\xi(t)) := \gamma\Big[\frac{J_\xi^T(t,\xi(t))}{\|J_\xi\|^2+\varepsilon}\Big]\Big[-\left|\Delta_{t-T}^{t}(\epsilon_y(\cdot,\xi(t)))\right| - \left[1+\bar{\phi}_\rho(t,\xi(t))\right]\sqrt{J}\Big] \quad (5.58)$$

since when injecting this expression in (5.58), one obtains:

$$\dot{J}(t) \leq -\Big[\frac{\gamma\|J_\xi\|^2}{\|J_\xi\|^2+\varepsilon} - 1\Big] \cdot \Big[\left|\Delta_{t-T}^{t}(\epsilon_y(\cdot,\xi(t)))\right| + \left[1+\bar{\phi}_\rho(t,\xi(t))\right]\sqrt{J}\Big] \quad (5.59)$$

This means that as long as:

$$\|J_\xi(t,\xi(t))\|^2 > \frac{\varepsilon}{\gamma-1} \quad (5.60)$$

the cost function J strictly decreases. Now using the inequality (5.47) with the above fact enables the following implication to be written:

$$\Big\{\Upsilon(J(t,\xi(t)) > \frac{\varepsilon}{\gamma-1}\Big\} \quad \Rightarrow \quad \Big\{\dot{J}(t,\xi(t)) < 0\Big\}. \quad (5.61)$$

This clearly shows that under the correction law (5.58), the set defined by:

$$\mathcal{A}_J := \Big\{(t,\xi) \quad | \quad J(t,\xi) \leq \Upsilon^{-1}\Big(\frac{\varepsilon}{\gamma-1}\Big)\Big\} \quad (5.62)$$

is an invariant and globally attractive set. But by the very definition of uniform observability (see definition 3), it can be inferred from (5.62) that the state estimation error $e = \xi(t) - x(t-T)$ satisfies the following asymptotic property:

$$\lim_{t\to\infty} \|\xi(t) - x(t-T)\| \leq \alpha^{-1} \circ \Upsilon^{-1}\Big(\frac{\varepsilon}{\gamma-1}\Big) \quad (5.63)$$

and by continuity of the system trajectories w.r.t the initial state, property (5.63) clearly implies:

$$\lim_{(\varepsilon/\gamma)\to 0} \Big[\lim_{t\to\infty} \|\hat{x}(t) - x(t)\|\Big] = 0. \quad (5.64)$$

The above discussion clearly proves the following result:

Proposition 1. *If the following conditions hold for the system (5.44)-(5.45):*

1. *The map f is continuously differentiable*
2. *The system is uniformly observable in the sense of definition 3*
3. *The uniform regularity assumption 1 is satisfied*
4. *It is possible to correctly estimate upper bounds of $y(\cdot)$ over past time intervals (see lemma 2)*

then for any a priori fixed desired precision $\eta > 0$ on the state estimation error, there is a sufficiently high ratio γ/ε such that the dynamic system given by:

$$\dot{\xi}(t) = f(t-T, \xi(t)) + c\big(t, \xi(t)\big) \tag{5.65}$$

$$\hat{x}(t) = X(t, t-T, \xi(t)) \tag{5.66}$$

where the correction term $c(t, \xi)$ is given by:

$$c(t, J) := \gamma\Big[\frac{J_\xi^T(t, \xi(t))}{\|J_\xi\|^2 + \varepsilon}\Big]\Big[-\big|\Delta_{t-T}^t(\epsilon_y(\cdot, \xi(t)))\big| - \big[1 + \bar{\phi}_\rho(t, \xi(t))\big]\sqrt{J}\Big] \tag{5.67}$$

leads to a state estimation error that asymptotically reaches the required precision η. ♡

It is worth noting that proposition 1 gives a receding-horizon observer that takes a rather classical form (differential equation built up with a correction term that is added to a copy of the system dynamic). There are two major differences however between this observer and classical analytic observers:

1. The first difference lies in the use of an integral norm J of the output prediction error in the correction term [see equation (5.67)] rather than its instantaneous value.
2. The second difference is the way the convergence is proved. While classical analytic observers investigate the evolution of the state estimation error which leads to the need for structural properties, here, the convergence proof is based on the convergence of J and this with the very definition of observability IMPLICITLY leads to the convergence of the state estimation error.

When compared to the classical moving-horizon observer scheme of section 5.3, the moving-horizon observer of proposition 1 contains apparently no optimization phase. Indeed, the optimization process is *embedded* in the dynamic of the internal state ξ. This dynamic

1. explicitly implements a gradient-based optimization process and
2. distribute the corresponding *iterations* over the system real life-time.

It is important to underline that the problem of local minima remains a common feature regardless the way the moving-horizon observer is implemented. In the context of proposition 1, this problem is hidden by the uniform global regularity assumption 1 (condition 3. of proposition 1). Note however that this assumption is not a constructive assumption in the sense that it is only needed for the convergence proof. The expression (5.67) of the observer dynamic is perfectly well defined even if this assumption is violated.

The real-time implementation of the observer equation (5.67) may face serious difficulties. This is because the computation of the gradient $J_\xi(t,\xi(t))$ is a quite involved task since it amounts to integrate a differential system of dimension $n(n+1)$ where n is the dimension of the state vector. This means that the time needed to perform the computation of the r.h.s of the observer equation, say τ_c can no more be neglected. This computation time represents naturally an upper bound on the sampling period $\tau_s(\geq \tau_c)$ that can be used to update the estimate of the state vector.

In the following section, a technical solution that is referred to as the post-stabilization technique is proposed in order to increase the sampling period while maintaining a good precision.

5.4.1 The Post Stabilization Technique

In order to simplify the expressions, in this section, the observer equations (5.65)-(5.66) are shortly re-written in the following compact form:

$$\dot{\xi}(t) = f_c(t,\xi(t),J_\xi(t)) \tag{5.68}$$
$$\hat{x}(t) = X(t,t-T,\xi(t)) \tag{5.69}$$

Proposition 1 states that observing the state of the system amounts to *integrate* the differential system (5.68). According to the discussion of the end of the preceding section, this integration has to be done using relatively high sampling period τ_s. In order to efficiently integrate the differential system (5.68) despite this fact, it is important to note that this system satisfies the following *nice property*:

property

> The sub-manifold $J(t,\xi(t)) = 0$ is invariant under the combined dynamic of the system and the observer equations (5.18) and (5.68).

In [18], an efficient integration scheme has been proposed that is dedicated to differential systems having invariant sub-manifolds. This technique is roughly depicted on figure 5.8. Namely, given the observer state $x(t_k)$ at instant $t_k = k\tau_s$, in order to obtain the next state $\xi(t_{k+1})$ at instant $t_{k+1} = (k+1)\tau_s$, the following steps are executed:

✓ First, the following differential system is integrated over $[t_k,t_{k+1}]$ starting from the initial condition $(t_k,\xi(t_k))$:

$$\dot{\xi}(t) = f_c(t,\xi(t),J_\xi(t_k)) \quad ; \quad t \in [t_k,t_{k+1}] \tag{5.70}$$

The corresponding solution at instant t_{k+1} is denoted by $\tilde{\xi}(t_{k+1})$ (see figure 5.8). Note that during the integration over $[t_k,t_{k+1}]$, the gradient $J_\xi(t_k)$ is kept equal to its initial value at instant t_k. Consequently, $\tilde{\xi}(t_{k+1})$ is a rough approximation of the exact integration of the observer equation.

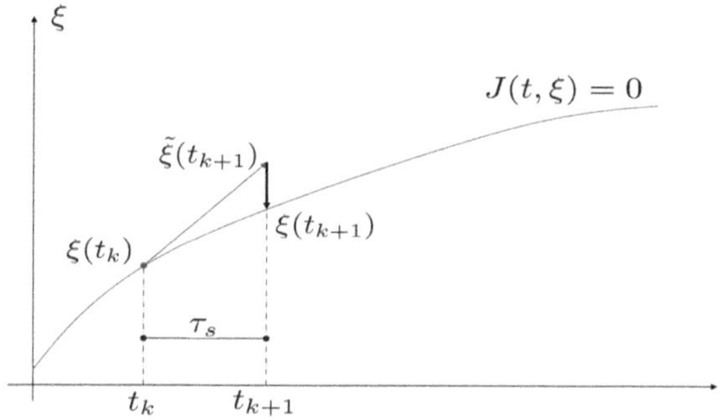

Fig. 5.8. The post-stabilization technique. To obtain the next observer state $\xi(t_{k+1})$, the observer equation is first integrated using constant gradient term $J_\xi(t_k, \xi(t_k))$, then the result $\tilde{\xi}(t_{k+1})$ is projected on the sub-manifold $J(t_{k+1}, \xi) = 0$.

✓ The second step in the post stabilization technique is to correct the rough approximation $\tilde{\xi}(t_{k+1})$ by projecting it on the manifold $J(t_{k+1}, \xi) = 0$. This is written as follows:

$$\xi(t_{k+1}) = \tilde{\xi}(t_{k+1}) - \frac{J_\xi(t_{k+1}, \tilde{\xi}(t_{k+1}))}{\|J_\xi(t_{k+1}, \tilde{\xi}(t_{k+1}))\|^2 + \nu} \cdot J\big(t_{k+1}, \tilde{\xi}(t_{k+1})\big) \quad (5.71)$$

where $\nu > 0$ is a regularization constant that is used to avoid numerical singularities close to the surface.

A detailed investigation on the consequence of the above mentioned post-stabilization technique is presented in [15]. In particular, it has been shown that when time invariant systems are considered the following asymptotic property holds:

$$\lim_{k \to \infty} J(\xi(t_k)) = O(\tau_s^4)$$

and this, regardless the order (≥ 1) of the integration scheme used to compute $\tilde{\xi}(t_{k+1})$.

5.4.2 Examples

In this section, two examples are given to illustrate the differential form of the moving-horizon observer presented in the preceding section. The first one (section 5.4.3) reports a successful industrial patented application [5] of this observer to the problem of the simultaneous estimation of the train velocity as well as the train position on a railways line. The second example (section 5.4.5) is a rather academic one that clearly shows the benefit from using the post-stabilization technique proposed in section 5.4.1 above. Another successful application of the differential moving-horizon observer can be found in [6] where this observation scheme has been applied to activated sludge processes used for waste-water treatment.

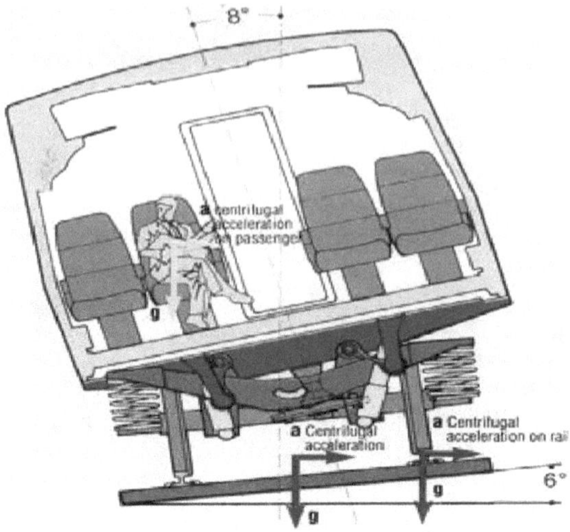

Fig. 5.9. Schematic view of a tilting train. The additional control-induced inclinaison of the compartment allows for a higher speed on existing rails while keeping the same comfort level (by maintaining the resulting felt acceleration normal to the compartment floor).

5.4.3 Nonlinear Observer for Tilting Trains

In this section, an industrial patented application [5] of the differential form of moving-horizon observer presented in this section is proposed[4]. The problem is first stated in the context of the control of tilting trains, then the need for an observer is explained and the performance of the proposed moving-horizon observer is shown.

Tilting Trains: The Control Problem

The problem of controlling tilting trains is schematically depicted on figure 5.9. Typically, when the train goes into a bend of curvature $\rho(r)$ at some curvilinear abscissa r on the rails, a passenger feels a centrifugal acceleration $V^2\rho(r)$. This acceleration when combined with the gravity gives a resulting acceleration that is not perpendicular to the compartment floor unless the rails present a by-construction inclinaison $\delta(r)$.

Consequently, the rails are inclined in accordance with some *nominal* optimal velocity V_{nom} by an angle $\bar{\delta}(r)$ satisfying:

$$\bar{\delta}(r) = \tan^{-1}\Big(\frac{1}{g} \cdot V_{nom}^2 \cdot \rho(r)\Big) \tag{5.72}$$

---

[4] This work has been achieved in a partnership context with the company ALSTOM-TRANSPORT (Villeurbanne, France). This partnership aimed to develop control algorithms for tilting trains. The work presented is described in details in the related patent [5].

Obviously, the curvature $\rho(\cdot)$ and the rails inclinaison $\delta(\cdot)$ become constant characteristics (profiles) of the rails. Now if the train follows these rails with a velocity that is significantly higher than the nominal velocity V_{nom} that has been used in the computation and the construction of the rails inclinaison profiles $\delta(\cdot)$, passenger would feel uncomfortable. The aim of the tilting train control is therefore to *compensate for the lack of rails inclinaison* by tilting the compartment using the dedicated jacks (see figure 5.9). Ideally, the additional inclinaison angle α_d is clearly given at instant t by:

$$\alpha_d(V(t), r(t)) := \tan^{-1}\Big(\frac{1}{g} \cdot V^2(t) \cdot \rho(r(t))\Big) - \bar{\delta}(r(t)) \tag{5.73}$$

Therefore, from a control point of view, the problem is to track a reference trajectory that depends on:

- The train's velocity $V(t)$
- The curvilinear abscissa of the train on the rails $r(t)$
- The geometric characteristics of the rails $\rho(\cdot)$ and $\bar{\delta}(\cdot)$

Remember that the origin of the control problem is related to the high velocities one aims to use that are higher than the nominal velocity V_{nom}. But the higher the velocity V is, the faster the set-point α_d changes since:

$$\dot{\alpha}_d = \frac{\partial \alpha_d}{\partial V}\dot{V} + \frac{\partial \alpha_d}{\partial r}V \approx \frac{\partial \alpha_d}{\partial r}(V, r)V \tag{5.74}$$

since the velocity of the train change slowly with time. This makes the tilting train control a very challenging problem that needs the use of advanced predictive control schemes enabling anticipating actions to be used. Indeed, a slight delay in the tracking may even give the inverse desired effect on the comfort level.

The Estimation Problem

Based on the above control problem description, it comes that anticipating the evolution of the desired set-point $\alpha_d(V(t), r(t))$ is a crucial issue. This means that the localization of the train on its rail is a key task in the overall control scheme. Note also that the estimation of a train velocity is a classical problem due to the need for a decentralized measurements for security reasons and due to the presence of slipping at the wheels level (see the patent [14] for more details on this critical issue).

Consequently, the estimation problem amounts to recover the evolution of both the curvilinear abscissa r and the error on the current estimation of the train velocity. To do this, the yaw angular velocity is available using a dedicated gyrometer that is fixed at the wheels level. Therefore, the dynamical system to be observed can be given as follows:

$$\dot{x}_1 = (1 + x_2) \cdot V_m(t) \tag{5.75}$$

$$\dot{x}_2 = 0 \tag{5.76}$$

$$y(t) = (1 + x_2(t))V_m(t) \cdot \rho(x_1(t)) \tag{5.77}$$

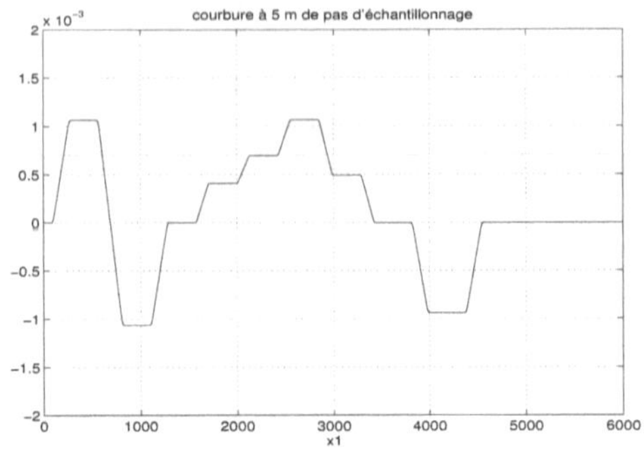

Fig. 5.10. Evolution of the curvature map $\rho(\cdot)$ on a portion of the Paris-Toulouse line. This map is used in the validating scenarios.

where $x_1 = r$ stands for the curvilinear abscissa of the train on the rail. x_2 is the relative error on the velocity, namely, the true velocity $V(t)$ is given by:

$$V(t) = (1 + x_2(t)) \cdot V_m(t)$$

The map $\rho(\cdot)$ is supposed to be available using dedicated series of measurements obtained during careful crossing of the line under consideration. The corresponding evolution of the curvature ρ as a function of the curvilinear abscissa is given on figure 5.10. This curve corresponds to the data characterizing a portion of the Paris-Toulouse line.

Note that in the above system model, the velocity measurement $V_m(\cdot)$ is supposed to be delivered by a dedicated velocity estimator or direct measurements. From the observation viewpoint, this signal is viewed as a known time varying signal over past intervals and can be handled using the estimation scheme through the time-varying character of the system model.

Based on the system model (5.75)-(5.77), the gradient $J_\xi(t,\xi)$ can be computed using the sensitivity matrix of the trajectory of the following system w.r.t initial conditions:

$$\begin{aligned}\dot{z}_1 &= (1 + z_2)V \\ \dot{z}_2 &= 0 \\ \dot{z}_3 &= \Big[(1 + z_2)V\rho(z_1) - y\Big]^2\end{aligned}$$

More precisely, one clearly has:

$$J_\xi(t,\xi) := \big(A_{31}(t)\ A_{32}(t)\big) \tag{5.78}$$

where the matrix $A(t) \in \mathbb{R}^{3\times 3}$ is the solution at instant t of the following differential system:

$$\dot{z}_1 = (1+z_2)V \tag{5.79}$$

$$\dot{z}_2 = 0 \qquad z(t-T) = \left(\xi^T \; 0\right)^T \tag{5.80}$$

$$\dot{z}_3 = \Big[(1+z_2)V\rho(z_1) - y(t)\Big]^2 \tag{5.81}$$

$$\dot{A}(\tau) = \begin{pmatrix} 0 & V(\tau) & 0 \\ 0 & 0 & 0 \\ \mathcal{X}_1(\tau) & \mathcal{X}_2(\tau) & 0 \end{pmatrix} A \quad ; \quad A(t-T) = \mathbb{I}_{3\times 3} \tag{5.82}$$

where the terms $\mathcal{X}_1$ and $\mathcal{X}_2$ are given by :

$$\mathcal{X}_1 = 2\Big[(1+z_2)V\cdot\rho(z_1) - y\Big](1+z_2)V\frac{\partial\rho}{\partial z_1}(z_1)$$

$$\mathcal{X}_2 = 2\Big[(1+z_2)V\cdot\rho(z_1) - y\Big]\cdot V\cdot\rho(z_1)$$

Note that by integrating the differential system (5.79)-(5.82) over $[t-T,t]$ one obtains simultaneously $J_\xi(t,\xi(t))$ by (5.78) but also $J(t,\xi(t))$ by:

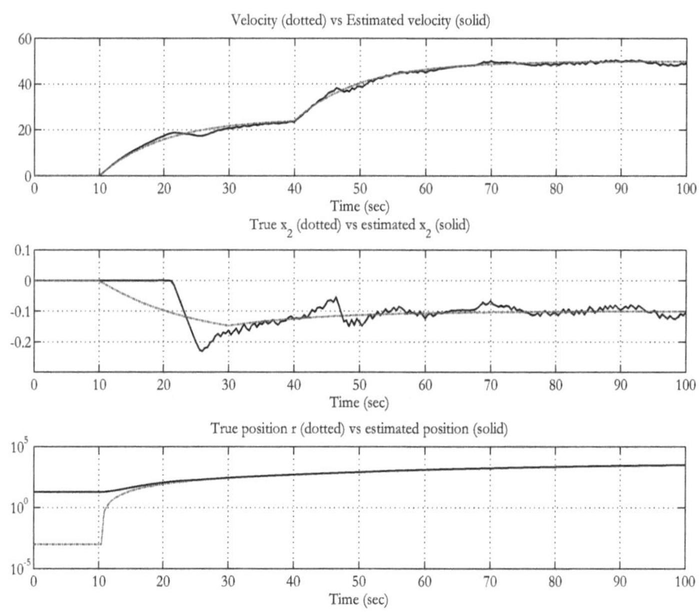

Fig. 5.11. Simulation of the differential moving-horizon observer when used to estimate the velocity and the position of a tilting train crossing a portion of the Paris-Toulouse line. Initial error on the position is equation de 20 m. The relative error on the velocity measurement varies form 0 to -15% during the first 20 seconds before it is settled to -10%. remember that the moving-horizon observer uses a constant evolution for this error when computing the cost function at each updating instant.

$$J(t, \xi(t)) = z_3(t).$$

Therefore, all that one needs to implement the differential moving horizon observer of proposition 1 can be obtained. It is worth noting that for this specific example, there is no need to integrate the 9th order differential system (5.79)-(5.82) since the structure of the system enables significant simplifications (see [5] for more details).

5.4.4 Simulations

Simulations are conducted using the portion of the Paris-Toulouse line (see the corresponding curvature on figure 5.10) with the following parameters:

$$T = 5\ sec \quad ; \quad \tau_s = 0.4\ sec \quad ; \quad \gamma = 0.2$$

Two simulations are proposed to illustrate the benefit from using the proposed observer. In the first (figure 5.11), an initial error on the position is introduced as well as a time varying relative error on the velocity measurement. In the second scenario, a different profil on the velocity measurement error is

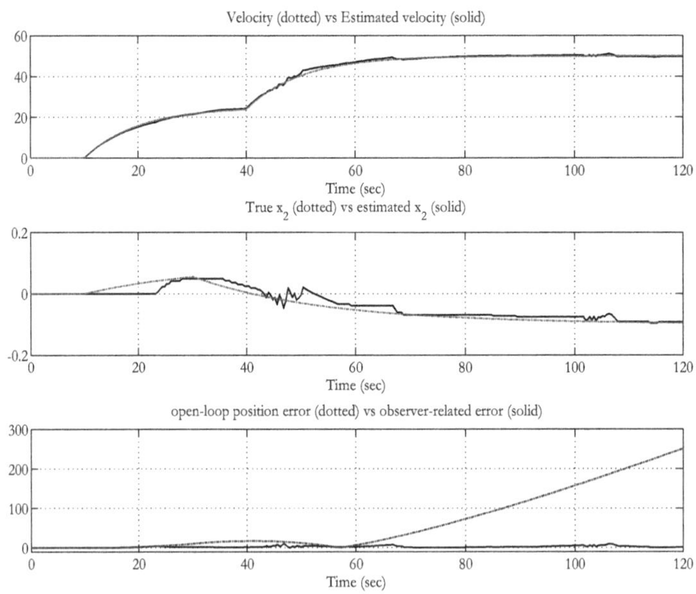

Fig. 5.12. Simulation of the differential moving-horizon observer when used to estimate the velocity and the position of a tilting train crossing a portion of the Paris-Toulouse line. Although there is no initial error on the position the bottom figure shows what would be the position error if no correction is made. The relative error on the velocity measurement varies form 0 to +5% and then to −10%. remember that the moving-horizon observer uses a constant evolution for this error when computing the cost function at each updating instant.

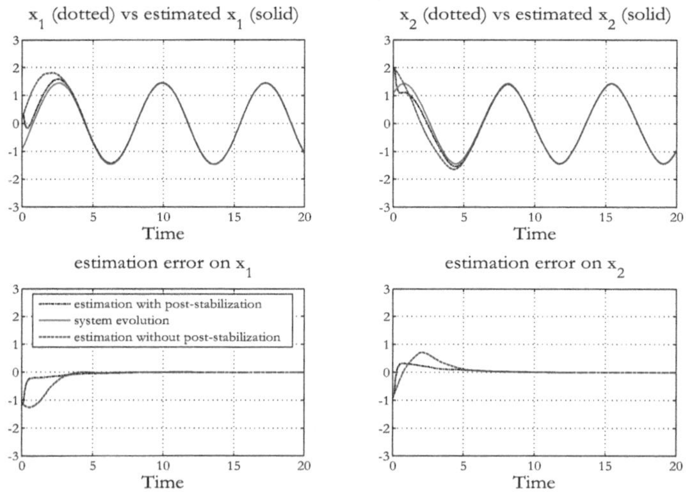

Fig. 5.13. Comparison between moving-horizon observer with (black dot-dashed line) and without (dotted) post-stabilization step. Here, the updating period is $\tau_s = 0.1$ s. This si sufficiently small to make the moving-horizon observer stabilizing even without the post stabilization step. Even in this case, note how the post stabilization step improves the quality of the estimation.

used without initial error on the train position. Despite the absence of initial error, figure 5.12 shows what would be the error on the estimated position if the velocity measurement were integrated without correction. The consequence of using such erroneous position on the overall tilting control loop would be clearly dramatic.

5.4.5 Illustrating the Benefit from Using the Post-stabilization Step

The aim of this section is to illustrate how the post-stabilization step proposed in section 5.4.1 enables the updating period to be increased leaving more time for computations. This is done using the following academic example:

$$\begin{aligned}\dot{x}_1 &= x_2\\ \dot{x}_2 &= -\sin(x_1) - 0.2x_1\cos(x_1x_2)\\ y &= x_1 + x_2\end{aligned}$$

Figures 5.13 and 5.14 show the behavior of the differential form of moving-horizon observer under two different updating periods $\tau_s = 0.1$ sec and $\tau_s = 0.4$ sec.

When $\tau_s = 0.1$ sec, the updating rate is sufficiently small for the observer to converge quite well even without the post stabilization step (see figure 5.13). However, when the updating period increases, the sampled observer fails to converge without the post stabilization step (see figure 5.14).

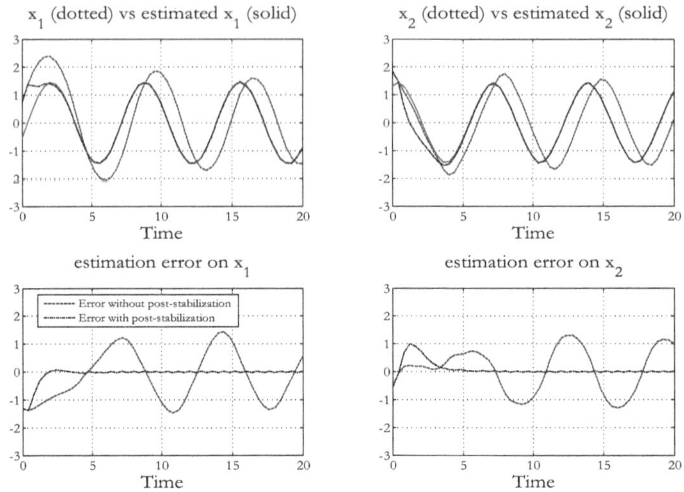

Fig. 5.14. Comparison between moving-horizon observer with (black dot-dashed line) and without (dotted) post-stabilization step. Here, the updating period is $\tau_s = 0.4$ s. Note how the post stabilization step enables the updating period to be increased while maintaining good precision.

It is worth noting that in both cases, it is the sampled version (5.70) of the differential moving-horizon observer that is implemented. This is precisely the reason for which a high updating period may destabilize the observer.

5.5 Moving Horizon Observers with Distributed Optimization

The differential form of the moving-horizon observer presented in section 5.4 tries to solve the optimization problem that underlines the state estimation problem using a gradient-based descent approach. Moreover, the iterations associated to this descent approach are distributed over the system life-time. The implementation of this approach needs however some regularity assumptions that guarantee the existence of all the partial derivatives of the cost function (the output prediction error).

In the present section, a more general viewpoint on the distributed-in-time optimization is adopted in order to get deeper insight on the resulting closed-loop behavior. The ideas developed here are closely connected to those *in the air* when real-time implementation of Model Predictive Control is addressed (see for instance [16, 1, 9, 8]). The main message of this section is that even when efficient and globally convergent optimizers are used, there is some optimal updating rate of the internal state of the observer. This optimal sampling rate corresponds to some optimal number of iterations of the optimizer between two successive updates. It goes without saying that the quantitative translation of this general result heavily depends on the system, the optimizer and the computational facilities and should be approached using a somehow *experimental way*.

In this section, the general *simulator* form (5.1) of the dynamic system is considered, namely:

$$x(t) = X(t, t_0, x_0),$$
$$y(t) = h(t, x(t)),$$

The measurement is assumed to be required with a sampling period τ_a. Note that τ_a defines the maximal frequency with which *additional new knowledge* is injected to any state estimation scheme. The acquisition period τ_a may be too small to be used as updating period[5] for the estimation. The updating period is considered here (without loss of generality) as a multiple of τ_a, namely, the updating period τ_u is defined by:

$$\tau_u = N_u \cdot \tau_a \quad \text{where } N_u \in \mathbb{N}$$

The resulting *updating instants* are therefore denoted by:

$$t_k = k \cdot \tau_u = k \cdot N_u \cdot \tau_a$$

The observation horizon T invoked in the above sections is here taken to be equal to an integer number N of acquisition periods, namely:

$$T := N \cdot \tau_a$$

that is, the observation horizon involves N past measurements. This enables the following cost function $J(t_k, \xi)$ to be defined at each updating instant t_k:

$$J(t_k, \xi) = \sum_{j=1}^{N} \left\| Y(t_k - j\tau_a, t_k - T, \xi) - y(t_k - j\tau_a) \right\|^2$$

Recall that minimizing $J(t_k, \xi)$ in the decision variable ξ amounts to look for the best estimate of the past state $x(t_k - T)$.

Following the same notations than those used in section 5.3, we assume that some iterative process $\mathcal{S}$ has been chosen to minimize $J(t, \xi)$ in the decision variable ξ, namely:

$$\xi^{(i+1)} = \mathcal{S}\Big(t, \xi^{(i)}, y_{t-T}^{t}\Big)$$

The result of n successive application of $\mathcal{S}$ for given t is denoted by:

$$\xi^{(i+n)} = \mathcal{S}^{(n)}\Big(t, \xi^{(i)}, y_{t-T}^{t}\Big)$$

Let us consider the following assumption about the efficiency of the iterative process $\mathcal{S}$:

[5] The precise meaning of an updating period is made clear later on.

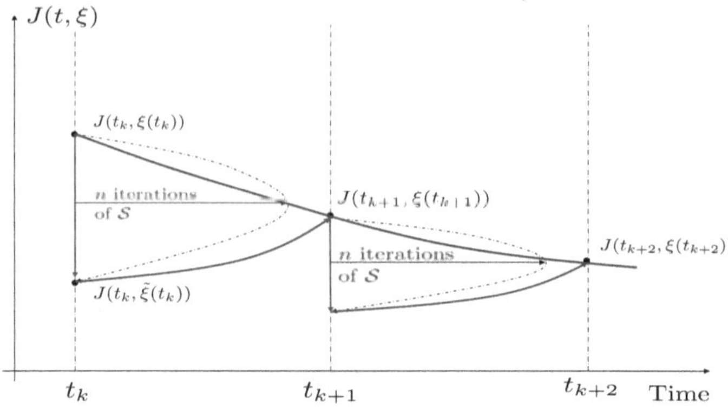

Fig. 5.15. Schematic view of the distributed-in-time optimization based observer. Note that the convergence of the overall estimation scheme is the result of a competition between a decreasing effect due to the optimizer and the increasing effect due to the natural divergence of open-loop state estimation.

Assumption 2 [Efficiency of the optimizer]
Iterative process $\mathcal{S}$ is efficient in the sense that there exists some efficiency map $\alpha_{eff} : \mathbb{N} \rightarrow [0,1[$ such that for all t and ξ, one has:

$$J\Big(t, \mathcal{S}^{(n)}(t,\xi,y_{t-T}^{t})\Big) \leq \alpha_{eff}(n)\cdot J(t,\xi) \tag{5.83}$$

where $\alpha(\cdot)$ is a decreasing function such that $\alpha(0)=1$. ♡

The state estimation algorithm studied in this section is defined by the following rules (see figure 5.15):

(1) **Initial conditions.** Given that the estimation is based on past measurements over some prediction horizon $T = N\cdot\tau_a$. The estimation begins as soon as N measurements have been acquired. Consider some integer $k = k_0$ such that $k_0\cdot N_u > N$. Assume that the current estimation of ξ is $\xi(t_k)$.

(2) **Updating ξ.** The computation of $\xi(t_{k+1})$ is done in two steps:

1. First n successive iterations are performed using the optimization process $\mathcal{S}$ to decrease the value of the cost function $J(t_k,\xi)$. This is written as follows

$$\tilde{\xi}(t_k) = \mathcal{S}^n(t_k,\xi(t_k),y_{t_k-T}^{t_k}) \tag{5.84}$$

Note that according to the assumption (5.83) on the optimizer's efficiency, one can write the following inequality:

$$J(t_k,\tilde{\xi}(t_k)) \leq \alpha_{eff}(n)\cdot J(t_k,\xi(t_k)) \tag{5.85}$$

2. Then the estimated value of $\xi(t_{k+1})$ is derived from $\tilde{\xi}(t_k)$ by integrating the system model:

$$\xi(t_{k+1}) = X(t_{k+1} - T, t_k - T, \tilde{\xi}(t_k)) \tag{5.86}$$

Note that when performing this *open-loop* updating over a time period of $N_u \tau_a = t_{k+1} - t_k$, some increase in the cost function have to be expected in general, this is stated by the following assumption:

Assumption 3 [open-loop behavior of the cost function]
When using open-loop prediction, the only inequality one can guarantee is given by:

$$J(t+\tau, X(t+\tau-T, t-T, \xi)) \leq \Big[J(t,\xi)\Big] \cdot \vartheta(\tau) \tag{5.87}$$

♡

using the inequality (5.87) with the following correspondances:

$$\xi = \tilde{\xi}(t_k) \quad ; \quad t = t_k \quad ; \quad \tau = N_u \cdot \tau_a$$

together with (5.86) enables to infer that when using the above estimation scheme, the inequality one can be sure of is the following:

$$J(t_{k+1}, \xi(t_{k+1})) \leq \Big[J(t_k, \tilde{\xi}(t_k))\Big] \cdot \vartheta(N_u \tau_a) \tag{5.88}$$

This with (5.85) enables the following inequality to be derived:

$$J(t_{k+1}, \xi(t_{k+1})) \leq \Big[\alpha_{eff}(n) \cdot \vartheta(N_u \tau_a)\Big] \cdot J(t_k, \xi(t_k)) \tag{5.89}$$

Note that the number of iterations n that may be performed during $N_u \tau_s$ time units is given by

$$n = E\Big(\frac{N_u \cdot \tau_a}{\tau_{iter}}\Big)$$

where τ_{iter} is the time needed for a single iteration.

Based on the above discussion, the following proposition can be derived:

Proposition 2 [Convergence of the distributed in time optimization based observers]
Under assumptions 2 and 3, the convergence of the distributed in time optimization based observer is guaranteed provided that the following inequality holds:

$$\varpi(N_u) := \alpha_{eff}\Big(E(\frac{N_u \tau_a}{\tau_{iter}})\Big) \cdot \vartheta(N_u \tau_a) < 1 \tag{5.90}$$

where

✓ τ_a *is the measurement acquisition period*
✓ $N_u\tau_a$ *is the updating period*
✓ τ_{iter} *is the time necessary to perform one iteration of the process* $\mathcal{S}$
✓ $\alpha_{eff}(\cdot)$ *is the optimizer efficiency map (see assumption 2)*
✓ $\vartheta(\cdot)$ *is the map characterizing the worst-case divergence rate of the open-loop prediction (see assumption 3)* ♠

Note that while condition (5.90) guarantees the convergence of the state estimation error. The corresponding *convergence time* (defined as the time needed for J to reach a value that is equal to 5% of its initial value) is still dependent on the value of N_u according to:

$$t_r(N_u) \approx \Big\lceil \frac{3N_u}{|\log\Big(\varpi(N_u)\Big)|} \Big\rceil \cdot \tau_a \tag{5.91}$$

It goes without saying that the exact expressions of the auxiliary functions $\alpha_{eff}(\cdot)$ and $\vartheta(\cdot)$ heavily depend on the system and the optimizer involved in the estimation scheme. however, in order to have concrete example showing the implication of the context on the *best* implementation parameters of the distributed-in-time optimization based state observers, let us consider the following structures for $\alpha_{eff}(\cdot)$ and $\vartheta(\cdot)$:

$$\alpha_{eff}(n) = \frac{D}{n^d + D} \quad ; \quad \vartheta(\tau) = \exp(\beta \cdot \tau) \tag{5.92}$$

Figures 5.16 and 5.17 give the evolution of the stability indicator ϖ and the settling time t_r as functions of the number of iterations N_u used to perform the observer internal state updating for two different sets of parameters used

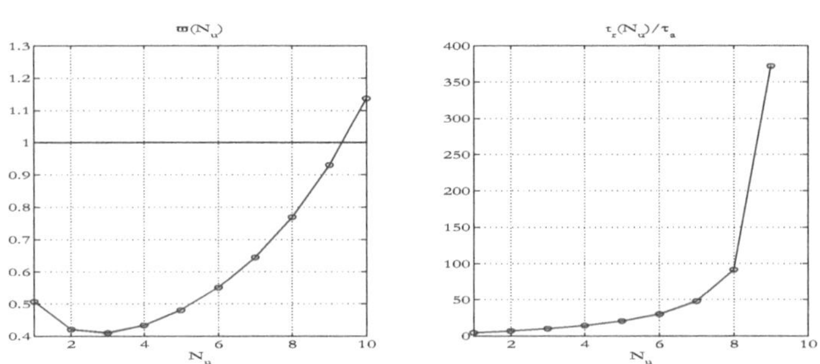

Fig. 5.16. Evolutions of the stability indicator $\varpi(N_u)$ and the settling time $t_r(N_u)$ vs the number of iterations N_u used to update the state estimation. The expressions (5.92) are used with the parameters $D = 3$, $d = 1$, $\beta \cdot \tau_a = 0.3$ and $\tau_a/\tau_{iter} = 5$. Under these conditions, stability cannot be guaranteed when more that 9 iterations are used. The optimal choice (in term of settling time) is the one where only one iteration is used to perform the updating.

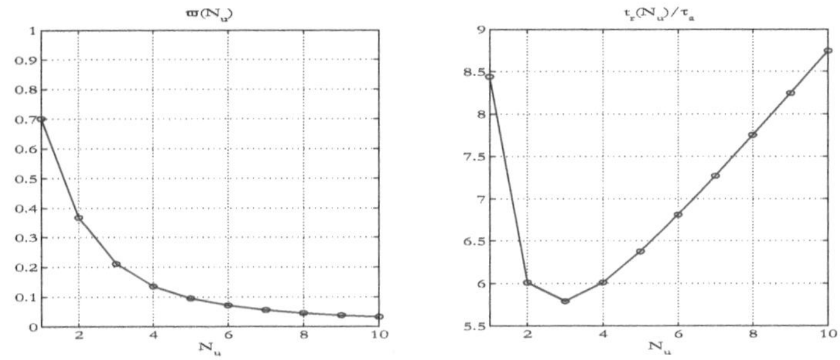

Fig. 5.17. Evolutions of the stability indicator $\varpi(N_u)$ and the settling time $t_r(N_u)$ vs the number of iterations N_u used to update the state estimation. The expressions (5.92) are used with the parameters $D = 50$, $d = 2$, $\beta \cdot \tau_a = 0.05$ and $\tau_a/\tau_{iter} = 5$. Under these conditions, while stability of the state estimation error seems guaranteed regardless the number of iterations used to perform the updating, the use of 3 iterations gives the best choice in term of settling time.

in (5.92). In the first case (figure 5.16), the instability rate of the open-loop estimation is high ($\beta \cdot \tau_a = 0.3$) leading to a maximum number of 9 iterations beyond which stability cannot be guaranteed. Moreover, the optimal choice in term of settling time is to use one single iteration before updating.

In the second case (figure 5.17), the instability rate is lower ($\beta \cdot \tau_a = 0.05$) and the efficiency of the iterations is higher ($d = 2$ rather than 1 in the first case). This leads to the stability being guaranteed for all number of iterations but with the optimal settling time corresponding to the use of 3 iterations.

5.6 Conclusion

The use of moving-horizon observers is intimately linked to the progress of nonlinear constrained optimization. However, the state estimation problem is not only an optimization problem. The way an optimization process can be used to result in a dynamic state estimator is to be carefully studied following (at least partially) some of the guidelines given in this chapter.

Another likely to be promising direction is to combine the partial use of analytic observer (on a part of the estimation problem) with the use of optimization process. This enables optimization to concentrate on those parts of the problem where no particular structural properties are available.

References

[1] Alvarado I. A., Findeisen R., Kuhl P., Allgower F., and Limon D. State estimation for repetitive processes using iteratively improving moving horizon observers. In *Proceedings of the IFAC World Congress*, Barcelona, Spain, 2002.

[2] Tornambè A. High gains observers for nonlinear systems. International Journal of Systems Science, 23, 1992.

[3] Rao C. V.; Rawlings J. B. and Mayne D. Q. Constrained state estimation for nonlinear discrete-time systems: Stability and moving horizon approximation. *IEEE Transactions on Automatic Control*, 48(2):246–258, 2003.
[4] Wang G. B., Peng S. S., and Huang H. A sliding observer for nonlinear process control. Chemical Engineering Science, 52, 1997.
[5] Alacoque J. C. and Alamir M. Method and device for locating a vehicle on a road. Patent EP1211152, Alstom-Transport, http://gauss.ffii.org/PatentView/EP1211152, 2002.
[6] Bassompierre C., Li S., and cadet C. Activated sludge processes state estimation: A nonlinear moving horizon observer approach. In *Proceedings of the American Control Conference*, Minneapolis, Minnesota, USA, 2006.
[7] Kiparissides C., Seferlis P., Mourikas G., and Morris A.J. Online optimization control of molecular weight properties in batch free-radical polymerization reactors. Industrial Engineering and Chemistry Research, 41, 2002.
[8] M. Diehl, R. Findeisen, and F. Allgöwer. A stabilizing real-time implementation of nonlinear model predictive control. In L. Biegler, O. Ghattas, M. Heinkenschloss, D. Keyes, and B. van Bloem Wanders, editors, *Real-Time PDE Optimization*, New York, 2006. Springer-Verlag. To appear.
[9] M. Diehl, R. Findeisen, Z. Nagy, H.G. Bock, J.P. Schlöder, and F. Allgöwer. Real-time optimization and nonlinear model predictive control of processes governed by differential-algebraic equations. *J. Proc. Contr.*, 4(12):577–585, 2002.
[10] Slotine J. E., Hedrick J. K., and Misawa E. A. On sliding observers for nonlinear systems. Journal of Dynamics Sysytems, Measurements and Control, 109, 1987.
[11] Michalska H. and Mayne D. Q. Moving-horizon observers and observer-based control. *IEEE Transactions on Automatic Control*, 40:995–1006, 1995.
[12] Buruaga I.S., Leiza J.R., and Asua J.M. Model-based control of emulsion terpolymerization based on calorimetric measurements. Polymer Reaction Engineering, 8(1), 2000.
[13] Calvillo-Corona L. and Alamir M. Towards the construction of robust optimal observers. In *Proceedings of the IFAC World Congress*, Barcelona, Spain, 2002.
[14] Thomas J. L., Latteux P., Alacoque J. C., Bornard G., and Alamir M. Method for estimating the velocity of a vehicle or an assembly of vehicles. Patent EP919814, Alcatel-Alstom Recherche, http://gauss.ffii.org/PatentView/EP919814, 2002.
[15] Alamir M. Optimization based nonlinear observers revisited. *International Journal of Control*, 72 (13):1204–1217, 1999.
[16] Alamir M. *Stabilization of Nonlinear Systems Using Receding-Horizon Control Schemes: A Parametrized Approach for Fast Systems.* Lecture Notes in Control and Information Sciences, Springer. Springer-Verlag, London, 2006.
[17] Alamir M., Sheibat-Othman N., and Othman S. On the use of nonlinear receding-horizon observers in batch terpolymerization processes. Submitted to International Journal of Modelling, Identification and Control, 2007.
[18] Ascher U. M. Stabilization of invariants of discretized differential systems. *Numerical Algorithms*, 14:1–24, 1997.
[19] Othman N., McKenna T.F., and Févotte G. On-line monitoring of emulsion terpolymerization processes. Polymer Reaction Engineering, 9(4), 2001.
[20] Gauthier J. P., Hammouri H., and Othman S. A simple observer for nonlinear systems – application to bio-reactors. IEEE Transactions on Automatic Control, 37, 1992.

[21] Aguilar-Lopez R. and Maya-Yescas R. State estimation for nonlinear systems under model uncertainties: a class of sliding-mode observers. Journal of Process Control, 15, 2005.
[22] BenAmor S., Doyle F.J., and McFarlane R. Polymer grade transition control using advanced real-time optimization sofware. Journal of Process Control, 14, 2004.
[23] Ahmed-Ali T. and Lamnhabi-Lagarrigue F. Sliding observer-controller design for uncertain triangular nonlinear systems. IEEE Transactions on Automatic Control, 44(6), 1999.
[24] Ohtsuka T. and Fujii H. A. Nonlinear receding horizon state estimation by realtime optiization technique. *J. Guidance, Control, Dynamics*, 19 (4):863–870, 1996.

6

Asymptotic Analysis and Observer Design in the Theory of Nonlinear Output Regulation

Alberto Isidori[1,2] and Lorenzo Marconi[2]

[1] DIS, Università di Roma "La Sapienza", Italy
isidori@ese.wustl.edu
[2] CASY-DEIS, Università of Bologna, Italy
lmarconi@deis.unibo.it

Summary. The purpose of these notes is to summarize a number on recent developments in the theory of output regulation for nonlinear systems. Cornerstones of these developments are the asymptotic analysis leading to a precise notion of steady state response for nonlinear systems and a number of concepts arising in the theory of nonlinear observers. The steady state analysis is the tool of choice for the identification of necessary conditions, which make it possible to express in simple terms a new nonlinear enhancement of the classical internal model principle. The theory of nonlinear observers, on the other hand, provides the appropriate ideas for the design of regulators for a fairly general class of nonlinear systems that satisfy a suitable minimum-phase assumption. The ideas in question are instrumental in the design of "asymptotic internal models", objects that serve the dual purpose of inducing a steady state in which the regulated variable vanishes and to make this steady state attractive.

6.1 Introduction

A central problem in control theory is the design of feedback controllers so as to have certain outputs of a given plant *to track* prescribed reference trajectories. In any realistic scenario, this control goal has to be achieved in spite of a good number of phenomena which would cause the system to behave differently than expected. These phenomena could be endogenous, for instance parameter variations, or exogenous, such as additional undesired inputs affecting the behavior of the plant.

In what follows, we address tracking problems that can be cast in the following terms. Consider a finite-dimensional, time-invariant, nonlinear system modelled by equations of the form

$$\begin{aligned} \dot{x} &= f(w, x, u) \\ e &= h(w, x) \\ y &= k(w, x)\,, \end{aligned} \tag{6.1}$$

in which $x \in \mathbb{R}^n$ is a vector of state variables, $u \in \mathbb{R}^m$ is a vector of inputs used for *control* purposes, $w \in \mathbb{R}^s$ is a vector of inputs which cannot be controlled and include *exogenous* commands, exogenous disturbances and model uncertainties, $e \in \mathbb{R}^p$ is a vector of *regulated* outputs which include tracking errors and any other variable that needs to be steered to 0, $y \in \mathbb{R}^q$ is a vector of outputs that

G. Besançon (Ed.): Nonlinear Observers and Applications, LNCIS 363, pp. 181–210, 2007.
springerlink.com

are available for *measurement* and hence used to feed the device that supplies the control action. The problem is to design a controller, which receives $y(t)$ as input and produces $u(t)$ as output, able to guarantee that, in the resulting closed-loop system, $x(t)$ remains bounded and

$$\lim_{t\to\infty} e(t) = 0\,, \tag{6.2}$$

regardless of what the exogenous input $w(t)$ actually is.

The ability to successfully address this problem very much depends on how much the controller is allowed to know about the exogenous disturbance $w(t)$. In the ideal situation in which $w(t)$ is available to the controller in real-time, the design problem indeed looks much simpler. This is, though, only an extremely optimistic situation which does not represent, in any circumstance, a realistic scenario. The other extreme situation is the one in which nothing is known about $w(t)$. In this, pessimistic, scenario the best result one could hope for is the fulfillment of some prescribed ultimate bound for $|e(t)|$, but certainly not a sharp goal such as (6.2). A more comfortable, intermediate, situation is the one in which $w(t)$ is only known *to belong to a fixed family* of functions of time, for instance the family of all solutions obtained from a fixed ordinary differential equation of the form

$$\dot{w} = s(w) \tag{6.3}$$

as the corresponding initial condition $w(0)$ is allowed to vary on a prescribed set. This situation is in fact sufficiently distant from the ideal but unrealistic case of perfect knowledge of $w(t)$ and from the realistic but conservative case of totally unknown $w(t)$. But, above all, this way of thinking at the exogenous inputs covers a number of cases of major practical relevance. There is, in fact, abundance of design problems in which parameter uncertainties, reference commands and/or exogenous disturbances can be modelled as functions of time that satisfy an ordinary differential equation.

The control law is to be provided by a system modelled by equations of the form

$$\begin{aligned} \dot{\xi} &= \varphi(\xi, y) \\ u &= \gamma(\xi, y) \end{aligned} \tag{6.4}$$

with state $\xi \in \mathbb{R}^\nu$. The initial conditions $x(0)$ of the *plant* (6.1), $w(0)$ of the *exosystem* (6.3) and $\xi(0)$ of the *controller* (6.4) are allowed to range over a fixed *compact* sets $X \subset \mathbb{R}^n$, $W \subset \mathbb{R}^s$ and, respectively $\Xi \subset \mathbb{R}^\nu$. All maps characterizing the of the controlled plant, of the exosystem and of the controller are assumed to be sufficiently differentiable.

The problem which will be studied, known as the *problem of output regulation* (or *generalized tracking problem* or also as *generalized servomechanism problem*) is to design a feedback controller of the form (6.4) so as to obtain a closed loop system in which all trajectories are bounded and the regulated output $e(t)$ asymptotically decays to 0 as $t \to \infty$. More precisely, it is required that the composition of (6.1), (6.3) and (6.4), that is the *autonomous* system

$$\begin{aligned} \dot{w} &= s(w) \\ \dot{x} &= f(w, x, \gamma(\xi, k(w, x))) \\ \dot{\xi} &= \varphi(\xi, k(w, x)) \end{aligned} \tag{6.5}$$

with output

$$e = h(w, x)\,,$$

be such that:

- the positive orbit of $W \times X \times \Xi$ is bounded,[1]
- $\lim_{t\to\infty} e(t) = 0$, uniformly in the initial condition.[2]

6.2 The Steady-State Response of a Nonlinear System

6.2.1 Background

The problem described in the previous section can be seen as the problem of forcing in the plant, by means of an appropriate control input $u(t)$, a response $x(t)$ that asymptotically compensates the effect, on the regulated variable $e(t)$, of the exogenous input $w(t)$. The classical way in which the problem is addressed for linear, time-invariant systems, when the exosystem is a neutrally stable linear system, is to seek a controller forcing in the associated closed-loop system (6.5) a (stable) "steady state" response entirely contained in the kernel of the map defining the tracking error e. Thus, it is natural to expect that a similar tool should also be effective in the more general setting considered here. It appears, though, that a rigorous investigation of the concept of "steady state", beyond the classical domain of linear system theory, had never been fully pursued. Thus, it seems natural, in a systematic analysis of the design problem outlined in the previous section, to begin with an attempt to offer a rigorous definition of the concept of "steady state" for a general nonlinear system.[3]

Thinking of a linear system, Gardner and Barnes [13], define the concept in question as follows: "A dynamical system is said to be in the *steady state* when the variables describing its behavior are either invariant with time, or are (sections of) periodic functions of time. A dynamical system is said to be in the *transient* (or unsteady) *state* when it is not in steady state."

One appealing feature of this definition is that no predetermined separation between inputs and outputs is sought, but the system is only analyzed in terms of how the variables that describe its behavior depend on time. Thus, the viewpoint applies to general dynamical systems and not necessarily to control systems with specified input and output; it is a precursor (at least as long as the notion of

[1] That is, the exists a bounded subset S of $\mathbb{R}^s \times \mathbb{R}^n \times \mathbb{R}^\nu$ such that, for any $(w_0, x_0, \xi_0) \in W \times X \times \Xi$, the integral curve $(w(t), x(t), \xi(t))$ of (6.5) passing through (w_0, x_0, ξ_0) at time $t = 0$ remains in S for all $t \geq 0$.

[2] That is, for every $\varepsilon > 0$ there exists a time $\bar{t}$, depending only on ε and *not on* (w_0, x_0, ξ_0) such that the integral curve $(w(t), x(t), \xi(t))$ of (6.5) passing through (w_0, x_0, ξ_0) at time $t = 0$ yields $\|e(t)\| \leq \varepsilon$ for all $t \geq \bar{t}$.

[3] The content of this section is a summary of results presented in [7].

steady state is concerned) of the modern *behavioral* viewpoint proposed by J.C. Willems [30]. The main restriction, though, is that this definition only applies to cases in which all relevant variables which describe the behavior of a dynamical systems are *periodic* (constant in particular) functions of time. This definition does not even cover the simple case in which the variables describing the behavior of a system are linear combinations of sinusoidal functions (of time) with at least two irrationally related angular frequencies, let alone the case in which the variables in question are "almost periodic".

Motivated by this *classical* idea of a steady state (extended to cover the case almost periodic functions of time) and by the fact that, in a stable linear system, any transient state asymptotically approaches a steady state, it is a common practice to regard a steady state as a kind of *limit* behavior. From this viewpoint, the steady state can be looked at either as the limit behavior which is approached when the *actual* time t tends to $+\infty$ or, respectively, as the limit behavior which is approached when the *initial* time t_0 tends to $-\infty$. The two alternatives are equivalent for a stable linear system. Adopting the first viewpoint, James, Nichols and Phillips [21] say that "the *transient response* of [a linear] filter is the difference between the actual output of the filter for $t > t_0$ and the asymptotic form that it approaches" and that "only when a filter is stable it is possible to speak with full generality of its response to an input that starts indefinitely far in the past." Adopting the other viewpoint, Zadeh and Desoer [31] define a "*ground state* of [the system], if it exists, [as] the limiting terminal state of [the system] when the zero input is applied, ... provided the limiting state γ is the same for all initial states" and afterward define "the *steady state response* of [the system] to an input $u_{(t_0,t]}$ [as] the limit, if it exists, of the ground-state response of [the system] to u as $t_0 \to -\infty$." [4]

6.2.2 Limit Sets

While the viewpoint of considering a system in steady state when its variables are almost periodic functions of time would suffice for a linear system, in a more general context it is inevitable to seek a definition which characterizes the steady state as a limiting behavior asymptotically approached as time increases (or decreases). Fundamental, in this respect, are certain concepts found in the classical 1927 essay by G.D.Birkhoff, in which he demonstrates that "with an arbitrary dynamical system ... there is associated always a closed set of 'central motions' which do possess this property of regional recurrence, towards which all other motions of the system in general tend asymptotically." [3, page 190]

The first, fundamental, ingredient in the process of characterizing the motions (of a dynamical system) which are asymptotically approached as time increases (or decreases), is the concept of ω-limit set of a given point. This concept is precisely defined as follows. Consider an *autonomous* ordinary differential equation

[4] Furthermore they observe, though, that "usually [the system] and u are such that, in [the expression of the response], γ can be replaced by an arbitrary initial state α without affecting the limiting value of the response as $t_0 \to -\infty$."

$$\dot{x} = f(x) \tag{6.6}$$

with $x \in \mathbb{R}^n$, $t \in \mathbb{R}$. It is well known that, if $f : \mathbb{R}^n \to \mathbb{R}^n$ is locally Lipschitz, for any $x_0 \in \mathbb{R}^n$, the solution of (6.6) with initial condition $x(0) = x_0$, denoted by $x(t, x_0)$, exists on some open interval of the point $t = 0$ and is unique.

Assume, in particular, that $x(t, x_0)$ is defined for all $t \geq 0$. A point x is said to be an ω-limit *point* of the motion $x(t, x_0)$ if there exists a sequence of times $\{t_k\}$, with $\lim_{k\to\infty} t_k = \infty$, such that

$$\lim_{k\to\infty} x(t_k, x_0) = x\,.$$

The ω-limit *set* of a point x_0, denoted $\omega(x_0)$, is *the union* of all ω-limit points of the motion $x(t, x_0)$.

It is obvious from this definition that an ω-limit point *is not* necessarily a limit of $x(t, x_0)$ as $t \to \infty$, as the solution in question may not admit any limit as $t \to \infty$. It happens though, that if the motion $x(t, x_0)$ is *bounded*, then $x(t, x_0)$ asymptotically approaches *the set* $\omega(x_0)$. This property is precisely described in what follows [3, page 198].

Lemma 1. Suppose there is a number M such that $\|x(t, x_0)\| \leq M$ for all $t \geq 0$. Then, $\omega(x_0)$ is a nonempty compact connected set, invariant under (6.6). Moreover, the distance of $x(t, x_0)$ from $\omega(x_0)$ tends to 0 as $t \to \infty$.

One of the remarkable features of $\omega(x_0)$, as indicated in this Lemma, is the fact that this set is *invariant* for (6.6). Invariance means that for any initial condition $\bar{x}_0 \in \omega(x_0)$ the solution $x(t, \bar{x}_0)$ of (6.6) exists *for all* $t \in (-\infty, +\infty)$ and that $x(t, \bar{x}_0) \in \omega(x_0)$ for all such t. Put in different terms, the set $\omega(x_0)$ is filled by motions of (6.6) which are bounded backward and forward in time. The other remarkable feature is that $x(t, x_0)$ *approaches* $\omega(x_0)$ as $t \to \infty$, in the sense that the distance *of the point* $x(t, x_0)$ (the value at time t of the solution of (6.6) starting in x_0 at time $t = 0$) *from the set* $\omega(x_0)$ tends to 0 as $t \to \infty$.

Since any motion $x(t, x_0)$ which is bounded in positive time asymptotically approaches the ω-limit set $\omega(x_0)$ as $t \to \infty$, one may be tempted to look, for a system (6.6) in which *all* motions are bounded in positive time, at the *union* of the limit sets of all points x_0, i.e. at the set

$$\Omega = \bigcup_{x_0 \in \mathbb{R}^n} \omega(x_0)$$

and to say that the system is in steady state if its state $x(t)$ evolves in the (invariant) set Ω. There is a major drawback, though, in taking this as a point of departure for the definition of "steady state" behavior of a nonlinear system: the convergence of $x(t, x_0)$ to Ω is not guaranteed to be *uniform* in x_0, even if the latter ranges over a compact set (see, e.g. [7]).

One of the main motivations for looking into the concept of steady state is the aim *to shape* the steady state response of a system to a given (or to a given family of) forcing inputs. But this motivation looses much of its meaning if the time needed to get within an ε-distance from the steady state may grow unbounded as the initial state changes (even when the latter is picked within a fixed *bounded* set). In other words, uniform convergence to the steady state (which is automatically guaranteed in the case of linear systems) is an indispensable feature to be required in a nonlinear version of this notion. The set Ω, the union of all ω-limit points of all points in the state space does not have this property of uniform convergence, but there is a larger set which does have this property. This larger set, known as the ω limit set *of a set*, is precisely defined as follows.

Consider again system (6.6), let B be a subset of $\mathbb{R}^n$ and suppose $x(t, x_0)$ is defined for all $t \geq 0$ and all $x_0 \in B$. The ω-limit set of B, denoted $\omega(B)$, is the set of all points x for which there exists a sequence of pairs $\{x_k, t_k\}$, with $x_k \in B$ and $\lim_{k\to\infty} t_k = \infty$ such that

$$\lim_{k\to\infty} x(t_k, x_k) = x\,.$$

It is clear from the definition that if B consists of only one single point x_0, all x_k's in the definition above are necessarily equal to x_0 and the definition in question reduces to the definition of ω-limit set of a point, given earlier. It is also clear form this definition that, if for some $x_0 \in B$ the set $\omega(x_0)$ is nonempty, all points of $\omega(x_0)$ are points of $\omega(B)$. In fact, all such points have the property indicated in the definition, if all the x_k's are taken equal to x_0. Thus, in particular, if all motions with $x_0 \in B$ are bounded in positive time,

$$\bigcup_{x_0 \in B} \omega(x_0) \subset \omega(B)\,.$$

However, the converse inclusion is not true in general.

The relevant properties of the ω-limit set of a set, which extend those presented earlier in Lemma 1, can be summarized as follows [16, page 8].

Lemma 2. *Let B be a nonempty bounded subset of $\mathbb{R}^n$ and suppose there is a number M such that $\|x(t, x_0)\| \leq M$ for all $t \geq 0$ and all $x_0 \in B$. Then $\omega(B)$ is a nonempty compact set, invariant under (6.6). Moreover, the distance of $x(t, x_0)$ from $\omega(B)$ tends to 0 as $t \to \infty$, uniformly in $x_0 \in B$. If B is connected, so is $\omega(B)$.*

Thus, as it is the case for the ω-limit set of a point, we see that the ω-limit set of a bounded set, being compact and invariant, is filled with motions which exist for all $t \in (-\infty, +\infty)$ and are bounded backward and forward in time.[5] But, above all, we see that the set in question is *uniformly* approached by motions with initial state $x_0 \in B$, a property that the set Ω does not have.

The set $\omega(B)$, as shown in the previous Lemma, asymptotically attracts, as $t \to \infty$, all motions that start in B. Since the convergence to $\omega(B)$ is uniform in x_0, it is also true that, whenever $\omega(B)$ is contained in the interior of B, the set $\omega(B)$ is *asymptotically stable*, in the sense of Lyapunov.

[5] The set of all such trajectories is a "behavior", in the sense of J.C. Willems [30].

In order to make this claim precise, recall that, for motions converging to a closed invariant set $\mathcal{A}$, the notion of asymptotic stability, a straightforward extension of the notion of asymptotic stability of an equilibrium, is defined as follows. Let $\mathcal{A} \subset \mathbb{R}^n$ be a *closed invariant set* for (6.6). The set $\mathcal{A}$ is asymptotically stable if the following hold:

(i) for every $\varepsilon > 0$, there exists $\delta > 0$ such that,

$$\text{dist}(x_0, \mathcal{A}) \le \delta \qquad \text{implies} \qquad \text{dist}(x(t, x_0), \mathcal{A}) \le \varepsilon \quad \text{for all } t \ge 0.$$

(ii) there exists a number $d > 0$ such that

$$\text{dist}(x_0, \mathcal{A}) \le d \qquad \text{implies} \qquad \lim_{t \to \infty} \text{dist}(x(t, x_0), \mathcal{A}) = 0\,.$$

As in the case of equilibria, for a closed invariant set $\mathcal{A}$ which is asymptotically stable for (6.6), the *domain of attraction* is the set of all x_0 for which $x(t, x_0)$ is defined for all $t \ge 0$ and $\text{dist}(x(t, x_0), \mathcal{A}) \to 0$ as $t \to \infty$.

With these definitions in mind, from the result of Lemma 2 it is possible to deduce the following important property.

Lemma 3. *Let B be a nonempty bounded subset of $\mathbb{R}^n$ and suppose there is a number M such that $\|x(t, x_0)\| \le M$ for all $t \ge 0$ and all $x_0 \in B$. Then $\omega(B)$ is a nonempty compact set, invariant under (6.6). Suppose also that $\omega(B)$ is contained in the interior of B. Then, $\omega(B)$ is asymptotically stable, with a domain of attraction that contains B.*

Finally, we conclude with another property, which is a straightforward consequence of the definitions.

Lemma 4. *Let B be a nonempty compact set, invariant under (6.6). Then $\omega(B) = B$.*

6.2.3 The Steady State Behavior of a Nonlinear System

Consider now again system (6.6), with initial conditions in a closed subset $X \subset \mathbb{R}^n$. Suppose the set X is a *positively invariant set*, which means that for any initial condition $x_0 \in X$, the solution $x(t, x_0)$ exists for all $t \ge 0$ and $x(t, x_0) \in X$ for all $t \ge 0$. The motions of this system are said to be *ultimately bounded* if there is a bounded subset B with the property that, for every compact subset X_0 of X, there is a time $T > 0$ such that $\|x(t, x_0)\| \in B$ for all $t \ge T$ and all $x_0 \in X_0$. In other words, if the motions of the system are ultimately bounded, every motion eventually enters and remains in the bounded set B.

Remark 1. Note that, since by hypothesis X is positively invariant, there is no loss of generality in assuming $B \subset X$ in the definition above. Note also that there exists a number M such that $\|x(t, x_0)\| \le M$ for all $t \ge 0$ and all $x_0 \in B$. In fact, let $\text{Cl}(B)$ denote the closure of B, which is a compact subset of X, and let M_1 the denote the maximum of $\|x\|$ as $x \in \mathrm{Cl}(B)$. By definition of ultimate boundedness, there is a time T such that $\|x(t,x_0)\| \le M_1$, for all $t \ge T$ and all $x_0 \in \mathrm{Cl}(B)$. Moreover, since $x(t,x_0)$ depends continuously on (t,x_0), there exists a number M_2 such that $\|x(t,x_0)\| \le M_2$ for all $0 \le t \le T$ and all x_0 in $\mathrm{Cl}(B)$. Thus, the property in question is fulfilled with $M = \max\{M_1, M_2\}$. By virtue of this property, one can conclude from Lemma 2 that the set $\omega(B)$ is nonempty and has all the properties indicated in the Lemma itself. Finally, note that, for a system whose motions are ultimately bounded, the set $\omega(B)$ is a unique well-defined set, regardless of how B is taken. In fact, let B' be any other bounded subset of X with the property indicated in the definition of ultimate boundedness. Then, it is not difficult to prove,[6] using the various definitions, that $\omega(B') \subset \omega(B)$. Reversing the role of the two sets shows that $\omega(B) \subset \omega(B')$, i.e. that the two sets in question are identical. ◁

For systems whose motions are ultimately bounded, the notion of steady state can be defined as follows.

Definition 1. *Suppose the motions of system (6.6), with initial conditions in a closed and positively invariant set X, are ultimately bounded. A* steady state motion *is any motion with initial condition in $x(0) \in \omega(B)$. The set $\omega(B)$ is the* steady state locus *of (6.6) and the* restriction *of (6.6) to $\omega(B)$ is the* steady state behavior *of (6.6).* ◁

The notion thus introduced recaptures the classical notion of steady state for linear systems and provides a new powerful tool to deal with similar issues in the case of nonlinear systems.

Example 1. In order to see how this notion includes the classical viewpoint, consider an n-dimensional, single-input, *asymptotically stable* linear system

$$\dot{x} = Fx + Gu \tag{6.7}$$

forced by the harmonic input $u(t) = u_0 \sin(\omega t + \phi_0)$. A simple method to determine the periodic motion of (6.7) consists in viewing the forcing input $u(t)$ as provided by an autonomous "signal generator" of the form

$$\begin{pmatrix} \dot{w}_1 \\ \dot{w}_2 \end{pmatrix} = \begin{pmatrix} 0 & \omega \\ -\omega & 0 \end{pmatrix} \begin{pmatrix} w_1 \\ w_2 \end{pmatrix}$$
$$u = w_1$$

and in analyzing the state state behavior of the associated "augmented" system

[6] Let $\bar{x}$ be a point of B'. By hypothesis, there exists a sequence $\{\bar{x}_k, \bar{t}_k\}$, with $\bar{x}_k \in B'$ and $\lim_{k\to\infty} \bar{t}_k = \infty$ such that $x(\bar{t}_k, \bar{x}_k)$ converges to $\bar{x}$ as $k \to \infty$. As all such $\bar{x}_k$'s are in a compact subset of X, by definition of B there exist a time $T > 0$ such that all points $x_k = x(T, \bar{x}_k)$ are points of B. Set $t_k = \bar{t}_k - T$ and consider the sequence $\{x_k, t_k\}$. Trivially $x(t_k, x_k)$, being equal to $x(\bar{t}_k, \bar{x}_k)$, converges to $\bar{x}$ as $k \to \infty$. Thus, $\bar{x}$ is a point of B also.

$$\begin{aligned} \dot{w} &= \begin{pmatrix} 0 & \omega \\ -\omega & 0 \end{pmatrix} w \\ \dot{x} &= Fx + G \begin{pmatrix} 1 & 0 \end{pmatrix} w . \end{aligned} \tag{6.8}$$

As a matter of fact, let Π be the unique solution of the Sylvester equation and observe that the graph of the linear map

$$\begin{aligned} \pi : \mathbb{R}^2 &\to \mathbb{R}^n \\ w &\mapsto \Pi w \end{aligned}$$

is an invariant subspace for the system (6.8). Since all trajectories of (6.8) approach this subspace as $t \to \infty$, the steady state behavior of (6.8) is determined by the restriction of its motion to this invariant subspace.

Revisiting this analysis from the viewpoint of the more general notion of steady state introduce above, let $W \subset \mathbb{R}^2$ be a set of the form

$$W = \{w \in \mathbb{R}^2 : \|w\| \leq c\} \tag{6.9}$$

in which c is a fixed number, and suppose the set of initial conditions for (6.8) is $W \times \mathbb{R}^n$. This is in fact the case when the problem of evaluating the periodic response of (6.7) to harmonic inputs whose amplitude does not exceed a fixed number c is addressed. The set W is compact and invariant for the upper subsystem of (6.8). Therefore, by Lemma 4, the ω-limit set of W under the motion of the upper subsystem of (6.8) is the subset W itself.

The set $W \times \mathbb{R}^n$ is closed and positively invariant for the full system (6.8) and, moreover, since the lower subsystem of (6.8) is a linear asymptotically stable system driven by a bounded input, it is immediate to check that the motions of system (6.8), with initial conditions taken in $W \times \mathbb{R}^n$, are ultimately bounded. As a matter of fact, any bounded set B of the form

$$B = \{(w, x) \in \mathbb{R}^2 \times \mathbb{R}^n : w \in W, \|x - \Pi w\| \leq d\}$$

in which d is any positive number, has the property indicated in the definition of ultimate boundedness. Note also that any of such B satisfies $B \subset W \times \mathbb{R}^n$. It is easy to check that

$$\omega(B) = \{(w, x) \in \mathbb{R}^2 \times \mathbb{R}^n : w \in W, x = \Pi w\},$$

i.e. $\omega(B)$ is the graph of the restriction of the map π to the set W. Note that $\omega(B)$ is independent of the choice of B (so long as B is a set with having the properties indicated in the definition of ultimate boundedness). The restriction of (6.8) to the invariant set $\omega(B)$ characterizes the steady state behavior of (6.7) under the family of all harmonic inputs of fixed angular frequency ω, and amplitude not exceeding c. ◁

Example 2. A similar result, namely the fact that the *steady state locus* is the *graph* of a map, can be reached if the "signal generator" is any nonlinear system, with initial conditions chosen in a compact invariant set W. More precisely, consider an augmented system of the form[7]

$$\begin{aligned} \dot{w} &= s(w) \\ \dot{x} &= Fx + Gq(w)\,, \end{aligned} \tag{6.10}$$

in which $w \in W \subset \mathbb{R}^r$, $x \in \mathbb{R}^n$, and assume that: (i) all eigenvalues of F have negative real part, (ii) the set W is a compact set, invariant for the the upper subsystem of (6.10).

As in the previous example, the ω-limit set of W under the motion of the upper subsystem of (6.10) is the subset W itself. Moreover, since the lower subsystem of (6.10) is a linear asymptotically stable system driven by the bounded input $u(t) = q(w(t, w_0))$, it is easy to check that the motions of system (6.10), with initial conditions taken in $W \times \mathbb{R}^n$, are ultimately bounded. As a matter of fact, so long as $w(0) \in W$, the input $q(w(t, w_0))$ to the lower subsystem of (6.10) is bounded by some fixed number U and standard arguments can be invoked to show that

$$\|x(t)\| \leq Ke^{-\lambda t}\|x(0)\| + LU$$

for all $t \geq 0$, in which K, λ and L are appropriate positive numbers. Thus, any bounded set B of the form

$$B = \{(w, x) \in \mathbb{R}^r \times \mathbb{R}^n : w \in W, \|x\| \leq 2LU\}$$

has the property indicated in the definition of ultimate boundedness.

Moreover, it is possible to show that, regardless of how B is taken, $\omega(B)$ is the graph of the map

$$\begin{aligned} \pi : W &\rightarrow \mathbb{R}^n \\ w &\mapsto \pi(w)\,, \end{aligned}$$

defined by

$$\pi(w) = \lim_{T \to \infty} \int_{-T}^{0} e^{-F\tau} Gq(w(\tau, w))d\tau\,. \tag{6.11}$$

The explanation of this fact reposes on the following arguments. First of all, observe that – since $q(w(t, w))$ is by hypothesis a bounded function of t and all eigenvalues of F have negative real part – the limit on the right-hand side of (6.11) exists and is finite. Then, a simple calculation shows that the graph of the map π is invariant for (6.10). To see why this is the case, pick any initial condition (w_0, x_0) on the graph of π and compute the solution $x(t)$ of the lower equation of (6.10) by means of the classical variation of constants formula, to obtain

$$x(t) = e^{Ft}x_0 + \int_0^t e^{F(t-\tau)} Gq(w(\tau, w_0))d\tau$$

[7] We retain, throughout, the assumption that both $s(w)$ and $q(w)$ are locally Lipschitz functions.

Since by hypothesis $x_0 = \pi(w_0)$, using (6.11) we obtain

$$\begin{aligned} x(t) &= e^{Ft}\int_{-\infty}^{0} e^{-F\tau}Gq(w(\tau,w_0))d\tau + \int_0^t e^{F(t-\tau)}Gq(w(\tau,w_0))d\tau \\ &= \int_{-\infty}^{t} e^{F(t-\tau)}Gq(w(\tau,w_0))d\tau = \int_{-\infty}^{0} e^{-F\theta}Gq(w(\theta+t,w_0))d\theta \\ &= \int_{-\infty}^{0} e^{-F\theta}Gq(w(\theta,w(t,w_0)))d\theta = \pi(w(t,w_0)) = \pi(w(t)) \end{aligned}$$

and this proves the invariance of the graph of π for (6.10). From this property, it is immediately concluded that any point of the graph of π is necessarily a point of $\omega(B)$. To complete the proof of the claim it remains to show that no other point of $W \times \mathbb{R}^n$ can be a point of $\omega(B)$. But this is a direct consequence of the fact that F has eigenvalues with negative real part. In fact, this assumption implies that all motions of (6.10) whose initial condition is not on the graph of π are unbounded in backward time and therefore cannot be contained in $\omega(B)$, which is know to be a bounded set. ◁

There are various ways in which the result discussed in the previous example can be generalized. For instance, it can be extended to describe the steady state response of a nonlinear system

$$\dot{x} = f(x,u) \tag{6.12}$$

in the neighborhood of a locally exponentially stable equilibrium point. To this end, suppose that $f(0,0) = 0$ and that the matrix

$$F = \left[\frac{\partial f}{\partial x}\right](0,0)$$

has all eigenvalues with negative real part. Then, it is well known (see e.g. [15, page 275]) that it is always possible to find a compact subset $X \subset \mathbb{R}^n$, which contains $x = 0$ in its interior and a number $\sigma > 0$ such that, if $\|x_0\| \in X$ and $\|u(t)\| \leq \sigma$ for all $t \geq 0$, the solution of (6.12) with initial condition $x(0) = x_0$ satisfies $\|x(t)\| \in X$ for all $t \geq 0$. Suppose that the input u to (6.12) is produced, as before, by a signal generator of the form

$$\begin{aligned} \dot{w} &= s(w) \\ u &= q(w) \end{aligned} \tag{6.13}$$

with initial conditions chosen in a compact invariant set W and, moreover, suppose that, $\|q(w)\| \leq \sigma$ for all $w \in W$. If this is the case, the set $X \times W$ is positively invariant for

$$\begin{aligned} \dot{w} &= s(w) \\ \dot{x} &= f(x,q(w)), \end{aligned} \tag{6.14}$$

and the motions of the latter are ultimately bounded, with $B = X \times W$. The set $\omega(B)$ may have a complicated structure but it is possible to show, by means arguments similar to those which are used in the proof of the Center Manifold theorem, that if X and B are small enough the set in question can still be expressed as the graph of a map $x = \pi(w)$. In particular, the graph in question is precisely the center manifold of (6.14) at $(0,0)$ if $s(0) = 0$ and the matrix

$$S = \left[\frac{\partial s}{\partial w}\right](0)$$

has all eigenvalues on the imaginary axis.

A common feature of the examples discussed above is the fact that the set $\omega(B)$ can be expressed as the graph of a map $x = \pi(w)$. This means that, so long as this is the case, a system of the form (6.12) has a *unique* well defined *steady state response* to the input $u(t) = q(w(t))$. As a matter of fact, the response in question is precisely $x(t) = \pi(w(t))$. Of course, in general, this may not be the case and *multiple* steady state responses to a given input may occur. In general, the following property holds.

Lemma 5. *Let W be a compact set, invariant under the flow of (6.13). Let X be a closed set and suppose that the motions of (6.14) with initial conditions in $W \times X$ are ultimately bounded. Then, the steady state locus of (6.14) is the graph of a set-valued map defined on the whole of W.*

Proof. By hypothesis, see Lemma 4, $\omega(W) = W$. As a consequence, for all $\bar{w} \in W$ there is a sequence $\{w_k, t_k\}$ with w_k in W for all k such that $\bar{w} = \lim_{k\to\infty} w(t_k, w_k)$. Set $p = \mathrm{col}(w, x)$ and let $\phi(t, p_0)$ denote the integral curve of (6.14) passing through p_0 at time $t = 0$. Pick *any* point $x_0 \in X$ and let $p_k = \mathrm{col}(w_k, x_0)$. If the motions of (6.14) are ultimately bounded, there is a bounded set B and a time $T > 0$ such that $\phi(t, p_k) \in B$ for all $t \geq T$ and all $k > 0$. Pick any integer h such that $t_h \geq T$, set $\bar{p}_k = \phi(t_h, p_k)$ and $\bar{t}_k = t_k - t_h$, for $k \geq h$, and observe that, by construction, $\phi(t_k, p_k) = \phi(\bar{t}_k, \bar{p}_k)$. The sequence $\{\phi(\bar{t}_k, \bar{p}_k)\}$ is bounded. Hence, there exists a subsequence $\{\phi(\hat{t}_k, \hat{p}_k)\}$ converging to a point $\hat{p} = \mathrm{col}(\hat{w}, \hat{x})$, which is a point of $\omega(B)$ because all $\bar{p}_k$'s are in B. Since system (6.14) is upper triangular, necessarily $\hat{w} = \bar{w}$. This shows that, for any point $\bar{w} \in W$, there is a point $\hat{x} \in X$ such that $(\bar{w}, \hat{x}) \in \omega(B)$, as claimed. ◁

6.3 Necessary Conditions for Output Regulation

Taking advantage of the notions introduced in the previous section, we are now in a position to highlight some general properties that any controller that solves a problem of output regulation must necessarily have. Recall that, as defined in section 6.1, the problem of output regulation is solved if, in the composite system (6.5):

- the positive orbit of $W \times X \times \Xi$ is bounded,
- $\lim_{t\to\infty} e(t) = 0$, uniformly in the initial condition.

The notions introduced in the previous section are instrumental to prove the following, elementary – but fundamental – result, which is a nonlinear enhancement of a Lemma of [11] on which all the theory of output regulation for linear systems is based.

Lemma 6. *Suppose the positive orbit of $W \times X \times \Xi$ is bounded. Then*

$$\lim_{t\to\infty} e(t) = 0$$

if and only if

$$\omega(W \times X \times \Xi) \subset \{(w,x,\xi) : h(w,x) = 0\}. \tag{6.15}$$

Proof. Set, for convenience, $B = W \times X \times \Xi$. Set $p = \mathrm{col}(w,x,\xi)$ and let $\phi(t,p_0)$ denote the integral curve of (6.5) passing through p_0 at time $t = 0$. By Lemma 2,

$$\lim_{t\to\infty} \mathrm{dist}(\phi(t,p_0), \omega(B)) = 0\,,$$

uniformly in p_0. Thus, condition (6.15) is sufficient.

To prove the converse, set $K = \{(w,x,\xi) : h(w,x) = 0\}$. Since $h(w,x)$ is continuous and the positive orbit of $W \times X \times \Xi$ is bounded, to say that $\lim_{t\to 0} e(t) = 0$, uniformly in p_0, is equivalent to say that, for any $\varepsilon > 0$, there exists $\bar{t}$ such that, for any $p_0 \in B$,

$$\mathrm{dist}(\phi(t,p_0), K) \le \varepsilon, \qquad \text{for all } t \ge \bar{t}.$$

Pick any point $p \in \omega(B)$. Then, there exist a sequence of pairs (p_k, t_k), with $p_k \in B$ and $t_k \to \infty$ as $k \to \infty$ with the following property: for any $\varepsilon > 0$, there exists $\bar{k}$ such that

$$\mathrm{dist}(\phi(t_k,p_k), p) \le \varepsilon, \qquad \text{for all } k \ge \bar{k}.$$

Without loss of generality, we can pick $\bar{k}$ such that $t_{\bar{k}} \ge \bar{t}$. Thus, for every $\varepsilon > 0$, $\mathrm{dist}(p, K) \le 2\varepsilon$, i.e. $\mathrm{dist}(p, K) = 0$. Since K is closed, $p \in K$ and (6.15) follows. ◁

It is seen from this simple result that the problem of output regulation, as defined in section 6.1, can be simply cast as the problem of *shaping the steady state locus of the closed loop system*, in such a way that property (6.15) holds.

To proceed with the analysis in a more concrete fashion, we consider from now on the special case in which the controlled plant (6.4) is modelled by equations *in normal form*

$$\begin{aligned}
\dot{z} &= f_0(w,z) + f_1(w,z,e_1)e_1 \\
\dot{e}_1 &= e_2 \\
&\;\;\vdots \\
\dot{e}_{r-1} &= e_r \\
\dot{e}_r &= q(w,z,e_1,\ldots,e_r) + b(w,z,e_1,\ldots,e_r)u \\
e &= e_1 \\
y &= \mathrm{col}(e_1,\ldots,e_r)\,,
\end{aligned} \tag{6.16}$$

with state $(z, e_1, \ldots, e_r) \in \mathbb{R}^{n-r} \times \mathbb{R}^r$, control input $u \in \mathbb{R}$, regulated output $e \in \mathbb{R}$, measured output $y \in \mathbb{R}^r$. The functions $f_0(\cdot), f_1(\cdot), q(\cdot), b(\cdot), s(\cdot)$ in (6.16) and (6.3) are assumed to be at least continuously differentiable. It is also assumed that

$$b(w, z, e_1, \ldots, e_r) \neq 0 \qquad \forall (w, z, e_1, \ldots, e_r).$$

The initial conditions of (6.16) range on a set $Z \times E$, in which Z is a fixed *compact* subset of $\mathbb{R}^{n-r}$ and $E = \{(e_1, \ldots, e_r) \in \mathbb{R}^r : |e_i| \leq c\}$, with c a fixed number.

Suppose that a controller of the form (6.4) solves the problem of output regulation. Then Lemma 2 applies and, since $e = e_1$, we deduce that the steady state locus of the closed loop system (6.5) is necessarily a subset of the set of all states in which $e_1 = 0$. This being the case, it is seen from the form of the equations (6.16) that, when the closed loop system (6.5) is in steady state, necessarily also

$$e_2 = e_3 = \cdots = e_r = 0.$$

As a consequence, the following conclusions hold:

- *The steady state locus $\omega(W \times Z \times E \times \Xi)$ of the closed-loop system is a subset of the set $W \times \mathbb{R}^{n-r} \times \{0\} \times \mathbb{R}^{\nu}$.*
- *The restriction of the closed-loop system to its steady state locus $\omega(W \times Z \times E \times \Xi)$ reduces to*

$$\begin{aligned} \dot{w} &= s(w) \\ \dot{z} &= f_0(w, z) \\ \dot{\xi} &= \varphi(\xi, 0). \end{aligned} \tag{6.17}$$

- *For each $(w, z, 0, \ldots, 0, \xi) \in \omega(W \times Z \times E \times \Xi)$*

$$0 = q(w, z, 0, \ldots, 0) + b(w, z, 0, \ldots, 0)\gamma(\xi, 0). \tag{6.18}$$

The prior analysis implicitly assumes that the positive orbit of W under the flow of exosystem is bounded, i.e. that the motions of the exosystem asymptotically approach the its own steady state locus $\omega(W)$. In principle, $\omega(W)$ may differ from W but there is no loss of generality in assuming from the very beginning that the two sets coincide. After all, the problem in question is a problem concerning how the closed-loop system behaves in steady state and there is no special interest in considering exosystems that are not "in steady state". We make this assumption precise as follows (see Lemma 4).

Assumption (i): the compact set W is invariant for (6.3). ◁

With this in mind we observe that, by Lemma 5, if the positive orbit of $W \times Z \times E \times \Xi$ under the flow of (6.5) is bounded, then $\omega(W \times Z \times E \times \Xi)$ is the graph of a (possibly set-valued) map defined on the whole of W. Consider now the set

$$\mathcal{A}_{\text{ss}} = \{(w, z) : (w, z, 0, \ldots, 0, \xi) \in \omega(W \times Z \times E \times \Xi), \text{ for some } \xi \in \mathbb{R}^{\nu}\}$$

and define the map

$$\begin{aligned} u_{\text{ss}} : \mathcal{A}_{\text{ss}} &\to \mathbb{R} \\ (w, z) &\mapsto -\frac{q(w, z, 0, \ldots, 0)}{b(w, z, 0, \ldots, 0)}. \end{aligned}$$

By construction, the set $\mathcal{A}_{ss}$ is the graph of a (possibly set-valued) map defined on the whole of W, which is invariant for the dynamics of

$$\begin{aligned} \dot{w} &= s(w) \\ \dot{z} &= f_0(w, z), \end{aligned} \tag{6.19}$$

that are precisely *the zero dynamics of the "augmented system"* (6.3) – (6.16), while the map $u_{ss}(\cdot)$ is the control that forces the motion of (6.3) – (6.16) to evolve on $\mathcal{A}_{ss}$.

With this in mind, the conclusions reached above can be rephrased in the following terms. Suppose that a controller of the form (6.4) solves the problem of output regulation for (6.16) with exosystem (6.3). Then, there exists a (possibly set-valued) map defined on the whole of W whose graph $\mathcal{A}_{ss}$ is invariant for the autonomous system (6.19). Moreover, for each $(w_0, z_0) \in \mathcal{A}_{ss}$ there is a point $\xi_0 \in \mathbb{R}^\nu$ such that the integral curve of (6.19) issued from (w_0, z_0) and the integral curve of

$$\dot{\xi} = \varphi(\xi, 0)$$

issued from ξ_0 satisfy

$$u_{ss}(w(t), z(t)) = \gamma(\xi(t)), \qquad \forall t \in \mathbb{R}.$$

This is a nonlinear version of the celebrated *internal model principle* of [12].

6.4 Sufficient Conditions for Output Regulation

6.4.1 The Control Structure

On the basis of the ideas presented in the previous section we proceed now with the construction of a controller that solves the problem of output regulation. The "steady state" features of this controller are those identified at the end of the section, namely this controller has to be able to "generate" all controls of the form $u_{ss}(w(t), z(t))$ for any "steady state" trajectory $w(t), z(t)$ of (6.19). The controller should incorporate a device that generates all such trajectories (the *internal model*), thus making sure that the "appropriate" state-state behavior takes place, and a device guaranteeing that convergence to this specific steady state behavior occurs. It is here that additional assumptions are needed.

Note that, since W is invariant for $\dot{w} = s(w)$, the closed cylinder

$$\mathcal{C} := W \times \mathbb{R}^{n-r}$$

is locally invariant for (6.19). Hence, it is natural regard (6.19) as a system defined on $\mathcal{C}$ and endow the latter with the subset topology.

Assumption (ii): there exists a bounded subset B of $\mathcal{C}$ which contains the positive orbit of the set $W \times Z$ under the flow of (6.19) and the resulting omega-limit set $\omega(W \times Z)$ satisfies

$$(w,z)\in\mathcal{C}\,, \quad |(w,z)|_{\omega(W\times Z)} \le d_0 \quad \Rightarrow \quad z\in Z \tag{6.20}$$

where d_0 is a positive number. ◁

While in the analysis of the necessity we have only identified the existence of a compact set (actually, the graph of a map defined on W) which is invariant for (6.19), the new assumption (ii) implies, in its first part, the existence of a compact set (still the graph of a map defined on W)

$$\mathcal{A} := \omega(W\times Z)$$

which is not only invariant but also uniformly attractive of all trajectories of (6.19) issued from points of $W\times Z$. The second part of the assumption, in turn, guarantees that this set is also stable in the sense of Lyapunov (see Lemma 3). In the next assumption we strengthen this property by also requiring the set $\mathcal{A}$ is locally exponentially stable (this assumption is useful to straighten the subsequent analysis, but is not essential).

Assumption (iii): there exist $M\ge 1$, $\lambda>0$ such that

$$(w_0,z_0)\in\mathcal{C}\,, \quad |(w_0,z_0)|_{\mathcal{A}} \le d_0 \quad \Rightarrow \quad |(w(t),z(t))|_{\mathcal{A}} \le Me^{-\lambda t}|(w_0,z_0)|_{\mathcal{A}}$$

in which $(w(t),z(t))$ denotes the solution of (6.19) passing through (w_0,z_0) at time $t=0$. ◁

System (6.16) being affine in the control input u, it seems natural to look for a controller having a similar structure, namely a controller of the form

$$\begin{aligned}\dot\xi &= \varphi(\xi)+\psi(\xi)v\\ u &= \gamma(\xi)+v\end{aligned} \tag{6.21}$$

with state $\xi\in\mathbb{R}^\nu$, in which v is a residual control input, to be eventually chosen as a function of the measured output y. Here $\varphi(\cdot)$, $\psi(\cdot)$ and $\gamma(\cdot)$ are functions to be determined.

As a matter of fact, it will be possible to show that, if the triplet $\{\varphi(\xi),\psi(\xi),\gamma(\xi)\}$ possesses what we will define as *asymptotic internal model* property, the choice of the residual control v in (6.21) as

$$v = ky$$

solves the problem of output regulation, provided that the gain coefficient k is appropriately chosen.

6.4.2 The Asymptotic Internal Model Property

To simplify the exposition, we address first the special case in which the controlled system (6.16) has relative degree 1, i.e. is modelled by equations of the form

$$\begin{aligned}\dot{z} &= f_0(w,z) + f_1(w,z,e)e\\ \dot{e} &= q(w,z,e) + b(w,z,e)u\\ y &= e\,.\end{aligned} \tag{6.22}$$

As a matter of fact, as sketched at the end of the section, the case of higher relative degree can easily be reduced to this one.

We begin by rewriting the zero dynamics of the augmented system (6.3) – (6.22), given by

$$\begin{aligned}\dot{w} &= s(w)\\ \dot{z} &= f_0(w,z)\,,\end{aligned} \tag{6.23}$$

in the more compact form

$$\dot{\mathbf{z}} = \mathbf{f}_0(\mathbf{z}) \tag{6.24}$$

where $\mathbf{z} := \mathrm{col}(w,z)$. Moreover, consistently with this notation, we rewrite the term $q(w,z,e) + b(w,z,e)u$ in (6.22) as

$$q(w,z,e) + b(w,z,e)u = \mathbf{q}_0(\mathbf{z})\mathbf{b}_0(\mathbf{z}) + \mathbf{q}_1(\mathbf{z},e)e + [\mathbf{b}_0(\mathbf{z}) + \mathbf{b}_1(\mathbf{z},e)e]u$$

in which

$$\mathbf{b}_0(\mathbf{z}) = b(w,z,0)\,, \qquad \mathbf{q}_0(\mathbf{z}) = \frac{q(w,z,0)}{b(w,z,0)}\,,$$

and we denote by $\mathbf{Z} := W \times Z$ the compact set where the initial condition $\mathbf{z}(0)$ is supposed to range. Observe also that, by assumption, $\mathbf{b}_0(\mathbf{z})$ is nowhere zero and thus it has a well defined sign. In view of this, the overall system (6.3)– (6.22) is rewritten as

$$\begin{aligned}\dot{\mathbf{z}} &= \mathbf{f}_0(\mathbf{z}) + \mathbf{f}_1(\mathbf{z},e)e\\ \dot{e} &= \mathbf{b}_0(\mathbf{z})[\mathbf{q}_0(\mathbf{z}) + u] + [\mathbf{q}_1(\mathbf{z},e) + \mathbf{b}_1(\mathbf{z},e)u]e\end{aligned} \tag{6.25}$$

where $\mathbf{f}_1(\mathbf{z},e) = \mathrm{col}(0, f_1(w,z,e))$ and the initial conditions $(\mathbf{z}(0), e(0))$ range in the set $\mathbf{Z} \times E$.

Controlling this system by means of (6.21) yields a closed-loop system

$$\begin{aligned}\dot{\mathbf{z}} &= \mathbf{f}_0(\mathbf{z}) + \mathbf{f}_1(\mathbf{z},e)e\\ \dot{e} &= \mathbf{b}_0(\mathbf{z})[\mathbf{q}_0(\mathbf{z}) + \gamma(\xi) + v] + [\mathbf{q}_1(\mathbf{z},e) + \mathbf{b}_1(\mathbf{z},e)(\gamma(\xi) + v)]e\\ \dot{\xi} &= \varphi(\xi) + \psi(\xi)v\end{aligned}$$

which, regarded as a system with input v and output e, has relative degree 1 and zero dynamics given by

$$\begin{aligned}\dot{\mathbf{z}} &= \mathbf{f}_0(\mathbf{z})\\ \dot{\xi} &= \varphi(\xi) - \psi(\xi)[\gamma(\xi) + \mathbf{q}_0(\mathbf{z})]\,.\end{aligned} \tag{6.26}$$

Thus, in view of well-known results, it is reasonable to expect that if the latter possesses a compact invariant set which attracts all trajectories with initial conditions in $\mathbf{Z} \times \Xi$, a high-gain control (on e) be able to steer this variable to arbitrary small values. Note that system (6.26) can be viewed as system (6.24), which by assumption already possesses an invariant set which attracts all initial conditions in $\mathbf{Z}$, driving a system of the form

$$\dot{\xi} = \varphi(\xi) - \psi(\xi)[\gamma(\xi) + \mathbf{q}_0(\mathbf{z})]\,. \tag{6.27}$$

Thus, the important thing needed, to guarantee that system (6.26) possesses a compact invariant set which attracts all trajectories with initial conditions in $\mathbf{Z} \times \Xi$, is to secure that system (6.27) possesses a bounded response to the input $\mathbf{z}$ provided by (6.24). It is here, again, that the notion of steady state turns out to be useful.

Definition 2. *The triplet* $\{\varphi(\xi), \psi(\xi), \gamma(\xi)\}$ *has the* asymptotic internal model property *if there exists a* C^1 *map* $\tau : W \times \mathbb{R}^{n-r} \to \mathbb{R}^d$ *such that:*

(i) the vector fields $\mathbf{f}_0|_{\mathcal{A}}$ *and* φ *are* τ*-related, namely*

$$\frac{\partial \tau(\mathbf{z})}{\partial \mathbf{z}} \mathbf{f}_0(\mathbf{z}) = \varphi(\tau(\mathbf{z})) \qquad \forall\, \mathbf{z} \in \mathcal{A}\,, \tag{6.28}$$

and

$$\mathbf{q}_0(\mathbf{z}) + \gamma \circ \tau(\mathbf{z}) = 0 \qquad \forall\, \mathbf{z} \in \mathcal{A}\,; \tag{6.29}$$

(ii) in the composite system

$$\begin{aligned} \dot{\mathbf{z}} &= \mathbf{f}_0(\mathbf{z}) \\ \dot{\xi} &= \varphi(\xi) - \psi(\xi)[\gamma(\xi) + \mathbf{q}_0(\mathbf{z})] \end{aligned} \tag{6.30}$$

the set

$$\mathrm{graph}(\tau|_{\mathcal{A}}) = \{(\mathbf{z}, \xi) : \mathbf{z} \in \mathcal{A},\ \xi = \tau(\mathbf{z})\}$$

uniformly and locally exponentially attracts $\mathbf{Z} \times \Xi$.

Remark 2. Note that conditions (6.28) and (6.29) simply express the property that the restriction to $\mathcal{A}$ of the autonomous system with output

$$\dot{\mathbf{z}} = \mathbf{f}_0(\mathbf{z}), \qquad y = -\mathbf{q}_0(\mathbf{z}) \tag{6.31}$$

is *immersed* into the system (see [18, page 406])

$$\dot{\xi} = \varphi(\xi), \qquad y = \gamma(\xi)\,. \triangleleft \tag{6.32}$$

The conditions indicated in (i) imply the invariance of the compact set $\mathrm{graph}(\tau|_{\mathcal{A}})$ under the flow of (6.30). If condition (ii) also holds, the set in question can be identified with $\omega(\mathbf{Z} \times \Xi)$, the limit set of $\mathbf{Z} \times \Xi$ under the flow of (6.30), as shown below in the proof of Theorem 1. In other words, if conditions (i) and (ii) hold, the set $\mathrm{graph}(\tau|_{\mathcal{A}})$ is the steady state locus of the composite system (6.26).

We defer to the next section the description of relevant cases in which a controller which possesses the asymptotic internal model property can be constructively designed. Now, we continue the main argument, that is we show why assumptions (i), (ii), (iii) and the asymptotic internal model property suffice solve the problem output regulation by means of a controller of the form (6.21). This is formally stated and proved in the next theorem (see [25]).

Theorem 1. *Pick compact sets* $\mathbf{Z}$*,* E *and* Ξ *for the initial conditions of the closed-loop system (6.3), (6.22), (6.21). Assume that (i)-(ii)-(iii) hold and that the triplet* $\{\varphi, \psi, \gamma\}$ *has the asymptotic internal model property. Assume, in addition, that the vector field* $\psi(\xi)$ *is complete. Then there exists* $k^\star > 0$ *such that for all* $k \geq k^\star$ *the controller (6.21) with* $v = -\text{sign}(\mathbf{b}_0)ke$ *solves the problem of output regulation.*

Proof. The first crucial step to prove the lemma is to show that the trajectories of (6.26) originating from $\mathbf{Z}\times\Xi$ are bounded and the resulting ω-limit set $\omega(\mathbf{Z}\times\Xi)$ is precisely $\text{graph}(\tau|_{\mathcal{A}})$. To this end note that boundedness of the trajectories is a consequence of requirement (ii) in the definition of the asymptotic internal model property. To show that $\omega(\mathbf{Z}\times\Xi) = \text{graph}(\tau|_{\mathcal{A}})$ note that, by the triangular structure of (6.26), it turns out that

$$\omega(\mathbf{Z} \times \Xi) \subset \mathcal{C}\,.$$

Furthermore, by requirement (i) in definition 1, it follows that $\text{graph}(\tau|_{\mathcal{A}})$ is an invariant set for (6.26) and thus $\text{graph}(\tau|_{\mathcal{A}}) \subset \omega(\mathbf{Z} \times \Xi)$. To prove that $\text{graph}(\tau|_{\mathcal{A}}) \equiv \omega(\mathbf{Z} \times \Xi)$ we proceed by contradiction. For, suppose that there exists $(\mathbf{z}_0', \xi_0') \in \omega(\mathbf{Z} \times \Xi)$ such that

$$|(\mathbf{z}_0', \xi_0')|_{\text{graph}(\tau|_{\mathcal{A}})} = c > 0 \tag{6.33}$$

and denote by $(\mathbf{z}'(t), \xi'(t))$ the solution of (6.26) at time t passing through $(\mathbf{z}_0', \xi_0')$ at time $t = 0$. As $\omega(\mathbf{Z} \times \Xi)$ is (backward) invariant and compact there exists a number $K_1 > 0$ such that

$$|(\mathbf{z}'(t), \xi'(t))|_{\text{graph}(\tau|_{\mathcal{A}})} \leq K_1 \qquad \text{for all } t \leq 0\,. \tag{6.34}$$

Now note that by uniform attractiveness in requirement (ii) of definition 1, it turns out that for all positive $K_2 \leq K_1$ there exists $T > 0$ such that for all $(\mathbf{z}_0, \xi_0) \in \mathbf{Z} \times \Xi$ satisfying

$$|(\mathbf{z}_0, \xi_0)|_{\text{graph}(\tau|_{\mathcal{A}})} \leq K_1 \tag{6.35}$$

the trajectory $(\mathbf{z}(t), \xi(t))$ of (6.30) passing through $(\mathbf{z}_0, \xi_0)$ at time $t = 0$ is such that

$$|(\mathbf{z}(T), \xi(T))|_{\text{graph}(\tau|_{\mathcal{A}})} \leq K_2\,. \tag{6.36}$$

Moreover local exponential stability in the second requirement of the previous definition implies the existence of positive d, M, λ such that for all $(\mathbf{z}_0, \xi_0)$ satisfying

$$|(\mathbf{z}_0, \xi_0)|_{\text{graph}(\tau|_{\mathcal{A}})} \leq d$$

the trajectory is such that

$$|(\mathbf{z}(t), \xi(t))|_{\text{graph}(\tau|_{\mathcal{A}})} \leq M e^{-\lambda t} |(\mathbf{z}_0, \xi_0)|_{\text{graph}(\tau|_{\mathcal{A}})}\,.$$

Combining the previous two properties with K_2 chosen so that $K_2 \leq d$ and T consequently, it is possible to check that

$$
\begin{aligned}
&|(\mathbf{z}(0), \xi(0))|_{\mathrm{graph}(\tau|_{\mathcal{A}})} \leq K_1 \qquad \Rightarrow \\
&|(\mathbf{z}(t), \xi(t))|_{\mathrm{graph}(\tau|_{\mathcal{A}})} \leq \bar{M} e^{-\lambda t} |(\mathbf{z}(0), \xi(0))|_{\mathrm{graph}(\tau|_{\mathcal{A}})}
\end{aligned}
$$

where $\bar{M} := \max\{M, K_1/e^{-\lambda T}\}$. From this, choosing T' such that $\bar{M}e^{-\lambda T'} K_1 \leq 0.5c$ and using (6.34), it turns out that

$$
\begin{aligned}
|(\mathbf{z}'_0, \xi'_0)|_{\mathrm{graph}(\tau|_{\mathcal{A}})} &\leq \bar{M} e^{-\lambda T'} |(\mathbf{z}'(-T'), \xi'(-T'))|_{\mathrm{graph}(\tau|_{\mathcal{A}})} \\
&\leq \bar{M} e^{-\lambda T'} K_1 \leq 0.5c
\end{aligned}
$$

which contradicts (6.33). This proves that $\mathrm{graph}(\tau|_{\mathcal{A}}) \equiv \omega(\mathbf{Z} \times \Xi)$.

Since the vector field $\psi(\xi)$ is complete, it is possible to replace the set of coordinates ξ by a set of new coordinates $\eta = F(\mathbf{z}, e, \xi)$, with $F(\mathbf{z}, e, \xi)$ such that

$$
\frac{\partial F}{\partial e}[\mathbf{b}_0(\mathbf{z}) + \mathbf{b}_1(\mathbf{z}, e)e] + \frac{\partial F}{\partial \xi}\psi(\xi) = 0 \,.
$$

Implementing this change of coordinates and setting $p := \mathrm{col}(\mathbf{z}, \eta)$, one can rewrite the resulting system in simplified form as

$$
\begin{aligned}
\dot{p} &= f(p) + \ell(p, e) \\
\dot{e} &= q(p) + r(p, e) + b(p, e)v
\end{aligned}
\tag{6.37}
$$

in which

$$
f(p) = \begin{pmatrix} \mathbf{f}_0(\mathbf{z}) \\ \varphi(\eta) - \psi(\eta)[\gamma(\eta) + \mathbf{q}_0(\mathbf{z})] \end{pmatrix}
$$

$$
q(p) = \mathbf{b}_0(\mathbf{z})[\gamma(\eta) + \mathbf{q}_0(\mathbf{z})] \qquad\qquad b(p, e) = \mathbf{b}_0(\mathbf{z}) + \mathbf{b}_1(\mathbf{z}, e)e
$$

and $\ell(p, e)$, $r(p, e)$ are suitably defined smooth functions of their arguments such that $\ell(p, 0) = r(p, 0) = 0$ for all p. In particular, from the first part of the proof, it turns out that the zero dynamics $\dot{p} = f(p)$ of (6.37) posses an uniformly attractive (locally exponentially) compact invariant set on which $q(p)$ is identically zero. From this and the choice $v = -\mathrm{sign}(\mathbf{b}_0)ke$ the claim of the Theorem follows by high-gain arguments such as the ones used in [24] (see Theorems 2 and 3 in the quoted reference). ◁

Remark 3. For completeness, we sketch how the case of higher relative degree is handled. Consider again system (6.16) and replace the variable e_r by

$$
\tilde{e} := e_r + g^{r-1}a_0 e_1 + g^{r-2}a_1 e_2 + \ldots + g a_{r-2} e_{r-1} \tag{6.38}
$$

where g is a positive parameter (to be determined later) and $a_0, a_1, \ldots, a_{r-2}$ is any fixed set of coefficients chosen in such a way that the polynomial

$$
\lambda^{r-1} + a_{r-2}\lambda^{r-2} + \ldots + a_1\lambda + a_0 = 0
$$

is Hurwitz. This change of variable transforms system (6.16) into a system of the form

$$\begin{aligned} \dot{\tilde{z}} &= \tilde{f}_0(w,\tilde{z}) + \tilde{f}_1(w,\tilde{z},\tilde{e})\tilde{e} \\ \dot{\tilde{e}} &= \tilde{q}(w,\tilde{z},\tilde{e}) + \tilde{b}(w,\tilde{z},\tilde{e})u\,, \end{aligned} \tag{6.39}$$

in which $\tilde{z} \in \mathbb{R}^{n-1}$ is defined as

$$\tilde{z} = \mathrm{col}(z, e_1, \ldots, e_{r-1})\,,$$

and where

$$\tilde{f}_0(w,\tilde{z}) + \tilde{f}_1(w,\tilde{z},\tilde{e})\tilde{e} = \begin{pmatrix} f(w,z,e_1) \\ e_2 \\ \cdots \\ e_{r-1} \\ -g^{r-1}a_0e_1 - g^{r-2}a_1e_2 - \ldots - ga_{r-2}e_{r-1} + \tilde{e} \end{pmatrix}$$

and

$$\begin{aligned} \tilde{q}(w,\tilde{z},\tilde{e}) &= q(w,z,e_1,\ldots,e_{r-1},-g^{r-1}a_0e_1 - \ldots - ga_{r-2}e_{r-1} + \tilde{e}) \\ &\quad - g^{r-1}a_0e_2 - \cdots - g^2a_{r-3}e_{r-1} - ga_{r-2}[-g^{r-1}a_0e_1 - \ldots - ga_{r-2}e_{r-1} + \tilde{e}]\,. \end{aligned}$$

System (6.39) is in all identical to system (6.22). Standard arguments (see e.g. [4]) can be invoked to show that, if g is sufficiently large, assumptions (ii) and (iii) hold. Thus, the problem of output regulation can be solved by means of a controller of the form (6.21) with $v = -\mathrm{sign}(b)k\tilde{e}$. Since $\tilde{e}$ is a linear combination of the components of the measured output y of (6.16), the result follows. ◁

Remark 4. If assumption (iii) does not hold, the proposed controller can still solve the problem of output regulation, but in this case a nonlinear control $v = \kappa(e)$ may be needed, where $\kappa(e)$ is a function which is not necessarily Lipschitz at the origin (see [24]). ◁

6.5 Achieving the Asymptotic Internal Model Property

Goal of this section is to present relevant cases, taken from existing literature, in which a controller satisfying the asymptotic internal model property can be constructed. Looking at the definition, the property in question is easily seen to be related to the capability of reproducing, by means of the output $\gamma(\xi)$ of the system $\dot{\xi} = \varphi(\xi) - \psi(\xi)(\gamma(\xi) + \mathbf{q}_0(\mathbf{z}))$, the *asymptotic* behavior of the output $\mathbf{q}_0(\mathbf{z})$ of the system $\dot{\mathbf{z}} = \mathbf{f}_0(\mathbf{z})$. This, in particular, shows up through the two requirements detailed in the definition: the first which asks for the existence of the invariant compact set graph$(\tau|_{\mathcal{A}})$ for the two systems on which the two outputs coincide, and the second in which this set is required to be (locally exponentially) attractive for the composite system (6.30). The importance of the first requirement in the design of regulators was highlighted, in the literature, as early as in [8] (see also [18, page 406]). If this requirement is met, the regulator is able to generate the appropriate steady state control. It is the second requirement, though, that makes it possible to render that particular steady state attractive.

The problem in question is closely related to the problem of designing *nonlinear observers*. As a matter of fact it is seen from item (i) of the definition that, for each $\mathbf{z}_0 \in \mathcal{A}$, the function of time

$$\hat{\xi}(t) = \tau(\mathbf{z}(t, \mathbf{z}_0))$$

which is defined (and bounded) for all $t \in \mathbb{R}$ satisfies

$$\frac{\mathrm{d}\hat{\xi}(t)}{\mathrm{d}t} = \varphi(\hat{\xi}(t)) \tag{6.40}$$

and, moreover

$$\gamma(\hat{\xi}(t)) = -\mathbf{q}_0(\mathbf{z}(t, \mathbf{z}_0)) .$$

In view of the latter, system (6.27) can be rewritten in the form

$$\dot{\xi} = \varphi(\xi) + \psi(\xi)[\gamma(\hat{\xi}) - \gamma(\xi)] \tag{6.41}$$

and interpreted as *a copy of the dynamics* (6.40) *of* $\hat{\xi}$ *corrected by an "innovation term"* $[\gamma(\hat{\xi}) - \gamma(\xi)]$ *weighted by an "output injection gain"* $\psi(\xi)$. This is the classical structure on an *observer* and the requirement in item (ii) of the definition precisely says that the difference $\xi(t) - \hat{\xi}(t)$ (the "observation error", in our interpretation) should asymptotically decay to zero (with ultimate exponential decay).

This interpretation is at the basis of a number of major recent advances in the design of regulators. In fact, in a number of recent papers, this interpretation has been pursued and, taking into consideration various approaches to the design of nonlinear observers, has lead to effective design methods. In the remaining part of this section, we summarize these results. More specifically, in the next two subsections we show how the theory of *nonlinear high gain observers* (see [14]) and, respectively, *nonlinear adaptive observers* (see [2], [26]) can be successfully employed in the construction of the triplet $\{\varphi, \psi, \gamma\}$. In doing this we follow design procedures which have been proposed respectively in [6] and [10]. The constructions in question rely upon the special technical assumption that the set of functions of time obtained by letting $\mathbf{z}_0$ vary over $\mathcal{A}$ in $\mathbf{q}_0(\mathbf{z}(t, \mathbf{z}_0))$ is a subset of the set of solution of a fixed (nonlinear) ordinary differential equation. This assumption has been weakened in [24], where the theory nonlinear developed in [1] is used to arrive at a characterization of a triplet $\{\varphi, \psi, \gamma\}$ under milder conditions, as described in the last part of the section.

6.5.1 Gauthier-Kupka's Internal Model (see [6])

Assume the existence of an integer $d > 0$, of a locally Lipschitz function $f : \mathbb{R}^d \to \mathbb{R}$ such that, for any $\mathbf{z} \in \mathcal{A}$, the solution $\mathbf{z}(t)$ of passing through $\mathbf{z}$ at time $t = 0$ is such that the function $\rho(t) := \mathbf{q}_0(\mathbf{z}(t))$ satisfies

$$\rho^{(d)}(t) = f(\rho(t), \rho^{(1)}(t), \ldots, \rho^{(d-1)}(t))$$

for all $t \in \mathbb{R}$.

Let $\tau' : B \to \mathbb{R}^d$ be the map defined as

$$\tau'(\mathbf{z}) := \mathrm{col}(\mathbf{q}_0(\mathbf{z}), L_{\mathbf{f}_0}\mathbf{q}_0(\mathbf{z}), \ldots, L_{\mathbf{f}_0}^{d-1}\mathbf{q}_0(\mathbf{z})) \tag{6.42}$$

and let $f_c : \mathbb{R}^d \to \mathbb{R}$ be a function with compact support which agrees with $f(\cdot)$ on $\tau'(\mathcal{A})$, namely

$$f_c|_{\tau'(\mathcal{A})} = f|_{\tau'(\mathcal{A})} \qquad \text{and} \qquad |f_c(s)| \le K < \infty \qquad \text{for all } s \in \mathbb{R}^d .$$

Then, it easy to check that the properties indicated in item (i) of the definition are fulfilled by choosing

$$\varphi(\xi) = \begin{pmatrix} \xi_2 \\ \vdots \\ \xi_d \\ f_c(\xi_1, \xi_2, \ldots, \xi_d) \end{pmatrix}, \qquad \gamma(\xi) = \xi_1, \tag{6.43}$$

with $\tau(\mathbf{z}) = \tau'(\mathbf{z})$. Comparing this construction with the remark after Definition 2 we observe, in particular, that system (6.31) is *immersed into a system which is uniformly observable*, in the sense of [14] (even though system (6.31) might not have had such a property). It is precisely this that makes it possible to choose $\psi(\xi)$ in such a way that also the property indicated in item (ii) of the definition can be achieved.

As a matter of fact, the property in question is achieved by choosing

$$\psi(\xi) = D_k \begin{pmatrix} c_0 \\ \vdots \\ c_{d-1} \end{pmatrix}$$

where $D_k = \mathrm{diag}(k, k^2, \cdots, k^d)$, k is a design parameter, and the c_i's are such that the polynomial $\lambda^d + c_0\lambda^{d-1} + \cdots + c_{d-1} = 0$ is Hurwitz, as formally proved in Lemmas 1 and 2 of [6] to which the interested reader is referred for details.

It is worth noting that the assumption in question clearly covers the interesting (and widely addressed in the recent past literature, see [17]) case in which the function $f(\cdot)$ is linear, namely the case in which (6.31) is immersed into a linear observable system. In this case, although the choice indicated above is clearly still valid, a more direct way of designing the regulator is to use $f(\cdot)$ instead of $f_c(\cdot)$ in the definition of $\varphi(\xi)$, to set $\psi(\xi) = G$ and simply choose G in such a way that $\dot{\xi} = \varphi(\xi) - G\gamma(\xi)$ is a stable linear system.

6.5.2 Bastin-Gevers's Internal Model (see [10])

Implicit in the setup of the problem of output regulation is the possibility that the vector w of exogenous inputs includes a set of uncertain constant parameters.

The latter can be uncertain parameters in the model of the controlled plant (6.16) but also uncertain parameters affecting the dynamics of some other exogenous inputs. In this case, in fact, one can still consider a set (w_1, w_2) of exogenous inputs obeying

$$\dot{w}_1 = s_1(w_1, w_2)$$
$$\dot{w}_2 = 0$$

in which $s_1(w_1, w_2)$ explicitly depends on w_2. If this is the case, it is unlikely that an assumption such as the one introduced at the beginning of the earlier section is going to be fulfilled, and different scenarios have to be considered. A an obvious option would be to assume the existence of a function $f : \mathbb{R}^d \times \mathbb{R}^q \to \mathbb{R}$ and of a map $\theta : \mathcal{A} \to \mathbb{R}^q$ such that, for any $\mathbf{z} \in \mathcal{A}$, the solution $\mathbf{z}(t)$ of passing through $\mathbf{z}$ at time $t = 0$ is such that the functions $\rho(t) := \mathbf{q}_0(\mathbf{z}(t))$ and $\theta(t) := \theta(\mathbf{z}(t))$ satisfy

$$\rho^{(d)}(t) = f(\rho(t), \rho^{(1)}(t), \ldots, \rho^{(d-1)}(t), \theta(t)) \quad \text{and} \quad \theta^{(1)}(t) = 0$$

for all $t \in \mathbb{R}$. In this case, though, while the immersion property (i) is easily fulfilled (exactly as in the previous case), it becomes quite hard to have property (ii) fulfilled. In order to make this possible, some extra (stringent) assumptions, on the function f, must be imposed.

Note that, if the hypothesis indicated above holds, system (6.31) is immersed into the $(d + q)$ – dimensional system

$$\dot{\eta} = \begin{pmatrix} \eta_2 \\ \vdots \\ \eta_d \\ f(\eta_1, \eta_2, \ldots, \eta_d, \theta) \end{pmatrix}, \qquad y = \eta_1 \tag{6.44}$$

$$\dot{\theta} = 0 \tag{6.45}$$

via the pair of maps

$$\eta = \tau'(\mathbf{z}), \qquad \theta = \theta(\mathbf{z}),$$

in which $\tau'(\mathbf{z})$ is the map defined in (6.42). The assumption that we make now is that there is a globally defined diffeomorphism $\tilde{\eta} = \Phi(\eta)$ that changes system (6.44) into a system in adaptive observability form, in the sense of [26], namely a system of the form

$$\dot{\tilde{\eta}} = A\tilde{\eta} + \phi(C\tilde{\eta}) + \Omega(C\tilde{\eta})\theta, \qquad y = C\tilde{\eta} \tag{6.46}$$

in which A, C is an observable pair, and $\phi : \mathbb{R} \to \mathbb{R}^d$ and $\Omega : \mathbb{R} \to \mathbb{R}^{d \times q}$ are smooth functions. Conditions under which this is possible are well-known and can be found, for instance, in [26]. Note that, if this assumption holds, the map $\tilde{\tau}(\mathbf{z}) := \Phi(\tau'(\mathbf{z}))$ satisfies

$$\frac{\partial \tilde{\tau}}{\partial \mathbf{z}} \mathbf{f}_0(\mathbf{z}) = A\tilde{\tau}(\mathbf{z}) + \phi(C\tilde{\tau}(\mathbf{z})) + \Omega(C\tilde{\tau}(\mathbf{z}))\theta(\mathbf{z}), \qquad \mathbf{q}_0(\mathbf{z}) = C\tilde{\tau}(\mathbf{z}). \tag{6.47}$$

This being said, we define now the triplet $\{\varphi(\xi), \psi(\xi), \gamma(\xi)\}$ as follows (see [10])

$$\xi = \mathrm{col}(\xi_1, \xi_2, \xi_3) \qquad \text{with } \xi_1 \in \mathbb{R}^d,\ \xi_2 \in \mathbb{R}^q,\ \xi_3 \in \mathbb{R}^{d-1} \times \mathbb{R}^q,$$

$$\varphi(\xi) = \begin{pmatrix} A\xi_1 + \phi_c(C\xi_1) + \Omega_c(C\xi_1)\xi_2 - M(\xi_3)\mathrm{dzv}_\ell(\xi_2) \\ -\mathrm{dzv}_\ell(\xi_2) \\ F\xi_3 + G\Omega_c(C\xi_1) \end{pmatrix},$$
$$\psi(\xi) = \begin{pmatrix} H(\xi_3, \xi_1) \\ \beta(\xi_3, \xi_1) \\ 0 \end{pmatrix}, \qquad \gamma(\xi) = C\xi_1 \tag{6.48}$$

in which $\phi_c(\cdot)$ and $\Omega_c(\cdot)$ denote functions with compact support which agree with $\phi(\cdot)$ and $\Omega(\cdot)$ on $C\tilde{\tau}(\mathcal{A})$, $F \in \mathbb{R}^{d-1 \times d-1}$ and $G \in \mathbb{R}^{d-1 \times d}$ are chosen as

$$F = \begin{pmatrix} -b_2 & 1 \cdots 0 & 0 \\ \cdot & \cdots\cdots & \cdot \\ -b_{d-1} & 0 \cdots 0 & 1 \\ -b_d & 0 \cdots 0 & 0 \end{pmatrix}, \qquad G = \begin{pmatrix} -b_2 & 1 \cdots 0 & 0 & 0 \\ \cdot & \cdots\cdots & \cdot & \cdot \\ -b_{d-1} & 0 \cdots 0 & 1 & 0 \\ -b_d & 0 \cdots 0 & 0 & 1 \end{pmatrix} \tag{6.49}$$

and $M(\cdot)$, $\beta(\cdot,\cdot)$, $H(\cdot,\cdot)$ as

$$M(\xi_3) = \begin{pmatrix} 0 \\ \xi_3 \end{pmatrix}, \qquad \begin{aligned} \beta^{\mathrm{T}}(\xi_3, \xi_1) &= CAM(\xi_3) + C\Omega_c(C\xi_1), \\ H(\xi_3, \xi_1) &= M(\xi_3)\beta(\xi_3, \xi_1) + K, \end{aligned}$$

where the b_i's, $i = 2, \cdots, d$, and K are design parameters. Finally, $\mathrm{dzv}_\ell(\cdot)$ is the vector-valued *dead-zone* function defined as

$$\mathrm{dzv}_\ell(\mathrm{col}(s_1, \ldots, s_d)) = \mathrm{col}(\mathrm{dz}_\ell(s_1), \ldots, \mathrm{dz}_\ell(s_d)) \tag{6.50}$$

in which $\mathrm{dz}_\ell(\cdot)$ is any continuously differentiable function satisfying

$$\mathrm{dz}_\ell(x) = \begin{cases} 0 & \text{if } |x| \le \ell \\ x & \text{if } |x| \ge \ell + 1\,. \end{cases}$$

Lengthy, but not difficult, computations can be used to check that if the coefficients b_i's are chosen so that the matrix F is Hurwitz and ℓ so that $\ell \ge \max_{\mathbf{z} \in \mathcal{A}} |\theta(\mathbf{z})|$, then the map

$$\tau(\mathbf{z}) = \mathrm{col}(\tilde{\tau}(\mathbf{z}),\, \theta(\mathbf{z}),\, \sigma(\mathbf{z})) \quad \text{where} \quad \sigma(\mathbf{z}) = \int_{-\infty}^{0} e^{-Fs} G\Omega(C\tilde{\tau}(\mathbf{z}(s, \mathbf{z})))ds \tag{6.51}$$

is such that $\mathrm{graph}(\tau|_{\mathcal{A}})$ is invariant for $\dot{\xi} = \varphi(\xi)$ and $\mathbf{q}_0|_{\mathcal{A}} = \gamma \circ \tau|_{\mathcal{A}}$ and thus the first requirement in the Definition 2 is fulfilled. Furthermore it can be proved that, if K is appropriately chosen, $\mathrm{graph}(\tau|_{\mathcal{A}})$ also uniformly (and locally exponentially) attracts $\mathbf{Z} \times \Xi$ under the flow of (6.30), namely that the triplet (6.48) also fulfills the second requirement of Definition 2. The result in question is presented in the next proposition, whose proof – which relies upon a persistence of excitation condition – can be found in [10] .

Proposition 1. *Fix* $(\varphi(\xi), \gamma(\xi), \psi(\xi))$ *as in (6.48) and* $\tau(\mathbf{z})$ *as in (6.51). Set* $b = \mathrm{col}(1, b_2, \ldots, b_{q_1})$ *and choose*

$$K = Ab + \lambda b$$

with λ *a design parameter. If for all* $\mathbf{z}_0 \in \mathcal{A}$ *the following implication is true (persistence of excitation condition)*

$$\varsigma^{\mathrm{T}} \beta\left(\sigma\left(\mathbf{z}(t, \mathbf{z}_0)\right), \tilde{\tau}\left(\mathbf{z}(t, \mathbf{z}_0)\right)\right) = 0 \quad \forall t \geq 0 \qquad \Rightarrow \qquad \varsigma \equiv 0,$$

then there exists $\lambda^\star > 0$ *such that for all* $\lambda \geq \lambda^\star$ *the set* $\mathrm{graph}(\tau|_{\mathcal{A}})$ *uniformly (locally exponentially) attracts* $\mathbf{Z} \times \Xi$ *under the flow of (6.30).*

It is interesting to note that the analysis discussed above covers also the particular case in which the exosystem state $\mathbf{z}$ includes a vector ϱ of *constant* uncertain parameters ranging in a compact set $P \subset \mathbb{R}^p$ and there exists a differentiable map $\tau' : B \to \mathbb{R}^d$ such that[8]

$$\begin{aligned} \frac{\partial \tau'(\mathbf{z})}{\partial \mathbf{z}} \mathbf{f}_0(\mathbf{z}) &= S(\varrho)\tau'(\mathbf{z}) \\ \mathbf{q}_0(\mathbf{z}) &= \Gamma(\varrho)\tau'(\mathbf{z}) \end{aligned} \tag{6.52}$$

in which $(S(\varrho), \Gamma(\varrho)) \in \mathbb{R}^{d\times d} \times \mathbb{R}^{1\times d}$ is an observable pair for all $\varrho \in \mathcal{P}$. In fact, note that, since the pair $(S(\varrho), \Gamma(\varrho))$ is observable for all ϱ, standard arguments can be used to show that there exist a nonsingular matrix $M(\varrho) \in \mathbb{R}^{d\times d}$, a column vector $L(\varrho) \in \mathbb{R}^{d\times 1}$, and an observable (parameter independent) pair $(A, C) \in \mathbb{R}^{d\times d} \times \mathbb{R}^{1\times d}$ such that

$$\begin{aligned} M(\varrho)S(\varrho)M(\varrho)^{-1} &= A + L(\varrho)C \\ \Gamma(\rho)M(\varrho)^{-1} &= C\,. \end{aligned}$$

From this, it turns out that relation (6.47) holds with $q = d$, $\phi(s) = 0$, $\Omega(s) = sI_d$, $\tau(\mathbf{z}) = M(\varrho)\tau'(\mathbf{z})$ and $\theta(\mathbf{z}) = L(\varrho)$. This, in particular, shows that general design procedure leading to chioce of the triplet $(\varphi(\xi), \gamma(\xi), \psi(\xi))$ in (6.48) can be successfully adopted. It is interesting to note, however, that in this particular case the general procedure detailed above can be simplified to obtain the triplet fulfilling the asymptotic internal model property in a more direct and effective way. How this is possible is explained in the following (see [10] for details).

Let $(F, G) \subset \mathbb{R}^{d\times d} \times \mathbb{R}^{1\times d}$ be an arbitrary controllable pair with F Hurwitz and let $T(\varrho)$ denote the unique nonsingular solution of the Sylvester equation

$$FT(\varrho) - T(\varrho)S(\varrho) = -G\Gamma(\varrho)$$

and $\Psi(\varrho)$ the row vector $\Psi(\varrho) = \Gamma(\varrho)T^{-1}(\varrho)$. By bearing in mind the definition (6.50), set $\xi = \mathrm{col}(\xi_1, \xi_2)$ with $\xi_1 \in \mathbb{R}^d$ and $\xi_2 \in \mathbb{R}^d$, and choose the triplet as

[8] This scenario is representative of the important case in which $\mathbf{q}_0(\mathbf{z}(t))$ is the sum of a finite number of periodic signals of uncertain amplitude, phase and frequency (see [28]).

$$\varphi(\xi) = \begin{pmatrix} (F + G\xi_2^{\mathrm{T}})\xi_1 \\ -\mathrm{dzv}_\ell(\xi_2) \end{pmatrix}, \qquad \gamma(\xi) = \xi_2^{\mathrm{T}}\xi_1, \qquad \psi(\xi) = \begin{pmatrix} G \\ \xi_1 \end{pmatrix}. \tag{6.53}$$

Simple, though lengthy, algebra can be used to show that if ℓ is chosen so that

$$\ell \geq \max_{\varrho \in P} |\Psi^{\mathrm{T}}(\varrho)|$$

then the first requirement of Definition 1 is satisfied by the triplet (6.53) through the map

$$\tau(\mathbf{z}) = \begin{pmatrix} T(\varrho)\,\tau'(\mathbf{z}) \\ \Psi^{\mathrm{T}}(\varrho) \end{pmatrix}, \tag{6.54}$$

in which, as also stressed above, the constant parameters ϱ can be though as trivial components of $\mathbf{z}$. Moreover also the second requirement of Definition 1 can be shown to be satisfied provided that a persistence of excitation condition, detailed in the next proposition, is fulfilled. For the proof of this proposition the interested reader is referred to [10].

Proposition 2. *Fix $(\varphi(\xi), \gamma(\xi), \psi(\xi))$ as in (6.53) and $\tau(\mathbf{z})$ as in (6.54). If there exist positive T and K such that*

$$\int_t^{t+T} \tau'(\mathbf{z}(s, \mathbf{z}_0))\, \tau'^{\mathrm{T}}(\mathbf{z}(s, \mathbf{z}_0))\, ds \geq KI$$

for all $t \geq 0$ and for all $\mathbf{z_0} \in \mathcal{A}$, then the set $\mathrm{graph}(\tau|_{\mathcal{A}})$ *uniformly (locally exponentially) attracts $\mathbf{Z} \times \Xi$ under the flow of (6.30).*

6.5.3 Andrieu-Praly's Internal Model (see [24])

In this section we follow the theory presented in [24] to show that, in order to achieve the asymptotic internal model property, assumptions such as those considered in the previous two subsections, namely the existence of an ordinary differential equation to be fulfilled by any of the functions $\rho(t)$, are not needed. As opposite to the frameworks discussed in the previous two subsections, though, this kind theory only guarantees the existence of a triplet having the internal model property, while its actual construction may be difficult.

Let $(F, G) \in \mathbb{R}^{d \times d} \times \mathbb{R}^{d \times 1}$ be a controllable pair and set

$$\varphi(\xi) = F\xi + G\gamma(\xi), \qquad \psi(\xi) = G$$

with $\gamma : \mathbb{R}^d \to \mathbb{R}$ a continuous function to be chosen in such a way that the proposed triplet has the required properties.

With this choice it turns out that the composite system (6.30) assumes the form

$$\begin{aligned} \dot{\mathbf{z}} &= \mathbf{f}_0(\mathbf{z}) \\ \dot{\xi} &= F\xi - G\mathbf{q}_0(\mathbf{z})\,. \end{aligned} \tag{6.55}$$

The first step in proving that the triplet in question can be made to satisfy the asymptotic internal model property is presented in the following proposition whose only requirement is that the matrix F is Hurwitz.

Proposition 3. *Consider system (6.55) under the assumption (i)-(ii)-(iii) in Section 6.2. There exists an $\ell > 0$ such that if the eigenvalues of F have real parts lower than $-\ell$, then the map*

$$\tau(\mathbf{z}) = \int_{-\infty}^{0} e^{-Fs} G\mathbf{q}_0(\mathbf{z}(s,\mathbf{z}))ds \tag{6.56}$$

is differentiable, satisfies

$$\frac{\partial \tau}{\partial \mathbf{z}} \mathbf{f}_0(\mathbf{z}) = F\tau(\mathbf{z}) - G\mathbf{q}_0(\mathbf{z}) \qquad \forall \mathbf{z} \in \mathcal{A}\,, \tag{6.57}$$

and it is such that the set graph($\tau|_{\mathcal{A}}$) is locally exponentially stable for (6.55) with a domain of attraction containing $\mathbf{Z} \times \Xi$.

The proof that, if ℓ is large enough, the map (6.56) is differentiable can be found in [24]. The proof of the other properties is a straightforward consequence of the arguments presented in remark 2. It is worth stressing that the requirement of choosing ℓ sufficiently large is only a technical assumption needed to guarantee differentiability of the function τ (see the proof of Proposition 2 in [24]). In this sense the assumption in question must be not confused with any "high gain" requirement on the choice of F. Note, moreover, that the function γ and the dimension d of the pair (F, G) do not play any role in establishing this result.

As such, the previous result only guarantees the fulfillment of the requirement (ii) in definition 1, namely the existence of the exponentially stable set graph($\tau|_{\mathcal{A}}$) for system (6.30) but it says nothing regarding requirement (i). In this respect it is easy to realize that also requirement (i) is fulfilled if a function γ can be found that renders (6.29) satisfied. As a matter of fact, bearing in mind that $\varphi(\xi) = F\xi + G\gamma(\xi)$, (6.28) reads as

$$\frac{\partial \tau}{\partial \mathbf{z}} \mathbf{f}_0(\mathbf{z}) = F\mathbf{z} - G\gamma \circ \tau(\mathbf{z}) \qquad \forall \mathbf{z} \in \mathcal{A}$$

which, if (6.29) holds, reduces to (6.57). This, together with Proposition 3, shows that a triplet having the asymptotic internal model property can be found if a function $\gamma(\cdot)$ exists which satisfies (6.29). It is here that the dimension d of the pair (F, G) plays a role, as formalized in the next proposition whose proof can be found in [24].

Proposition 4. *Suppose*

$$d \geq 2(s + n - r) + 2\,.$$

Then for almost all[9] choices of a controllable pair (F, G), with F satisfying the condition indicated in Proposition 3, the map (6.56) satisfies

$$\tau(\mathbf{z}_1) = \tau(\mathbf{z}_2) \qquad \Rightarrow \qquad \mathbf{q}_0(\mathbf{z}_1) = \mathbf{q}_0(\mathbf{z}_2)\,.$$

As a consequence there exist a continuous map $\gamma : \tau(\mathcal{A}) \to \mathbb{R}$ fulfilling (6.29).

[9] See [24] for details.

Remark 5. The map $\tau(\cdot)$ in (6.56) is defined only on $\mathcal{A}$, but is not difficult to extend it to a C^1 map defined on the whole set $W \times \mathbb{R}^{n-r}$, as shown in [24]. Also the map $\gamma(\cdot)$ that makes (6.29) true can be extended to the whole $\mathbb{R}^d$, but this extension is only known to be continuous. The problem of determining when a C^1 extension exists is under current investigation. ◁

References

[1] Andrieu V, Praly L (2006) On the existence of a Kazantis-Kravaris/Luenberger observer, SIAM Journal on Control and Optimization, 45: 432–456

[2] Bastin G, Gevers M R (1988) Stable adaptive observers for non-linear time varying systems, IEEE Transaction on Automatic Control, 33: 650–657

[3] Birkhoff G D (1927) *Dynamical Systems* , American Mathematical Society.

[4] Byrnes C I, Isidori A (1991) Asymptotic Stabilization of Minimum Phase Nonlinear Systems, IEEE Trans. Automatic Control, 36: 1122-1137

[5] Byrnes C I, Isidori A (2003) Limit Sets, Zero Dynamics and Internal Models in the Problem of Nonlinear Output Regulation, IEEE Transaction on Automatic Control, 48: 1712–1723

[6] Byrnes C I, Isidori A (2004) Nonlinear Internal Models for Output Regulation, IEEE Transaction on Automatic Control, 49: 2244–2247

[7] Byrnes C I, Isidori A (2004) The Steady-State Response of a Nonlinear System: Ideas, Tools and Applications, *preprint.*

[8] Byrnes C I, Delli Priscoli F, Isidori A, Kang W (1997) Structurally Stable Output Regulation of Nonlinear Systems, Automatica, 33: 369–385

[9] Chen Z, Huang J (2004) Global robust servomechanism problem of lower triangular systems in the general case, Systems and Control Letters, 52: 209–220

[10] Delli Priscoli F, Marconi L, Isidori A (2006) A New Approach to Adaptive Nonlinear Regulation, SIAM Journal on Control and Optimization, 45: 829–855

[11] Francis B A (1977) The Linear Multivariable Regulator Problem, SIAM Journal on Control and Optimization, 14: 486–505

[12] Francis BA, Wonham W M (1976) The Internal Model Principle of Control Theory, Automatica, 12: 457–465

[13] Gardner M F, Barnes J L (1942) *Transients in Linear Systems*, Wiley.

[14] Gauthier J P, Kupka I (2001) *Deterministic Observation Theory and Applications*, Cambridge University Press.

[15] W. Hahn, *Stability of Motions*, Springer Verlag, 1967.

[16] Hale J K, Magalhães L T, Oliva W M (2002) *Dynamics in Infinite Dimensions*, Springer Verlag.

[17] Huang J, Lin C F (1994) On a Robust Nonlinear Multivariable Servomechanism Problem, IEEE Transaction on Automatic Control, 39: 1510–1513

[18] Isidori A *Nonlinear Control Systems* (1995), Springer-Verlag.

[19] Isidori A, Marconi L, Serrani A (2003) *Robust Autonomous Guidance: An Internal Model-based Approach*, Springer Verlag.

[20] Isidori A, Byrnes C I (1990) Output Regulation of Nonlinear Systems, IEEE Transaction on Automatic Control, 25: 131–140

[21] James H M, Nichols N B, Phillips R S (1947) *Theory of Servomechanisms*, McGraw-Hill.

[22] Kazantzis K, Kravaris C (1998) Nonlinear observer design using Lyapunov's auxiliary theorem, Systems and Control Letters, 34: 241-247

[23] Khalil H (1994) Robust Servomechanism Output Feedback Controllers for Feedback Linearizable Systems Automatica, 30: 587–1599
[24] Marconi L, Praly L, Isidori A (2006) Output Stabilization via Nonlinear Luenberger Observers, SIAM Journal on Control and Optimization, 45: 2277-2298.
[25] Marconi L, Isidori A (2006) A Unifying Approach to the Design of Nonlinear Output Regulators, in *Advances in Control Theory and Applications* (Bonivento *et al.* eds), Springer Verlag, LNCIS vol. 353. 185–200
[26] Marino R, Tomei P (1992) Global Adaptive Observers for Nonlinear Systems via Filtered Transformations, IEEE Transaction on Automatic Control, 37: 1239–1245
[27] Serrani A, Isidori A, Marconi L (2000) Semiglobal Output Regulation for Minimum-phase Systems, International Journal on Robust and Nonlinear Control, 10, pp. 379–396
[28] Serrani A, Isidori A, Marconi L (2001) Semiglobal Nonlinear Output Regulation with Adaptive Internal Model, IEEE Transaction on Automatic Control , 46: 1178-1194
[29] Teel A R, Praly L (1995) Tools for Semiglobal Stabilization by Partial State and Output Feedback, SIAM Journal on Control and Optimization, 33: 1443–1485
[30] Willems J C (1991) Paradigms and Puzzles in the Theory of Dynamical Systems, IEEE Trans. on Automatic Control, 36: 259–294.
[31] Zadeh L A, Desoer C A (1963) *Linear System Theory*, McGraw-Hill.

7

Parameter/Fault Estimation in Nonlinear Systems and Adaptive Observers

Gildas Besançon

Control Systems Department (former *Laboratoire d'Automatique de Grenoble*), GIPSA-lab. ENSIEG BP 46, 38402 Saint-Martin d'Hères, France
Gildas.Besancon@inpg.fr

7.1 Introduction and Problem Statement

Modeling and monitoring processes are clearly part of an overall control problem, as well as they can be considered by themselves, and each of them can be seen from an observer viewpoint: for the first one indeed, whenever the model structure is given, the problem amounts to that of estimating the model parameters. Even if this problem has been widely studied in the framework of *identification* [15, 13, ...], it can be recast in an observer formulation, by simply considering unknown parameters as constant state variables. For the second one, it has clearly also been very widely studied, in the community of *fault detection and diagnosis* [20, 7, ...]. But one can also use an observer to detect possible faults, for instance by comparing an observer output to the corresponding measured one. When taking into account possible faults through parameter changes in a model, fault detection (and isolation) can even be solved via parameter estimation.

For those reasons, the present chapter is dedicated to the observer problem in the presence of unknown parameters. In front of this one can clearly extend the state vector by including the unknown parameters in it, and thus be brought to an observer problem for an extended system. For such an approach the reader is referred to the other chapters of the present book discussing various possible observer techniques (see for instance the overview of chapter 1). An alternative approach is to rely on some possible observer design for the system assuming known parameters, and try to find some appropriate adaptation law for the unknown parameters so as to keep the observer convergence in the presence of those unknown parameters: this makes the observer a so-called *adaptive observer*. When the state vector can be estimated in this way, this actually corresponds to some *robust* design (w.r.t. parameter uncertainties) - but will be called adaptive *state* observer in the sequel (following the *adaptive* terminology), while when the parameter can be further estimated or the state vector can be reconstructed only together with the parameters, we actually solve a *joint parameter and state estimation* problem.

The present chapter will discuss some results on such *adaptive observers* for nonlinear systems. Section 7.2 will first briefly restate problems of parameter

G. Besançon (Ed.): Nonlinear Observers and Applications, LNCIS 363, pp. 211–222, 2007.
springerlink.com

estimation or fault detection, while section 7.3 will present appropriate forms for adaptive observers together with some possible corresponding designs.

Notice that the results here reported follow various previous works on this topic and with various co-workers (see [3, 26, 6, 4, 1] for instance).

7.2 Fault Diagnosis and Parameter Estimation

7.2.1 Fault Diagnosis

In this section, the problem of fault diagnosis is very briefly recalled, focusing on observer-based approaches, on the basis of [11]. This means that we here rely on some analytical model.

In a quite general approach, one can assume in that respect that the process to be monitored can be described by a state-space representation of the following form:

$$\begin{aligned} \dot{x}(t) &= f(x(t), u(t), d(t), f(t)) \\ y(t) &= h(x(t), u(t), d(t), f(t)) \end{aligned} \tag{7.1}$$

where $x \in I\!R^n$ is the state vector, $y \in I\!R^p$ is the measurement vector, $u \in I\!R^m$ is the known input vector, while $d \in I\!R^\nu$ denotes a vector of unknown disturbance inputs, and $f \in I\!R^q$ the vector of possible faults.

The problem of fault diagnosis is then that of detecting and isolating faults affecting the process, from the knowledge of model (7.1) together with input u, and the measurements of y.

Various approaches have been studied in this context, considering more or less specific functions f and h, and which can roughly be divided into 'parity space' methods on the one hand, and 'observer-based' methods on the other hand. Here we will just recall some features of the second ones (some relationships between both can actually be found [9]).

In such an approach, the basic idea is to use the output error between the measured output and the observer output as a detector for fault occurrences: after the transient due to initial conditions indeed, such a quantity should vanish in the absence of any fault, while it might be driven away from zero in the presence of some fault. Such an indicator is called a 'residual', and thus the fault detection problem amounts to a problem of residual generator (*Fundamental Problem of Residual Generation* FPRG [19]). In the case of multi-outputs systems or various fault models, one can use benches of observers to detect and isolate faults, by trying to design observers structurally insensitive to some faults and sensitive to other ones.

Notice that such an approach usually requires an additional stage for an actual decision, aiming at determining whether a residual significantly differs from zero and where the faults are most likely located, which will not be discussed here.

About the FPRG, it can be stated in a general form as follows:

For each fault f, find an auxiliary system :

$$\begin{aligned} \dot{z} &= \varphi(z, y, u) \\ r &= \psi(z, y, u) \end{aligned}$$

such that r in short is affected by f, not affected by d, and asymptotically decays to zero when f is identically zero.

If one can find a transformation of the original system so as to get a subsystem which is only affected by f, the FPRG amounts to an observer problem for this new representation.

A weaker formulation of the FPRG is that r must be *more affected by f than by d*, and some 'observer-based' residual generator can be obtained in some cases that attenuates the effect of disturbances and not that of faults (see e.g. [21, 22, 2]).

In a case of a subsystem not affected by disturbances (or when ignoring disturbances) the possible faults can also be detected by direct estimation: considering some a priori given fault models, this boils down to a problem of parameter estimation. Few words are added in that respect in next subsection.

7.2.2 Parameter Estimation

A possible approach to fault detection is to rely on parameter estimation algorithms, which is possible whenever faults can be modeled through parameter changes.

The problem here can be described on the basis of a state-space representation of the following form:

$$\begin{aligned} \dot{x}(t) &= f(\theta, x(t), u(t), t) \\ y(t) &= h(x(t)) \end{aligned} \tag{7.2}$$

with $x(t) \in \mathbb{R}^n$ the state vector, $y(t) \in \mathbb{R}^p$, the measurement vector, $u(t) \in \mathbb{R}^m$, the known input vector, and $\theta \in \mathbb{R}^q$ a vector of unknown variables assumed to be constant (which can stand for unknown model parameters or fault parameters).

Beyond all the methodologies developed within the identification community, a direct observer approach can be described as follows:

Consider the extended state vector $X := \begin{pmatrix} x \\ \theta \end{pmatrix}$ and the resulting extended representation:

$$\begin{aligned} \dot{X}(t) &= F(X(t), u(t), t) \\ y(t) &= H(X(t)) \end{aligned} \tag{7.3}$$

with $\dot{\theta} = 0$.

Then try to design an observer for this system relying on available techniques for nonlinear systems.

Notice that this approach does not specifically take advantage of θ being constant. Moreover, it might 'destroy' some appropriate structure of the original system (7.2) for observer design: for instance in the simple case of a *linear time-invariant* system including unknown parameters in its state space representation, the system is turned into a *nonlinear* one when following the above procedure.

An *adaptive* approach instead is based on a possible design for the original system, modified by an appropriate adaptation law.

As an example, consider the system of the following form:

$$\begin{aligned} \dot{x}_1(t) &= x_2(t) + \theta u(t) \\ \dot{x}_2(t) &= 2\theta u(t) \\ y(t) &= x_1(t) \end{aligned} \tag{7.4}$$

This system is of the (linear) form: $\dot{x}(t) = Ax(t) + B\theta u(t),\ y(t) = Cx(t)$ and it is clearly observable whenever θ is known.

From this, if θ was known, an observer could be classically designed as: $\dot{\hat{x}}(t) = A\hat{x}(t) + B\theta u(t) - K(C\hat{x}(t) - y(t))$.

When θ is unknown, we have to use an estimate $\hat{\theta}$ and find an adaptation law for this estimate so that the observer still works.

With $K = (3\ 2)^T$ and $\dot{\hat{\theta}} = -ku(C\hat{x} - y),\ k > 0$ it can here be shown that for bounded inputs, $\|\hat{x}(t) - x(t)\| \to 0$. Moreover, one can further characterize $u(t)$ so that we also have $|\hat{\theta}(t) - \theta| \to 0$ (see definition 2 and proposition 2 below).

This example yields the following remarks:

- The resulting adaptive scheme is indeed an observer achieving state estimation. A specificity is that under additional conditions on the input, it can further achieve parameter estimation;
- The extended system here is also of a specific form (state affine) for which an observer could have been designed, with on-line computation of a time-varying gain (Kalman-like design);
- The gain $(K^T,\ ku)^T$ in the above 'adaptive' design is explicitly obtained off-line;
- System (7.4) actually admits a particular (*passivity*) property between input θ and output y (see definition 1 below).

Hence this example shows how in some cases, appropriate properties allow to get an observer for state and parameter estimation. This is further discussed in next section.

7.3 Adaptive Observers

Let us consider here again systems described by a state-space representation of the form (7.2).

The observer problem which can be then discussed is that of obtaining an estimate of $x(t)$ from known $y(t), u(t), f, h$ in spite of unknown θ, or estimating both $x(t)$ and θ from known $y(t), u(t), f, h$.

Although this can be seen as a problem of *robust* observer design, it is usually referred to as an *adaptive observer* problem.

In order to be more precise w.r.t. the two possible 'versions' of the problem previously mentioned, we can use here two terminologies: *adaptive state observer* for the first one, and *joint state and parameter observer* for the second one.

As already highlighted in example (7.4), it will be seen that *adaptive state estimation* is possible under some 'passivity-like' condition w.r.t. parameters as inputs - as this commonly happens in adaptive systems [12, 14, 23], while parameter estimation additionally requires some 'persistent excitation' condition, as usual in identification problems.

Let us thus here recall the formal statement of those notions:

Definition 1 (Passivity). *[10]*
A system $\dot{\xi} = f(\xi, u)$, $y = h(\xi, u)$ *is* strictly state passive *[10] in short if there exists a storage function (positive semi-definite) V together with a positive definite function γ such that:*

$$u^T y \geq \frac{\partial V}{\partial x} f(x, u) + \gamma(x).$$

Definition 2 (Persistent Excitation). *[23]*
A signal $g : I\!R^+ \rightarrow I\!R^r$ *(or even* $I\!R^{r\times\rho}$*) satisfies the property of persistent excitation if there exist* $T, k_1, k_2 > 0$ *such that for all* $t \geq 0$*:*

$$k_1 I_r \geq \int_t^{t+T} g(\tau)g^T(\tau)d\tau \geq k_2 I_r \tag{7.5}$$

Notice that when considering bounded signals g, the upper bound in (7.5) can be omitted.

Notice also that this corresponds to the fact that system:

$$\begin{aligned}\dot{\theta} &= 0\\ y &= g^T(t)\theta\end{aligned}$$

satisfies a *uniform complete observability* property (see chapter 1).

When considering *simultaneous state and parameter* estimation, the persistent excitation condition comes at first, together with an appropriate structure for a possible observer design.

Obviously *adaptive state observers* can be recast in a framework of *joint state and parameter estimation* problem, but in order to emphasize possible cases when state and parameter estimation problems can be somehow decoupled, we will first focus on adaptive state observers and then on observers for joint state and parameter estimation.

7.3.1 Adaptive State Estimation

Most of available adaptive observers, in the sense of *state observers* have been proposed for systems which are linear in the unknown parameter vector θ. Let us thus consider systems of the following form:

$$\begin{aligned}\dot{x}(t) &= f(y(t), z(t), v(t)) + g(y(t), z(t), v(t))\theta\\ y(t) &= \begin{pmatrix} I_p & 0\end{pmatrix} x(t) \quad \text{with}\end{aligned} \tag{7.6}$$

with $x(t)=\begin{pmatrix} y(t)\\ z(t)\end{pmatrix} \in I\!R^n, y(t)\in I\!R^p, v(t)\in I\!R^m, \theta\in I\!R^q.$

In the adaptive observer approach the idea is to rely on some possible observer design when all parameters are known: so let us assume that there exists an observer giving an estimate $\hat{x}_\theta$ for x when θ is known, together with a Lyapunov function V for the dynamics of the observation error in $\hat{x}_\theta - x$ (see e.g. appendix of chapter 1 for a brief recall on Lyapunov results).

Then most of the available results on adaptive observers can be summarized by the following formulation [1]:

Proposition 1. *Consider a system (7.6) with an observer giving an estimate $\hat{x}_\theta$ for x when θ is known, together with a Lyapunov function V for $\hat{x}_\theta - x$. Then we have:*

(I) If:

$$\frac{\partial V}{\partial e} g(y, \sigma, v) = \varphi([I_p \ 0]e, y, \sigma, v)$$

for some function φ,
with g globally bounded and f, g globally Lipschitz w.r.t. z, uniformly w.r.t. (u, y, t)
then $lim_{t\to\infty}\|\hat{x}_{\hat{\theta}}(t) - x(t)\| = 0$ when $\dot{\hat{\theta}} = -\Lambda\varphi^T(\hat{y} - y, y, \hat{z}, v)$, $\Lambda > 0$.

(II) If moreover g is persistently exciting and $\dot{g}$ is bounded then we also have $lim_{t\to\infty}\|\hat{\theta}(t) - \theta\| = 0$.

This can be established by considering a modified Lyapunov function $V + \varepsilon^T\Lambda^{-1}\varepsilon$ with $\varepsilon := \hat{\theta} - \theta$ together with a convergence property resulting from persistent excitation condition (II) [1].

Notice that such conditions are typically satisfied in results of [16, 17, 8]: in those works indeed, $f(y, z, v) = Ax + B(y, v)$ or even $Ax + B(y, z, v)$, while $g(y, z, v) = G\psi(y, v)$ or even $G\psi(y, z, v)$, with in both cases appropriate conditions for the existence of a Lyapunov function for the observation error e of the form $V(e) = e^T Pe$ such that $PG = C^T$, which indeed guarantees condition (I). This is what happens for system (7.4) as well.

Notice also that condition (I) can be interpreted as a passivity-like condition as follows:

In view of definition 1, condition (I) can be replaced by the condition that the error system obtained with the observer and disturbed by a parameter error is strictly state passive for this parameter error as an input and some appropriate output function only depending on the measured output error and measured further signals $(y, \hat{z}, u, t\ )$.

On the other hand, condition (II) stands for some classical condition of 'persistent excitation'.

As a re-interpretation of proposition 1, it can be added that systems for which an adaptive state observer can be designed can actually be written under a specific form underlining this possible design. This specific form can be expressed as follows [1]:

Definition 3. *A system of the form:*

$$\begin{aligned} \dot{y} &= \alpha(y, \zeta, v) + \beta(y, \zeta, v)\theta; \\ \dot{\zeta} &= Z(y, \zeta, v) \end{aligned} \tag{7.7}$$

with $y \in I\!R^p, \zeta \in I\!R^r, u \in I\!R^m, \theta \in I\!R^q$
is said to be in nonlinear adaptive observer form *if:*

1. *y is the measured output;*

2. *There exists a proper decresent positive-definite C^1 function $V(t,e)$, such that for any initial condition for system (7.7), any admissible input u, any corresponding output solution of (7.7) $y(t)$, any $z, e \in I\!R^r$, and any $t \geq 0$, one has:*

$$\frac{\partial V}{\partial t}(t,e) + \frac{\partial V}{\partial e}(Z(y(t), e+z, v(t)) - Z(y(t), z, v(t))) \leq -\kappa(e),$$

   for some κ positive-definite;
3. *For the same conditions as in the above item, and with $x(t)$ denoting the corresponding state solution of (7.7), one has:*

$$(i)\ \|\alpha(y(t), e+z, v(t)) - \alpha(y(t), z, v(t))\| \leq \gamma_\alpha \sqrt{\kappa(e)};\ \gamma_\alpha > 0$$

$$\|\beta(y(t), e+z, v(t)) - \beta(y(t), z, v(t))\| \leq \gamma_\beta \sqrt{\kappa(e)};\ \gamma_\beta > 0$$

$$(ii)\ \|\beta(y(t), \zeta(t), v(t))\| \leq b;\ b > 0.$$

The terminology of 'adaptive observer form' is borrowed from [18], and is here motivated by the fact that structure (7.7) clearly guarantees a possible adaptive state observer design. Adding a condition of 'persistent excitation' further yields asymptotic estimation of θ. This is summarized as follows [1]:

Proposition 2. *Given a system (7.7), then*

$$\begin{aligned}
\dot{\hat{y}} &= \alpha(y, \hat{\zeta}, v) + \beta(y, \hat{\zeta}, v)\hat{\theta} - k_y(\hat{y} - y);\ k_y > 0 \\
\dot{\hat{\zeta}} &= Z(y, \hat{\zeta}, v) \\
\dot{\hat{\theta}} &= -k_\theta \beta^T(y, \hat{\zeta}, v)(\hat{y} - y)^T;\ k_\theta > 0
\end{aligned}$$

is an adaptive state observer in the sense that $\|\hat{y}(t) - y(t)\|$, $\|\hat{\zeta}(t) - \zeta(t)\|$ go to zero as t goes to infinity.

Furthermore if β is persistently exciting and $\dot{\beta}$ is bounded

then $\|\hat{\theta}(t) - \theta\|$ also decays to zero.

Structure (7.7) is also some kind of a *canonical form* for adaptive observer design, since it can be checked that systems of proposition 1 with a *quadratic* Lyapunov function for the observer error system can be turned into (7.7) by a simple change of coordinates [1].

In the simple example of system (7.4), it can easily be checked that $z_1 = x_1$, $z_2 = x_2 - 2x_1$ turn the system into the form (7.7): in this case, $\beta = u$ and thus the additional condition so as to get parameter estimation is that u should be bounded, as well as $\dot{u}$, and satisfy the persistent excitation condition.

More generally, it is clear now that for systems which can be turned into the form (7.7) by change of coordinates, the state can be estimated in spite of unknown parameters. It is also clear how those parameters can further be estimated too under appropriate excitation condition, yielding simultaneous state and parameter estimation. This can be recast in a more general way including recent results on adaptive observers for linear time-varying systems of [24] and subsequent developments.

7.3.2 Joint State and Parameter Estimation

In the previous subsection a specific form allowing adaptive observer design has been pointed out, and as long as this form can be obtained by change of state coordinates, it allows for adaptive *state* estimation.

Now by further considering changes of variables also possibly depending on *unknown parameters*, one can end up with a more general form allowing for adaptive observer design simultaneously estimating state and parameter vectors (which is actually required so as to be able to get estimates for the original state variables). Generalizing the approach sketched in [25], this can be expressed as follows:

Proposition 3. *Considering a system of the form (7.6), if it can be turned by a change of coordinates possibly depending on time and parameters* $z = \Phi(x, \theta, t)$ *- with* $x = \Psi(z, \theta, t)$ *bounded w.r.t.* t *- into the following form:*

$$\begin{aligned} \dot{z} &= Z(z, y, u, t) \\ y &= H(z, u, t) + D(z, u, t)\theta, \quad y \in \mathbb{R}^p, z \in \mathbb{R}^n, u \in \mathbb{R}^m, \theta \in \mathbb{R}^q \end{aligned} \tag{7.8}$$

such that:
[i] There exists a proper decrescent positive-definite C^1 *function* $V(t, e)$*, such that for any initial condition for system (7.8), any admissible input* u*, any* $z, e \in \mathbb{R}^n$*,* $y \in \mathbb{R}^p$*, and any* $t \geq 0$*, one has:*

$$\frac{\partial V}{\partial t}(t, e) + \frac{\partial V}{\partial e}[Z(e + z, y, u(t), t) - Z(z, y, u(t), t)] \leq -\gamma(e) \tag{7.9}$$

[ii] For any admissible input u *and corresponding trajectories* $z(t)$*,* D *is persistently exciting, and uniformly bounded by some* d*, with* $\|D(e + z, u(t), t) - D(z, u(t), t)\| \leq \gamma_D \sqrt{\gamma(e)}$ *and* $\|H(e + z, u(t), t) - H(z, u(t), t)\| \leq \gamma_H \sqrt{\gamma(e)}$ *for any* e, z *and some* $\gamma_D, \gamma_H > 0$*, then an adaptive observer for simultaneous estimation of state* x *and parameter* θ *can be designed as:*

$$\begin{aligned} \dot{\hat{z}} &= Z(\hat{z}, y, u, t) \\ \dot{\hat{\theta}} &= -\lambda D^T(\hat{z}, u, t)(D(\hat{z}, u, t)\hat{\theta} + H(\hat{z}, u, t) - y), \ \lambda > 0 \\ \hat{x} &= \Psi(\hat{z}, \hat{\theta}, t) \end{aligned} \tag{7.10}$$

This clearly follows from the structure of the error equations $e_z := \hat{z} - z$, $e_\theta = \hat{\theta} - \theta$:

$$\begin{aligned}\dot{e}_z &= Z(e_z + z, y, u, t) - Z(z, y, u, t)\\ \dot{e}_\theta &= -\lambda D^T(\hat{z}, u, t) D(\hat{z}, u, t) e_\theta + \Delta(z, \hat{z}, u, t)\end{aligned}$$

where Δ depends on errors $D(\hat{z}, u, t) - D(z, u, t)$ and $H(\hat{z}, u, t) - H(z, u, t)$. From this indeed, e_z vanishes, and thus so does Δ, while 0 is a globally exponentially stable equilibrium of the undisturbed equation in e_θ (obtained for $\Delta = 0$) from persistency of $D(\hat{z}, u, t)$ (which results from that of $D(z, u, t)$, since $D(\hat{z}, u, t) - D(z, u, t)$ is clearly $\mathcal{L}^2$).

The proof can be formally established by considering P such that:

$$\dot{P} + PD^T(\hat{z}, u, t)D(\hat{z}, u, t) + D^T(\hat{z}, u, t)D(\hat{z}, u, t)P = -I,$$

and $V(t, e_z) + \varepsilon e_\theta^T P e_\theta$ as a Lyapunov function for the overall error equation, with ε small enough.

Notice that the previously considered adaptive observer form (7.7) can clearly be turned into (7.8) by a simple transformation: $z_1 := y - \Gamma\theta$ with Γ such that $\dot{\Gamma} = -k\Gamma + \beta(y, \zeta, v)$ for any $k > 0$. Clearly indeed, this yields:

$$\dot{z}_1 = -kz_1 + \alpha + ky$$

while the equation in ζ remains unchanged, and $y = z_1 + \Gamma\theta$.

It can also easily be checked that conditions allowing an adaptive observer design for linear systems proposed in [24] ensure at the same time a possible transformation into a form (7.8) (such a transformation being actually instrumental in the proof). Those conditions indeed can be summarized as follows: considering a system described by

$$\begin{aligned}\dot{x}(t) &= A(t)x(t) + B(t)u(t) + \Psi(t)\theta\\ y(t) &= C(t)x(t)\end{aligned}$$

one basically needs the existence of a matrix $K(t)$ such that:

(a) $\xi = 0$ is an exponentially stable equilibrium of $\dot{\xi}(t) = [A(t) - K(t)C(t)]\xi(t)$;
(b) The solution of: $\dot{\Gamma}(t) = [A(t) - K(t)C(t)]\Gamma(t) + \Psi(t)$ is such that $C\Gamma(t)$ satisfies the condition of persistent excitation.

From this, the simple change of coordinates $z = x - \Gamma\theta$ yields:

$$\dot{z}(t) = [A(t) - K(t)C(t)]z(t) + K(t)y(t); \quad y(t) = C(t)z(t) + C(t)\Gamma(t)\theta$$

which satisfies conditions of proposition 3.

Notice finally that a specificity of structure (7.8) is that only y depends on θ and in a linear way.

However, by following the same idea of parameter/time-dependent change of coordinates, adaptive observers have also been obtained for classes of nonlinear systems whose *nominal* structures (with known parameters) admit an observer, but for which the parameter-linear form (7.8) cannot be obtained. This is for instance the case of systems admitting high gain observer designs [5]:

Proposition 4. *Let us consider a system of the form:*

$$\begin{aligned} \dot{x}(t) &= A_0 x(t) + \varphi(x(t), u(t)) + \psi(x(t), u(t))\theta \\ y(t) &= C_0 x(t) \end{aligned} \tag{7.11}$$

with $x \in I\!R^n, u \in I\!R^m, y \in I\!R, \theta \in I\!R^q$.

Let us assume that:

[A1] A_0, C_0, φ, ψ *satisfy:*

$$A_0 = \begin{pmatrix} 0 & 1 & & 0 \\ & & \ddots & \\ & & & 1 \\ 0 & & & 0 \end{pmatrix},$$
$$C_0 = \begin{pmatrix} 1 & 0 & \cdots & 0 \end{pmatrix}$$

and

$$\varphi(x, u) = \begin{pmatrix} \varphi_1(x_1, u) \\ \varphi_2(x_1, x_2, u) \\ \vdots \\ \varphi_n(x, u) \end{pmatrix}; \quad \psi(x, u) = \begin{pmatrix} 0 & \cdots & 0 \\ \vdots & & \vdots \\ 0 & \cdots & 0 \\ \psi_{n1}(x, u) & \cdots & \psi_{nq}(x, u) \end{pmatrix}$$

[A2] φ, ψ *are smooth functions w.r.t. their arguments, and* u *is bounded generating bounded states* $\|x(t)\| \leq X$ *while* $\|\theta\| \leq \Theta$.

[A3] Given K_0 *such that* $A_0 - K_0 C_0$ *is stable, inputs* u *are such that the state vector satisfies the following property:*

for any $x(0)$, *and any* $\Gamma(0) \in I\!R^{n \times q}$, *the solution* $\Gamma(t)$ *of:*

$$\dot{\Gamma}(t) = \lambda(A_0 - K_0 C_0)\Gamma(t) + \lambda\psi(x(t), u(t)) \tag{7.12}$$

is such that for some $t_0 \geq 0$:

$$\begin{aligned} &\exists \alpha, T \text{ independent of } \lambda \;:\; \forall t \geq t_0, \\ &\text{and for } \lambda \text{ large enough,} \\ &\int_t^{t+T} \Gamma(\tau)^T C_0^T C_0 \Gamma(\tau) d\tau \geq \alpha I \end{aligned} \tag{7.13}$$

Then for λ *large enough, the system below is an asymptotic observer for (7.11), in the sense that for any initial condition* $x(0)$ *and any* $\hat{\theta}(0)$, $\hat{x}(0)$ *respectively bounded by* Θ *and* X, $\|\hat{x}(t) - x(t)\|$ *and* $\|\hat{\theta}(t) - \theta\|$ *exponentially go to zero:*

$$\begin{aligned} \dot{\hat{\Gamma}}(t) &= \lambda(A_0 - K_0 C_0)\hat{\Gamma}(t) + \lambda\psi(\tilde{x}(t), u(t)) \\ \dot{\hat{x}}(t) &= A_0 \hat{x}(t) + \varphi(\tilde{x}(t), u(t)) + \psi(\tilde{x}(t), u(t))\tilde{\theta}(t) \\ &\quad + \Lambda(\lambda)^{-1}[\lambda K_0 + \hat{\Gamma}(t)\hat{\Gamma}^T(t) C_0^T][y(t) - C_0 \hat{x}(t)] \\ \dot{\hat{\theta}}(t) &= \lambda^n \hat{\Gamma}(t)^T C_0^T [y(t) - C_0 \hat{x}(t)] \\ \tilde{x} &= \hat{x} \; if \; \|\hat{x}\| \leq X, \; \tfrac{\hat{x}}{\|\hat{x}\|} X \; otherwise, \\ \tilde{\theta} &= \hat{\theta} \; if \; \|\hat{\theta}\| \leq \Theta, \; \tfrac{\hat{\theta}}{\|\hat{\theta}\|} \Theta \; otherwise. \end{aligned} \tag{7.14}$$

where $\Lambda(\lambda)$ *is a diagonal matrix whose entry* i *is given by* λ^{i-1} *(as in high gain observers).*

Typically here, assumptions $A1$ and $A2$ mean that the nominal system admits a possible high gain observer design, while assumption $A3$ is roughly met whenever $\psi(x(t), u(t))$ satisfies the persistent excitation condition.

7.4 Conclusions

In this chapter, a brief overview on possible *adaptive* observer designs has been given. Such designs are obviously of interest for state estimation in spite of unknown parameters, but also for parameter identification in continuous-time, process monitoring (and fault detection), as well as adaptive control for instance. They allow to take advantage of some appropriate structure for observer design of the 'nominal system'. But it can happen that the adaptive approach yields very similar result to that of 'state extension', as this has been shown for the case of state affine systems [3]. On the other hand, considering that adaptive designs anyway correspond in the end to observers for extended systems (with an extended state vector including unknown parameters) can be of interest for new nonlinear observer designs, for instance as proposed in [4].

Acknowledgement. The author would like to particularly thank Prof M. Kinnaert from the 'Université Libre de Bruxelles' for his overview on nonlinear observer-based fault detection, as well as Dr Q. Zhang from INRIA-IRISA for fruitful cooperations on the topic of this chapter.

References

[1] G. Besançon. Remarks on nonlinear adaptive observer design. *Systems & Control Letters*, 41:271–280, 2000.
[2] G. Besançon. High gain observation with disturbance attenuation and application to robust fault detection. *Automatica*, 39:1095–1102, 2003.
[3] G. Besançon, J. De Leon-Morales, and O. Huerta-Guevara. On adaptive observers for state affine systems. *Int. Journal of Control*, 79(6):581–591, 2006.
[4] G. Besançon and Q. Zhang. Further developments on nonlinear adaptive observers with application to fault detection. In *IFAC World Congress, Barcelona, Spain*, 2002.
[5] G. Besançon, Q. Zhang, and H. Hammouri. High gain observer based simultaneous state and parameter estimation in nonlinear systems. In *IFAC Symp Nonlinear Control Syst., Germany*, 2004.
[6] G. Besançon, Q. Zhang, and H. Hammouri. High gain observer based state and parameter estimation in nonlinear systems. In *IFAC Symp on Nonlinear Control Systems, Stuttgart, Germany*, 2004.
[7] M. Blanke, M. Kinnaert, J. Lunze, and M. Staroswiecki. *Diagnosis and Fault-tolerant Control (2nd Ed)*. Springer, 2006.
[8] Y.M. Cho and R. Rajamani. A systematic approach to adaptive observer synthesis for nonlinear systems. *IEEE Trans. on Automatic Control*, 42(4):534–537, 1997.

[9] V. Cocquempot. On the equivalence between observer-based and parity space approaches for fdi in non-linear systems. In *IFAC Safeprocess Budapest, Hungary*, 2000.

[10] H. Khalil. *Nonlinear Systems (2nd Ed)*. Prentice Hall, Upper Saddle River, NJ, 1996.

[11] M. Kinnaert. Fault diagnosis based on analytical models for linear and nonlinear models - A tutorial. In *IFAC Safeprocess, Washington, USA*, 2003.

[12] I.D. Landau. *Adaptive control - The model reference approach*. Marcel Dekker Inc., 1979.

[13] I.D. Landau and A. Voda (Eds). *System Identification (*in French*)*. Hermes Science Europe, 2001.

[14] I.D. Landau, R. Lozano, and M. M'Saad. *Adaptive control*. Springer, 1997.

[15] L. Ljung. *System Identification, Theory for the User (2nd Ed.)*. Prentice Hall, Upper Saddle River, NJ, 1999.

[16] R. Marino. Adaptive observers for single output nonlinear systems. *IEEE Trans. on Automatic Control*, 35(9):1054–1058, 1990.

[17] R. Marino and P. Tomei. Global adaptive observers for nonlinear systems via filtered transformations. *IEEE Trans. on Automatic Control*, 37(8):1239–1245, 1992.

[18] R. Marino and P. Tomei. *Nonlinear control design - Geometric, adaptive, robust*. Prentice Hall Int., 1995.

[19] M.A. Massoumnia, G.C. Verghese, and A.S. Willsky. Failure detection and identification. *IEEE Trans. on Automatic Control*, 34:316–321, 1989.

[20] R.J. Patton, P.M. Frank, and R.N. Clark (Eds). *Diagnosis in dynamics systems: theory and applications*. Prentice Hall, NJ, 1989.

[21] R.J. Patton and M. Hou. H_∞ estimation and robust fault detection. In *Proc. 4th European Control Conf., Brussels, Belgium*, 1997.

[22] C. De Persis and A. Isidori. On the design of fault detection filter with game-theoretic optimal sensitivity. *Int. J. Robust and Nonlinear Control*, 12:729–747, 2002.

[23] S. Sastry and M. Bodson. *Adaptive control - stability, convergence and robsutness*. Prentice Hall Int., NJ, 1989.

[24] Q. Zhang. Adaptive observer for MIMO linear time-varying systems. *IEEE Trans. on Automatic Control*, 47(3):525–29, 2002.

[25] Q. Zhang. Revisiting different adaptive observers through a unified formulation. In *joint IEEE Conf Decision and Control - Europ. Control Conf, Sevilla, Spain*, 2005.

[26] Q. Zhang and G. Besançon. Sensor fault estimation in nonlinear systems. In *IFAC World Congress, Prague, Czech Rep.*, 2005.

Index

Printing: Mercedes-Druck, Berlin
Binding: Stein + Lehmann, Berlin

Lecture Notes in Control and Information Sciences

Edited by M. Thoma, M. Morari

Further volumes of this series can be found on our homepage:
springer.com

Vol. 363: Besançon G. (Ed.)
Nonlinear Observers and Applications
224 p. 2007 [978-3-540-73502-1]

Vol. 362: Tarn T.-J.; Chen S.-B.;
Zhou C. (Eds.)
Robotic Welding, Intelligence and Automation
562 p. 2007 [978-3-540-73373-7]

Vol. 361: Méndez-Acosta H.O.; Femat R.;
González-Álvarez V. (Eds.):
Selected Topics in Dynamics and Control of Chemical and Biological Processes
320 p. 2007 [978-3-540-73187-0]

Vol. 360: Kozlowski K. (Ed.)
Robot Motion and Control 2007
452 p. 2007 [978-1-84628-973-6]

Vol. 359: Christophersen F.J.
Optimal Control of Constrained
Piecewise Affine Systems
190 p. 2007 [978-3-540-72700-2]

Vol. 358: Findeisen R.; Allgöwer
F.; Biegler L.T. (Eds.): Assessment and Future Directions of Nonlinear
Model Predictive Control
642 p. 2007 [978-3-540-72698-2]

Vol. 357: Queinnec I.; Tarbouriech
S.; Garcia G.; Niculescu S.-I. (Eds.):
Biology and Control Theory: Current Challenges
589 p. 2007 [978-3-540-71987-8]

Vol. 356: Karatkevich A.:
Dynamic Analysis of Petri Net-Based Discrete Systems
166 p. 2007 [978-3-540-71464-4]

Vol. 355: Zhang H.; Xie L.:
Control and Estimation of Systems with
Input/Output Delays
213 p. 2007 [978-3-540-71118-6]

Vol. 354: Witczak M.:
Modelling and Estimation Strategies for Fault
Diagnosis of Non-Linear Systems
215 p. 2007 [978-3-540-71114-8]

Vol. 353: Bonivento C.; Isidori A.; Marconi L.;
Rossi C. (Eds.)
Advances in Control Theory and Applications
305 p. 2007 [978-3-540-70700-4]

Vol. 352: Chiasson, J.; Loiseau, J.J. (Eds.)
Applications of Time Delay Systems
358 p. 2007 [978-3-540-49555-0]

Vol. 351: Lin, C.; Wang, Q.-G.; Lee, T.H., He, Y.
LMI Approach to Analysis and Control of
Takagi-Sugeno Fuzzy Systems with Time Delay
204 p. 2007 [978-3-540-49552-9]

Vol. 350: Bandyopadhyay, B.; Manjunath, T.C.;
Umapathy, M.
Modeling, Control and Implementation of Smart
Structures 250 p. 2007 [978-3-540-48393-9]

Vol. 349: Rogers, E.T.A.; Galkowski, K.;
Owens, D.H.
Control Systems Theory
and Applications for Linear
Repetitive Processes 482
p. 2007 [978-3-540-42663-9]

Vol. 347: Assawinchaichote, W.; Nguang, K.S.;
Shi P.
Fuzzy Control and Filter Design
for Uncertain Fuzzy Systems
188 p. 2006 [978-3-540-37011-6]

Vol. 346: Tarbouriech, S.; Garcia, G.; Glattfelder,
A.H. (Eds.)
Advanced Strategies in Control Systems
with Input and Output Constraints
480 p. 2006 [978-3-540-37009-3]

Vol. 345: Huang, D.-S.; Li, K.; Irwin, G.W. (Eds.)
Intelligent Computing in Signal Processing
and Pattern Recognition
1179 p. 2006 [978-3-540-37257-8]

Vol. 344: Huang, D.-S.; Li, K.; Irwin, G.W. (Eds.)
Intelligent Control and Automation
1121 p. 2006 [978-3-540-37255-4]

Vol. 341: Commault, C.; Marchand, N. (Eds.)
Positive Systems
448 p. 2006 [978-3-540-34771-2]

Vol. 340: Diehl, M.; Mombaur, K. (Eds.)
Fast Motions in Biomechanics and Robotics
500 p. 2006 [978-3-540-36118-3]

Vol. 339: Alamir, M.
Stabilization of Nonlinear Systems Using
Receding-horizon Control Schemes
325 p. 2006 [978-1-84628-470-0]

Vol. 338: Tokarzewski, J.
Finite Zeros in Discrete Time Control Systems
325 p. 2006 [978-3-540-33464-4]

Vol. 337: Blom, H.; Lygeros, J. (Eds.)
Stochastic Hybrid Systems
395 p. 2006 [978-3-540-33466-8]

Vol. 336: Pettersen, K.Y.; Gravdahl, J.T.; Nijmeijer, H. (Eds.)
Group Coordination and Cooperative Control
310 p. 2006 [978-3-540-33468-2]

Vol. 335: Kozłowski, K. (Ed.)
Robot Motion and Control
424 p. 2006 [978-1-84628-404-5]

Vol. 334: Edwards, C.; Fossas Colet, E.; Fridman, L. (Eds.)
Advances in Variable Structure and Sliding Mode Control
504 p. 2006 [978-3-540-32800-1]

Vol. 333: Banavar, R.N.; Sankaranarayanan, V.
Switched Finite Time Control of a Class of Underactuated Systems
99 p. 2006 [978-3-540-32799-8]

Vol. 332: Xu, S.; Lam, J.
Robust Control and Filtering of Singular Systems
234 p. 2006 [978-3-540-32797-4]

Vol. 331: Antsaklis, P.J.; Tabuada, P. (Eds.)
Networked Embedded Sensing and Control
367 p. 2006 [978-3-540-32794-3]

Vol. 330: Koumoutsakos, P.; Mezic, I. (Eds.)
Control of Fluid Flow
200 p. 2006 [978-3-540-25140-8]

Vol. 329: Francis, B.A.; Smith, M.C.; Willems, J.C. (Eds.)
Control of Uncertain Systems: Modelling, Approximation, and Design
429 p. 2006 [978-3-540-31754-8]

Vol. 328: Loría, A.; Lamnabhi-Lagarrigue, F.; Panteley, E. (Eds.)
Advanced Topics in Control Systems Theory
305 p. 2006 [978-1-84628-313-0]

Vol. 327: Fournier, J.-D.; Grimm, J.; Leblond, J.; Partington, J.R. (Eds.)
Harmonic Analysis and Rational Approximation
301 p. 2006 [978-3-540-30922-2]

Vol. 326: Wang, H.-S.; Yung, C.-F.; Chang, F.-R.
H_∞ Control for Nonlinear Descriptor Systems
164 p. 2006 [978-1-84628-289-8]

Vol. 325: Amato, F.
Robust Control of Linear Systems Subject to Uncertain
Time-Varying Parameters
180 p. 2006 [978-3-540-23950-5]

Vol. 324: Christofides, P.; El-Farra, N.
Control of Nonlinear and Hybrid Process Systems
446 p. 2005 [978-3-540-28456-7]

Vol. 323: Bandyopadhyay, B.; Janardhanan, S.
Discrete-time Sliding Mode Control
147 p. 2005 [978-3-540-28140-5]

Vol. 322: Meurer, T.; Graichen, K.; Gilles, E.D. (Eds.) Control and Observer Design for Nonlinear Finite and Infinite Dimensional Systems
422 p. 2005 [978-3-540-27938-9]

Vol. 321: Dayawansa, W.P.; Lindquist, A.; Zhou, Y. (Eds.)
New Directions and Applications in Control Theory
400 p. 2005 [978-3-540-23953-6]

Vol. 320: Steffen, T.
Control Reconfiguration of Dynamical Systems
290 p. 2005 [978-3-540-25730-1]

Vol. 319: Hofbaur, M.W.
Hybrid Estimation of Complex Systems
148 p. 2005 [978-3-540-25727-1]

Vol. 318: Gershon, E.; Shaked, U.; Yaesh, I.
H_∞ Control and Estimation of State-multiplicative Linear Systems
256 p. 2005 [978-1-85233-997-5]

Vol. 317: Ma, C.; Wonham, M.
Nonblocking Supervisory Control of State Tree Structures
208 p. 2005 [978-3-540-25069-2]

Vol. 316: Patel, R.V.; Shadpey, F.
Control of Redundant Robot Manipulators
224 p. 2005 [978-3-540-25071-5]

Vol. 315: Herbordt, W.
Sound Capture for Human/Machine Interfaces: Practical Aspects of Microphone Array Signal Processing
286 p. 2005 [978-3-540-23954-3]

Vol. 314: Gil', M.I.
Explicit Stability Conditions for Continuous Systems
193 p. 2005 [978-3-540-23984-0]

Vol. 313: Li, Z.; Soh, Y.; Wen, C.
Switched and Impulsive Systems
277 p. 2005 [978-3-540-23952-9]

Vol. 312: Henrion, D.; Garulli, A. (Eds.)
Positive Polynomials in Control
313 p. 2005 [978-3-540-23948-2]

Vol. 311: Lamnabhi-Lagarrigue, F.; Loría, A.; Panteley, E. (Eds.)
Advanced Topics in Control Systems Theory
294 p. 2005 [978-1-85233-923-4]

Vol. 310: Janczak, A.
Identification of Nonlinear Systems Using Neural Networks and Polynomial Models
197 p. 2005 [978-3-540-23185-1]

Vol. 309: Kumar, V.; Leonard, N.; Morse, A.S. (Eds.)
Cooperative Control
301 p. 2005 [978-3-540-22861-5]

SPRINGER NATURE

GPSR Compliance

The European Union's (EU) General Product Safety Regulation (GPSR) is a set of rules that requires consumer products to be safe and our obligations to ensure this.

If you have any concerns about our products, you can contact us on ProductSafety@springernature.com

In case Publisher is established outside the EU, the EU authorized representative is:

Springer Nature Customer Service Center GmbH
Europaplatz 3
69115 Heidelberg, Germany

Kolibri 360 GmbH
Ferdinand-Jühlke-Straße 7
99095 Erfurt, Deutschland
produktsicherheit@kolibri360.de